Year 5C
A Guide to Teaching for Mastery

Series Editor: Tony Staneff
Lead author: Josh Lury

Contents

Introduction to the author team	4
What is *Power Maths*?	5
What's different in the new edition?	6
Your *Power Maths* resources	7
The *Power Maths* teaching model	10
The *Power Maths* lesson sequence	12
Using the *Power Maths* Teacher Guide	15
Power Maths Year 5, yearly overview	16
Mindset: an introduction	21
The *Power Maths* characters	22
Mathematical language	23
The role of talk and discussion	24
Assessment strategies	25
Keeping the class together	27
Same-day intervention	28
The role of practice	29
Structures and representations	30
Variation helps visualisation	31
Practical aspects of *Power Maths*	32
Working with children below age-related expectation	34
Providing extra depth and challenge with *Power Maths*	36
Using *Power Maths* with mixed age classes	38
List of practical resources	39
Getting started with *Power Maths*	41

Unit 12 – Geometry: properties of shapes — 42

Understand and use degrees	44
Measure acute angles	48
Measure angles up to 180°	52
Draw lines and angles accurately	56
Calculate angles around a point	60
Calculate angles on a straight line	64
Lengths and angles in shapes	68
Regular and irregular polygons	72
Parallel lines	76
Perpendicular lines	80
Investigate lines	84
3D shapes	88
End of unit check	92

Unit 13 – Geometry: position and direction — 94

Read and plot coordinates	96
Problem solving with coordinates	100
Translate shapes	104
Translate points	108
Reflection	112
Reflection in horizontal and vertical lines	116
End of unit check	120

Unit 14 – Decimals — 122
Add and subtract decimals within 1 (1) — 124
Add and subtract decimals within 1 (2) — 128
Complements to 1 — 132
Add and subtract decimals across 1 — 136
Add decimals with the same number of decimal places — 140
Subtract decimals with the same number of decimal places — 144
Add decimals with a different number of decimal places — 148
Subtract decimals with a different number of decimal places — 152
Problem solving with decimals (1) — 156
Problem solving with decimals (2) — 160
Decimal sequences — 164
Multiply by 10 — 168
Multiply by 10, 100 and 1,000 — 172
Divide by 10 — 176
Divide by 10, 100 and 1,000 — 180
End of unit check — 184

Unit 15 – Negative numbers — 186
Understand negative numbers — 188
Count through zero — 192
Compare and order negative numbers — 196
Find the difference — 200
End of unit check — 204

Unit 16 – Measure – converting units — 206
Kilograms and kilometres — 208
Millimetres and millilitres — 212
Convert units of length — 216
Imperial units of length — 220
Imperial units of mass — 224
Imperial units of capacity — 228
Convert units of time — 232
Timetables – calculating — 236
Problem solving – units of measure (1) — 240
Problem solving – units of measure (2) — 244
End of unit check — 248

Unit 17 – Measure – volume — 250
Cubic centimetres — 252
Compare volumes — 256
Estimate volume — 260
End of unit check — 264

Introduction to the author team

Power Maths arises from the work of maths mastery experts who are committed to proving that, given the right mastery mindset and approach, **everyone can do maths**. Based on robust research and best practice from around the world, *Power Maths* was developed in partnership with a group of UK teachers to make sure that it not only meets our children's wide-ranging needs but also aligns with the National Curriculum in England.

Power Maths – White Rose Maths edition

This edition of *Power Maths* has been developed and updated by:

Tony Staneff, Series Editor and Author

Vice Principal at Trinity Academy, Halifax, Tony also leads a team of mastery experts who help schools across the UK to develop teaching for mastery via nationally recognised CPD courses, problem-solving and reasoning resources, schemes of work, assessment materials and other tools.

Josh Lury, Lead Author

Josh is a specialist maths teacher, author and maths consultant with a passion for innovative and effective maths education.

The first edition of *Power Maths* was developed by a team of experienced authors, including:

- **Tony Staneff and Josh Lury**
- **Trinity Academy Halifax** (Michael Gosling CEO, Emily Fox, Kate Henshall, Rebecca Holland, Stephanie Kirk, Stephen Monaghan and Rachel Webster)
- **David Board, Belle Cottingham, Jonathan East, Tim Handley, Derek Huby, Neil Jarrett, Stephen Monaghan, Beth Smith, Tim Weal, Paul Wrangles** – skilled maths teachers and mastery experts
- **Cherri Moseley** – a maths author, former teacher and professional development provider
- **Professors Liu Jian and Zhang Dan**, Series Consultants and authors, and their team of mastery expert authors: **Wei Huinv, Huang Lihua, Zhu Dejiang, Zhu Yuhong, Hou Huiying, Yin Lili, Zhang Jing, Zhou Da and Liu Qimeng**

    Used by over 20 million children, Professor Liu Jian's textbook programme is one of the most popular in China. He and his author team are highly experienced in intelligent practice and in embedding key maths concepts using a C-P-A approach.

- **A group of 15 teachers and maths co-ordinators**

    We consulted our teacher group throughout the development of *Power Maths* to ensure we are meeting their real needs in the classroom.

What is *Power Maths*?

Created especially for UK primary schools, and aligned with the new National Curriculum, *Power Maths* is a whole-class, textbook-based mastery resource that empowers every child to understand and succeed. *Power Maths* rejects the notion that some people simply 'can't do' maths. Instead, it develops growth mindsets and encourages hard work, practice and a willingness to see mistakes as learning tools.

Best practice consistently shows that mastery of small, cumulative steps builds a solid foundation of deep mathematical understanding. *Power Maths* combines interactive teaching tools, high-quality textbooks and continuing professional development (CPD) to help you equip children with a deep and long-lasting understanding. Based on extensive evidence, and developed in partnership with practising teachers, *Power Maths* ensures that it meets the needs of children in the UK.

Power Maths and Mastery

Power Maths makes mastery practical and achievable by providing the structures, pathways, content, tools and support you need to make it happen in your classroom.

To develop mastery in maths, children must be enabled to acquire a deep understanding of maths concepts, structures and procedures, step by step. Complex mathematical concepts are built on simpler conceptual components and when children understand every step in the learning sequence, maths becomes transparent and makes logical sense. Interactive lessons establish deep understanding in small steps, as well as effortless fluency in key facts such as tables and number bonds. The whole class works on the same content and no child is left behind.

Power Maths

- Builds every concept in small, progressive steps
- Is built with interactive, whole-class teaching in mind
- Provides the tools you need to develop growth mindsets
- Helps you check understanding and ensure that every child is keeping up
- Establishes core elements such as intelligent practice and reflection

The *Power Maths* approach

Everyone can!
Founded on the conviction that every child can achieve, *Power Maths* enables children to build number fluency, confidence and understanding, step by step.

Child-centred learning
Children master concepts one step at a time in lessons that embrace a concrete-pictorial-abstract (C-P-A) approach, avoid overload, build on prior learning and help them see patterns and connections. Same-day intervention ensures sustained progress.

Continuing professional development
Embedded teacher support and development offer every teacher the opportunity to continually improve their subject knowledge and manage whole-class teaching for mastery.

Whole-class teaching
An interactive, whole-class teaching model encourages thinking and precise mathematical language and allows children to deepen their understanding as far as they can.

What's different in the new edition?

If you have previously used the first editions of *Power Maths*, you might be interested to know how this edition is different. All of the improvements described below are based on feedback from *Power Maths* customers.

Changes to units and the progression

- The order of units has been slightly adjusted, creating closer alignment between adjacent year groups, which will be useful for mixed age teaching.
- The flow of lessons has been improved within units to optimise the pace of the progression and build in more recap where needed. For key topics, the sequence of lessons gives more opportunities to build up a solid base of understanding. Other units have fewer lessons than before, where appropriate, making it possible to fit in all the content.
- Overall, the lessons put more focus on the most essential content for that year, with less time given to non-statutory content.
- The progression of lessons matches the steps in the new White Rose Maths schemes of learning.

Lesson resources

- There is a Quick recap for each lesson in the Teacher Guide, which offers an alternative lesson starter to the Power Up for cases where you feel it would be more beneficial to surface prerequisite learning than general number fluency.
- In the **Discover** and **Share** sections there is now more of a progression from 1 a) to 1 b). Whereas before, 1 b) was mainly designed as a separate question, now 1 a) leads directly into 1 b). This means that there is an improved whole-class flow, and also an opportunity to focus on the logic and skills in more detail. As a teacher, you will be using 1 a) to lead the class into the thinking, then 1 b) to mould that thinking into the core new learning of the lesson.
- In the **Share** section, for KS1 in particular, the number of different models and representations has been reduced, to support the clarity of thinking prompted by the flow from 1 a) into 1 b).
- More fluency questions have been built into the guided and independent practice.
- Pupil pages are as easy as possible for children to access independently. The pages are less full where this supports greater focus on key ideas and instructions. Also, more freedom is offered around answer format, with fewer boxes scaffolding children's responses; squared paper backgrounds are used in the Practice Books where appropriate. Artwork has also been revisited to ensure the highest standards of accessibility.

New components

480 Individual Practice Games are available in *ActiveLearn* for practising key facts and skills in Years 1 to 6. These are designed in an arcade style, to feel like fun games that children would choose to play outside school. They can be accessed via the Pupil World for homework or additional practice in school – and children can earn rewards. There are Support, Core and Extend levels to allocate, with Activity Reporting available for the teacher. There is a Quick Guide on *ActiveLearn* and you can use the Help area for support in setting up child accounts.

There is also a new set of lesson video resources on the Professional Development tile, designed for in-school training in 10- to 20-minute bursts. For each part of the *Power Maths* lesson sequence, there is a slide deck with embedded video, which will facilitate discussions about how you can take your *Power Maths* teaching to the next level.

Your *Power Maths* resources

Pupil Textbooks

Discover, **Share** and **Think together** sections promote discussion and introduce mathematical ideas logically, so that children understand more easily.

Using a Concrete-Pictorial-Abstract approach, clear mathematical models help children to make connections and grasp concepts.

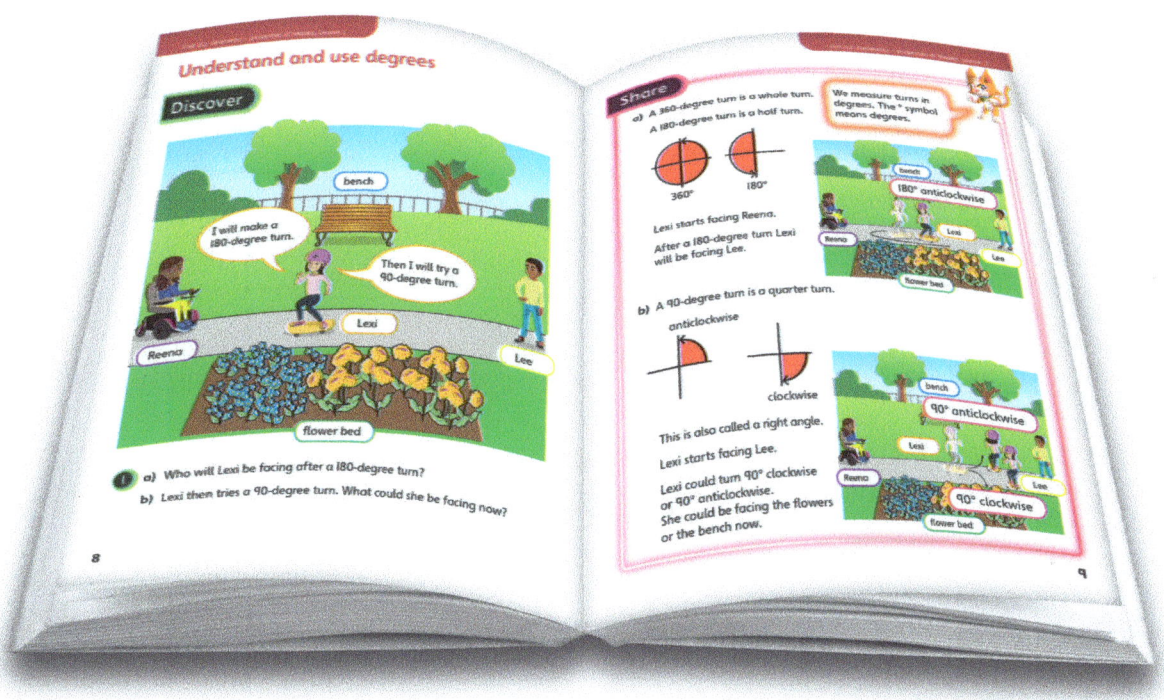

Appealing scenarios stimulate curiosity, helping children to identify the maths problem and discover patterns and relationships for themselves.

Friendly, supportive characters help children develop a growth mindset by prompting them to think, reason and reflect.

To help you teach for mastery, *Power Maths* comprises a variety of high-quality resources.

The coherent *Power Maths* lesson structure carries through into the vibrant, high-quality textbooks. Setting out the core learning objectives for each class, the lesson structure follows a carefully mapped journey through the curriculum and supports children on their journey to deeper understanding.

Pupil Practice Books

The Practice Books offer just the right amount of intelligent practice for children to complete independently in the final section of each lesson.

Practice questions are finely tuned to move children forward in their thinking and to reveal misconceptions.

The practice questions are for everyone – each question varies one small element to move children on in their thinking.

Calculations are connected so that children think about the underlying concept.

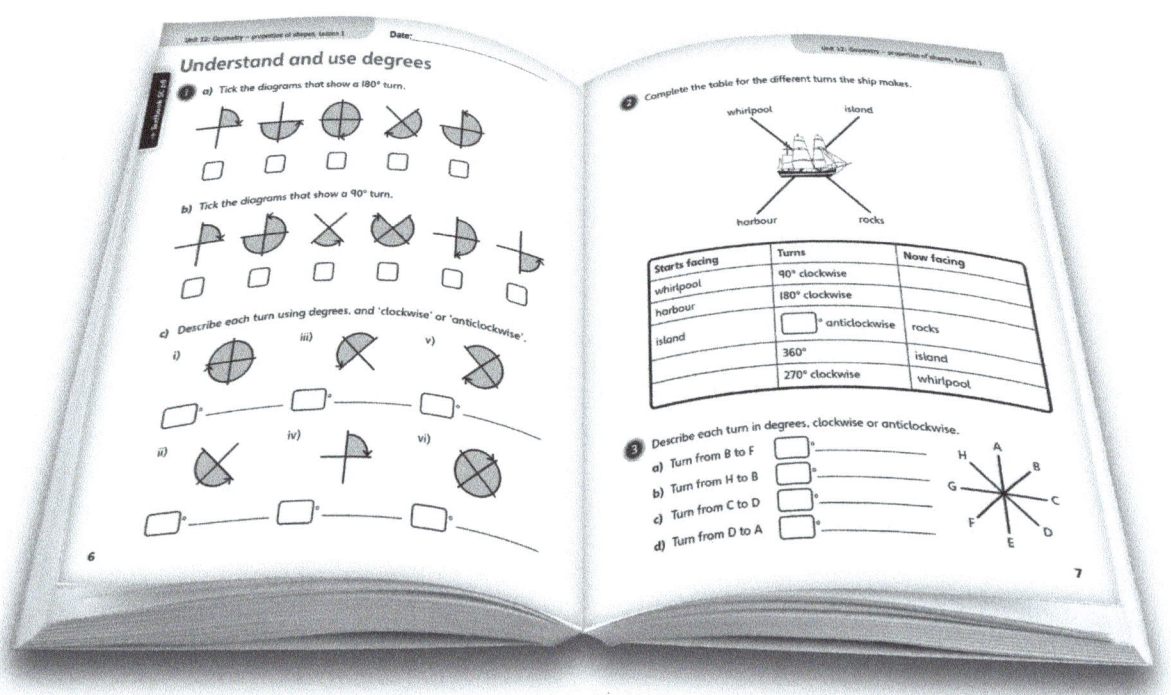

The *Power Maths* characters support and encourage children to think and work in different ways.

Challenge questions allow children to delve deeper into a concept.

Think differently questions encourage children to use reasoning as well as their mathematical knowledge to reach a solution.

Reflect questions reveal the depth of each child's understanding before they move on.

Online subscription

The online subscription will give you access to additional resources and answers from the Textbook and Practice Book.

eTextbooks

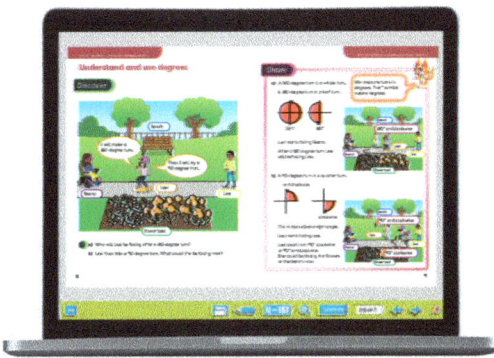

Digital versions of *Power Maths* Textbooks allow class groups to share and discuss questions, solutions and strategies. They allow you to project key structures and representations at the front of the class, to ensure all children are focusing on the same concept.

Teaching tools

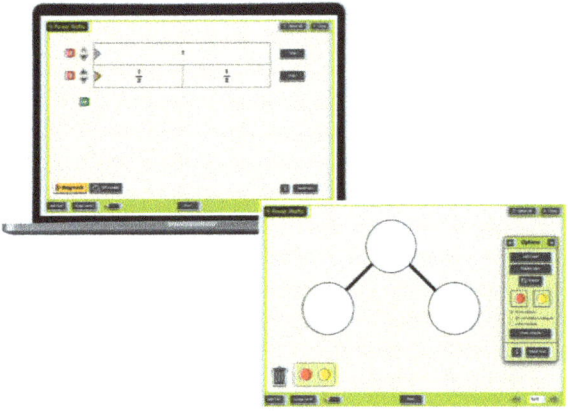

Here you will find interactive versions of key *Power Maths* structures and representations.

Power Ups

Use this series of daily activities to promote and check number fluency.

Online versions of Teacher Guide pages

PDF pages give support at both unit and lesson levels. You will also find help with key strategies and templates for tracking progress.

Unit videos

Watch the professional development videos at the start of each unit to help you teach with confidence. The videos explore common misconceptions in the unit, and include intervention suggestions as well as suggestions on what to look out for when assessing mastery in your students.

End of unit Strengthen and Deepen materials

The Strengthen activity at the end of every unit addresses a key misconception and can be used to support children who need it. The Deepen activities are designed to be low ceiling/high threshold and will challenge those children who can understand more deeply. These resources will help you ensure that every child understands and will help you keep the class moving forward together. These printable activities provide an optional resource bank for use after the assessment stage.

Individual Practice Games

These enjoyable games can be used at home or at school to embed key number skills (see page 6).

Professional Development videos and slides

These slides and videos of *Power Maths* lessons can be used for ongoing training in short bursts or to support new staff.

The *Power Maths* teaching model

At the heart of *Power Maths* is a clearly structured teaching and learning process that helps you make certain that every child masters each maths concept securely and deeply. For each year group, the curriculum is broken down into core concepts, taught in units. A unit divides into smaller learning steps – lessons. Step by step, strong foundations of cumulative knowledge and understanding are built.

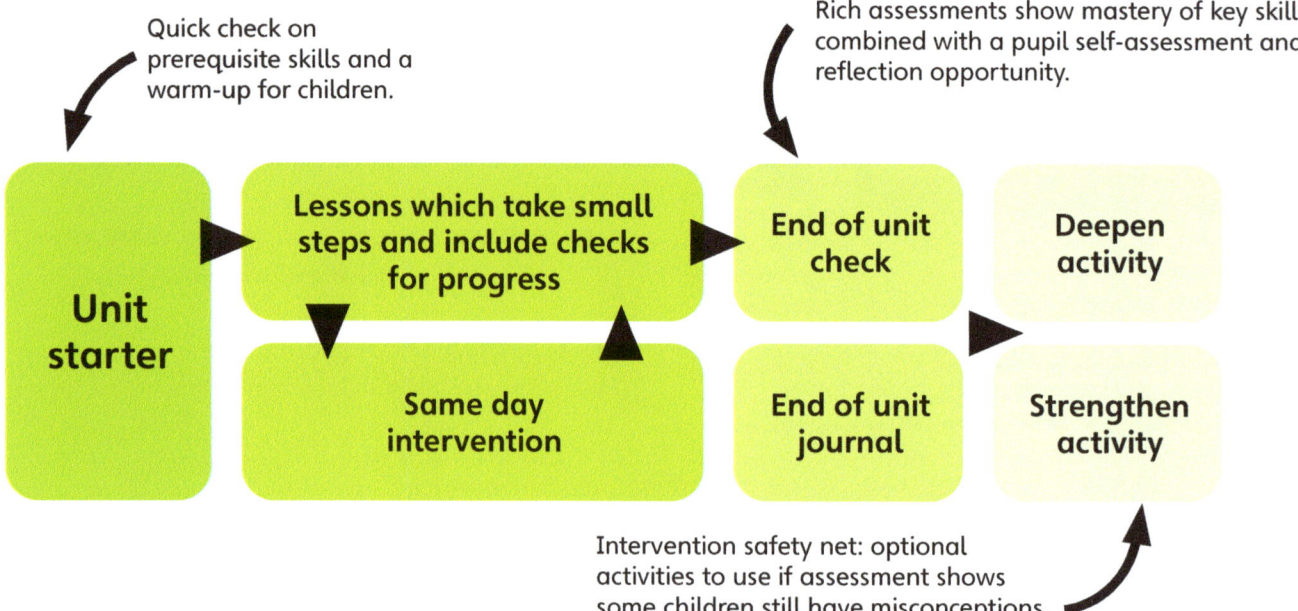

Unit starter

Each unit begins with a unit starter, which introduces the learning context along with key mathematical vocabulary and structures and representations.

- The Textbooks include a check on readiness and a warm-up task for children to complete.
- Your Teacher Guide gives support right from the start on important structures and representations, mathematical language, common misconceptions and intervention strategies.
- Unit-specific videos develop your subject knowledge and insights so you feel confident and fully equipped to teach each new unit. These are available via the online subscription.

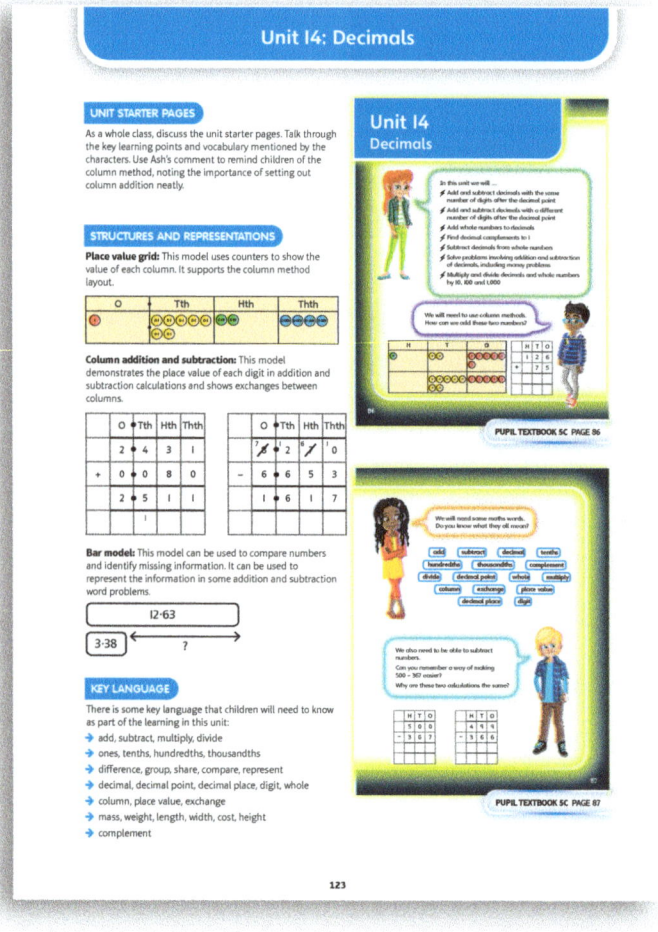

Lesson

Once a unit has been introduced, it is time to start teaching the series of lessons.

- Each lesson is scaffolded with Textbook and Practice Book activities and begins with a Power Up activity (available via online subscription) or the Quick recap activity in the Teacher Guide (see page 15).
- *Power Maths* identifies lesson by lesson what concepts are to be taught.
- Your Teacher Guide offers lots of support for you to get the most from every child in every lesson. As well as highlighting key points, tricky areas and how to handle them, you will also find question prompts to check on understanding and clarification on why particular activities and questions are used.

Same-day intervention

Same-day interventions are vital in order to keep the class progressing together. This can be during the lesson as well as afterwards (see page 28). Therefore, *Power Maths* provides plenty of support throughout the journey.

- Intervention is focused on keeping up now, not catching up later, so interventions should happen as soon as they are needed.
- Practice section questions are designed to bring misconceptions to the surface, allowing you to identify these easily as you circulate during independent practice time.
- Child-friendly assessment questions in the Teacher Guide help you identify easily which children need to strengthen their understanding.

End of unit check and journal

For each unit, the End of unit check in the Textbook lets you see which children have mastered the key concepts, which children have not and where their misconceptions lie. The Practice Books also include an End of unit journal in which children can reflect on what they have learned. Each unit also offers Strengthen and Deepen activities, available via the online subscription.

The Teacher Guide offers different ways of managing the End of unit assessments as well as giving support with handling misconceptions.

The End of unit check presents multiple-choice questions. Children think about their answer, decide on a solution and explain their choice.

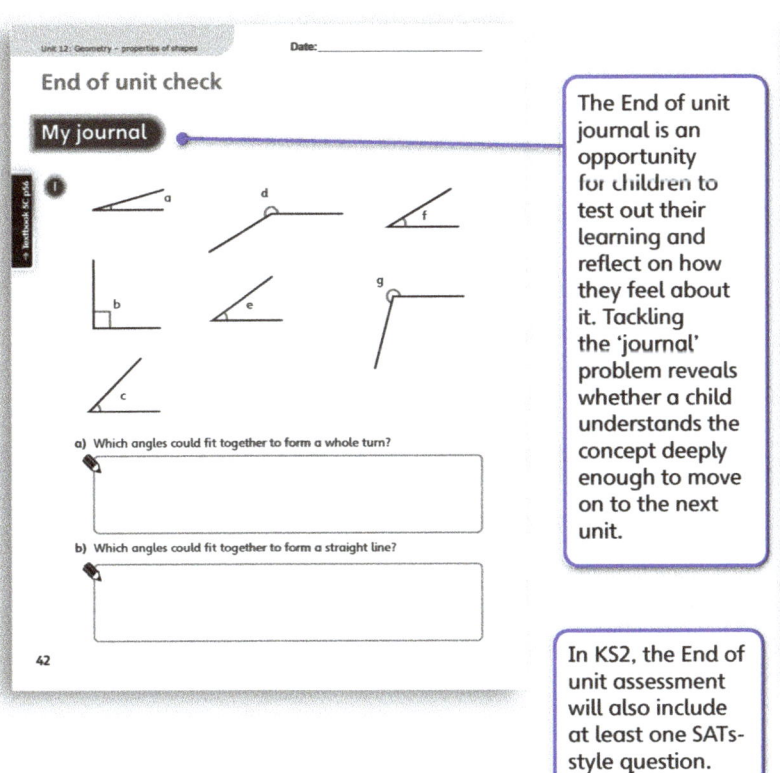

The End of unit journal is an opportunity for children to test out their learning and reflect on how they feel about it. Tackling the 'journal' problem reveals whether a child understands the concept deeply enough to move on to the next unit.

In KS2, the End of unit assessment will also include at least one SATs-style question.

The *Power Maths* lesson sequence

At the heart of *Power Maths* is a unique lesson sequence designed to empower children to understand core concepts and grow in confidence. Embracing the National Centre for Excellence in the Teaching of Mathematics' (NCETM's) definition of mastery, the sequence guides and shapes every *Power Maths* lesson you teach.

Flexibility is built into the *Power Maths* programme so there is no one-to-one mapping of lessons and concepts and you can pace your teaching according to your class. While some children will need to spend longer on a particular concept (through interventions or additional lessons), others will reach deeper levels of understanding. However, it is important that the class moves forward together through the termly schedules.

Power Up — 5 minutes

Each lesson begins with a Power Up activity (available via the online subscription) which supports fluency in key number facts.

The whole-class approach depends on fluency, so the Power Up is a powerful and essential activity.

The Quick recap is an alternative starter, for when you think some or all children would benefit more from revisiting pre-requisite work (see page 15).

TOP TIP
If the class is struggling with the task, revisit it later and check understanding.

Power Ups reinforce the two key things that are essential for success: times-tables and number bonds.

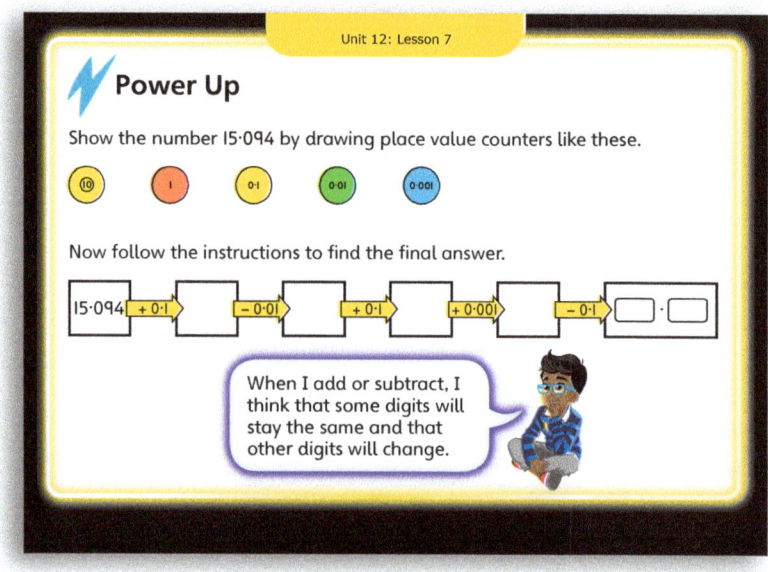

Discover — 10 minutes

A practical, real-life problem arouses curiosity. Children find the maths through story telling.

A real-life scenario is provided for the **Discover** section but feel free to build upon these with your own examples that are more relevant to your class, or get creative with the context.

TOP TIP
Discover works best when run at tables, in pairs with concrete objects.

Question ① a) tackles the key concept and question ① b) digs a little deeper. Children have time to explore, play and discuss possible strategies.

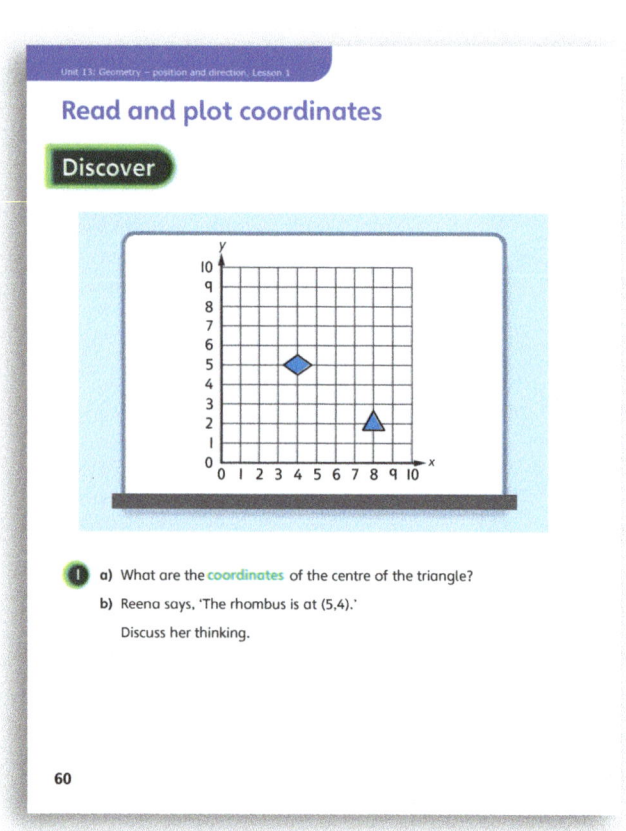

Share ⏱ 10 minutes

Teacher-led, this interactive section follows the **Discover** activity and highlights the variety of methods that can be used to solve a single problem.

TOP TIP
Pairs sharing a textbook is a great format for **Share**!

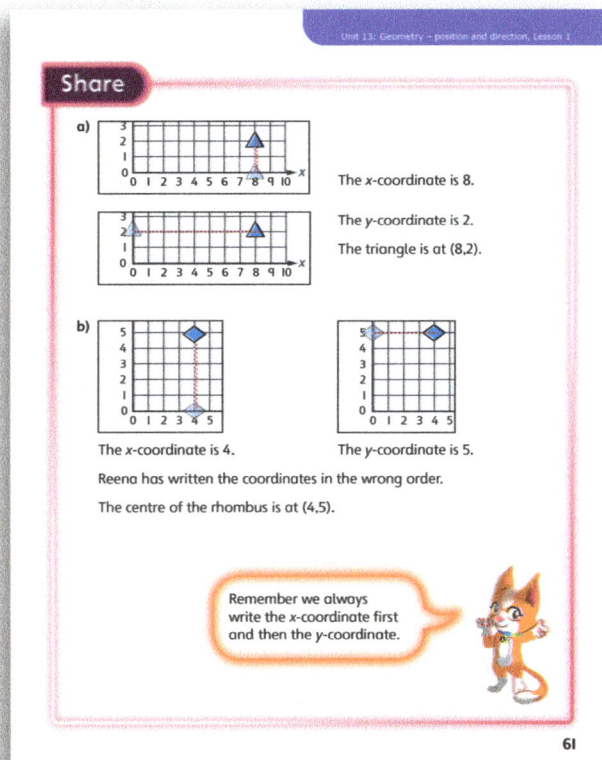

Your Teacher Guide gives target questions for children. The online toolkit provides interactive structures and representations to link concrete and pictorial to abstract concepts.

Bring children to the front to share and celebrate their solutions and strategies.

Think together

⏱ 10 minutes

Children work in groups on the carpet or at tables, using their textbooks or eBooks.

TOP TIP
Make sure children have mini whiteboards or pads to write on if they are not at their tables.

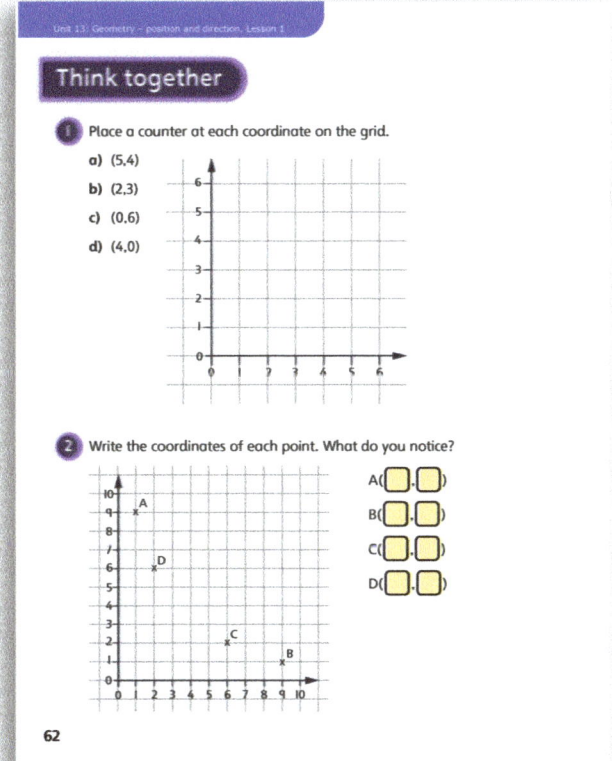

Using the Teacher Guide, model question ① for your class.

Question ② is less structured. Children will need to think together in their groups, then discuss their methods and solutions as a class.

In question ③ children try working out the answer independently. The openness of the **Challenge** question helps to check depth of understanding.

Practice ⏱ 15 minutes

Using their Practice Books, children work independently while you circulate and check on progress.

Questions follow small steps of progression to deepen learning.

TOP TIP
Some children could work separately with a teacher or assistant.

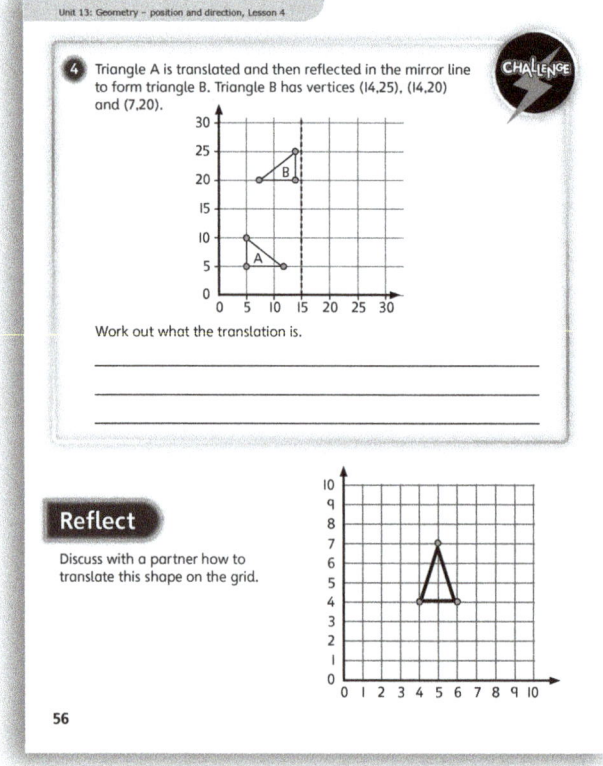

Are some children struggling? If so, work with them as a group, using mathematical structures and representations to support understanding as necessary.

There are no set routines: for real understanding, children need to think about the problem in different ways.

Reflect ⏱ 5 minutes

'Spot the mistake' questions are great for checking misconceptions.

The **Reflect** section is your opportunity to check how deeply children understand the target concept.

The Practice Books use various approaches to check that children have fully understood each concept.

Looking like they understand is not enough! It is essential that children can show they have grasped the concept.

Using the *Power Maths* Teacher Guide

Think of your Teacher Guides as *Power Maths* handbooks that will guide, support and inspire your day-to-day teaching. Clear and concise, and illustrated with helpful examples, your Teacher Guides will help you make the best possible use of every individual lesson. They also provide wrap-around professional development, enhancing your own subject knowledge and helping you to grow in confidence about moving your children forward together.

There is a Teacher Guide per year group for every term, with unit and lesson level guidance and support.

Never feel stuck! You will find ideas for introducing every unit and lesson and questions to encourage teacher reflection before and after each lesson.

Tips and advice on key elements such as C-P-A approaches, misconceptions, language, modelling growth mindsets and same day intervention.

Annotations for every Textbook and Practice Book page, providing prompts for key questions to ask to expose understanding and explanations as to why key questions have been chosen.

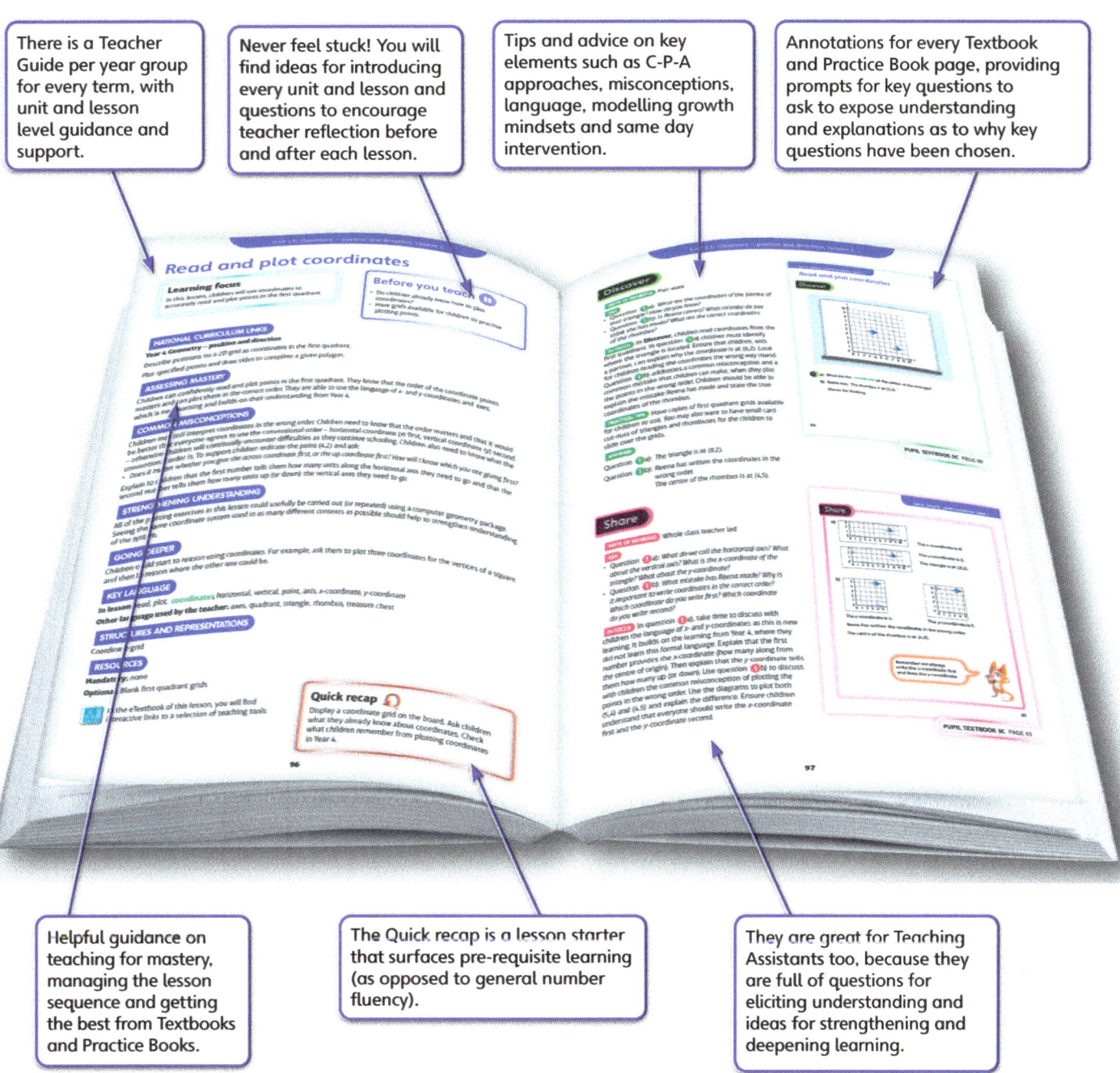

Helpful guidance on teaching for mastery, managing the lesson sequence and getting the best from Textbooks and Practice Books.

The Quick recap is a lesson starter that surfaces pre-requisite learning (as opposed to general number fluency).

They are great for Teaching Assistants too, because they are full of questions for eliciting understanding and ideas for strengthening and deepening learning.

At the end of each unit, your Teacher Guide helps you identify who has fully grasped the concept, who has not and how to move every child forward. This is covered later in the Assessment strategies section.

Power Maths Year 5, yearly overview

Textbook	Strand	Unit		Number of lessons
Textbook A / Practice Workbook A (Term 1)	Number – number and place value	1	Place value within 1,000,000 (1)	8
	Number – number and place value	2	Place value within 1,000,000 (2)	6
	Number – addition and subtraction	3	Addition and subtraction	12
	Number – multiplication and division	4	Multiplication and division (1)	10
	Number – fractions (including decimals and percentages)	5	Fractions (1)	8
	Number – fractions (including decimals and percentages)	6	Fractions (2)	11
Textbook B / Practice Workbook B (Term 2)	Number – multiplication and division	7	Multiplication and division (2)	10
	Number – fractions (including decimals and percentages)	8	Fractions (3)	7
	Number – fractions (including decimals and percentages)	9	Decimals and percentages	15
	Measurement	10	Measure – perimeter and area	8
	Statistics	11	Graphs and tables	6
Textbook C / Practice Workbook C (Term 3)	Geometry – properties of shapes	12	Geometry – properties of shapes	12
	Geometry – position and direction	13	Geometry – position and direction	6
	Number – fractions (including decimals and percentages)	14	Decimals	15
	Number – number and place value	15	Negative numbers	4
	Measurement	16	Measure – converting units	10
	Measurement	17	Measure – volume	3

Power Maths Year 5, Textbook 5C (Term 3) overview

Strand	Unit		Lesson number	Lesson title	NC Objective 1	NC Objective 2
Geometry – properties of shapes	12	Geometry – properties of shapes	1	Understand and use degrees	Know angles are measured in degrees: estimate and compare acute, obtuse and reflex angles	Identify: – angles at a point and one whole turn (total 360°) – angles at a point on a straight line and $\frac{1}{2}$ a turn (total 180°) – other multiples of 90°
Geometry – properties of shapes	12	Geometry – properties of shapes	2	Measure acute angles	Know angles are measured in degrees: estimate and compare acute, obtuse and reflex angles	
Geometry – properties of shapes	12	Geometry – properties of shapes	3	Measure angles up to 180°	Know angles are measured in degrees: estimate and compare acute, obtuse and reflex angles	Draw given angles, and measure them in degrees (°)
Geometry – properties of shapes	12	Geometry – properties of shapes	4	Draw lines and angles accurately	Draw given angles, and measure them in degrees (°)	
Geometry – properties of shapes	12	Geometry – properties of shapes	5	Calculate angles around a point	Identify: – angles at a point and one whole turn (total 360°) – angles at a point on a straight line and $\frac{1}{2}$ a turn (total 180°) – other multiples of 90°	
Geometry – properties of shapes	12	Geometry – properties of shapes	6	Calculate angles on a straight line	Identify: – angles at a point and one whole turn (total 360°) – angles at a point on a straight line and $\frac{1}{2}$ a turn (total 180°) – other multiples of 90°	
Geometry – properties of shapes	12	Geometry – properties of shapes	7	Lengths and angles in shapes	Use the properties of rectangles to deduce related facts and find missing lengths and angles	
Geometry – properties of shapes	12	Geometry – properties of shapes	8	Regular and irregular polygons	Distinguish between regular and irregular polygons based on reasoning about equal sides and angles	
Geometry – properties of shapes	12	Geometry – properties of shapes	9	Parallel lines	Identify horizontal and vertical lines and pairs of perpendicular and parallel lines (Year 3)	
Geometry – properties of shapes	12	Geometry – properties of shapes	10	Perpendicular lines	Identify horizontal and vertical lines and pairs of perpendicular and parallel lines (Year 3)	
Geometry – properties of shapes	12	Geometry – properties of shapes	11	Investigate lines	Identify horizontal and vertical lines and pairs of perpendicular and parallel lines (Year 3)	
Geometry – properties of shapes	12	Geometry – properties of shapes	12	3D shapes	Identify 3D shapes, including cubes and other cuboids, from 2D representations	
Geometry – position and direction	13	Geometry – position and direction	1	Read and plot coordinates	Describe positions on a 2D grid as coordinates in the first quadrant (Year 4)	Plot specified points and draw sides to complete a given polygon (Year 4)
Geometry – position and direction	13	Geometry – position and direction	2	Problem solving with coordinates	Describe positions on a 2D grid as coordinates in the first quadrant (Year 4)	Plot specified points and draw sides to complete a given polygon (Year 4)
Geometry – position and direction	13	Geometry – position and direction	3	Translate shapes	Identify, describe and represent the position of a shape following a reflection or translation, using the appropriate language, and know that the shape has not changed	
Geometry – position and direction	13	Geometry – position and direction	4	Translate points	Identify, describe and represent the position of a shape following a reflection or translation, using the appropriate language, and know that the shape has not changed	

Strand	Unit		Lesson number	Lesson title	NC Objective 1	NC Objective 2
Geometry – position and direction	13	Geometry – position and direction	5	Reflection	Identify, describe and represent the position of a shape following a reflection or translation, using the appropriate language, and know that the shape has not changed	
Geometry – position and direction	13	Geometry – position and direction	6	Reflection in horizontal and vertical lines	Identify, describe and represent the position of a shape following a reflection or translation, using the appropriate language, and know that the shape has not changed	
Number – fractions (including decimals and percentages)	14	Decimals	1	Add and subtract decimals within 1 (1)	Solve problems involving number up to three decimal places	
Number – fractions (including decimals and percentages)	14	Decimals	2	Add and subtract decimals within 1 (2)	Solve problems involving number up to three decimal places	
Number – fractions (including decimals and percentages)	14	Decimals	3	Complements to 1	Solve problems involving number up to three decimal places	
Number – fractions (including decimals and percentages)	14	Decimals	4	Add and subtract decimals across 1	Solve problems involving number up to three decimal places	
Number – fractions (including decimals and percentages)	14	Decimals	5	Add decimals with the same number of decimal places	Solve problems involving number up to three decimal places	
Number – fractions (including decimals and percentages)	14	Decimals	6	Subtract decimals with the same number of decimal places	Solve problems involving number up to three decimal places	
Number – fractions (including decimals and percentages)	14	Decimals	7	Add decimals with a different number of decimal places	Solve problems involving number up to three decimal places	
Number – fractions (including decimals and percentages)	14	Decimals	8	Subtract decimals with a different number of decimal places	Solve problems involving number up to three decimal places	
Number – fractions (including decimals and percentages)	14	Decimals	9	Problem solving with decimals (1)	Solve problems involving number up to three decimal places	
Number – fractions (including decimals and percentages)	14	Decimals	10	Problem solving with decimals (2)	Solve problems involving number up to three decimal places	
Number – fractions (including decimals and percentages)	14	Decimals	11	Decimal sequences	Read, write, order and compare numbers with up to three decimal places	

Strand	Unit		Lesson number	Lesson title	NC Objective 1	NC Objective 2
Number – fractions (including decimals and percentages)	14	Decimals	12	Multiply by 10	Recognise and use thousandths and relate them to tenths, hundredths and decimal equivalents	Solve problems involving number up to three decimal places
Number – fractions (including decimals and percentages)	14	Decimals	13	Multiply by 10, 100 and 1,000	Recognise and use thousandths and relate them to tenths, hundredths and decimal equivalents	Solve problems involving number up to three decimal places
Number – fractions (including decimals and percentages)	14	Decimals	14	Divide by 10	Recognise and use thousandths and relate them to tenths, hundredths and decimal equivalents	Solve problems involving number up to three decimal places
Number – fractions (including decimals and percentages)	14	Decimals	15	Divide by 10, 100 and 1,000	Recognise and use thousandths and relate them to tenths, hundredths and decimal equivalents	Solve problems involving number up to three decimal places
Number – number and place value	15	Negative numbers	1	Understand negative numbers	Interpret negative numbers in context, count forwards and backwards with positive and negative whole numbers, including through zero	
Number – number and place value	15	Negative numbers	2	Count through zero	Interpret negative numbers in context, count forwards and backwards with positive and negative whole numbers, including through zero	
Number – number and place value	15	Negative numbers	3	Compare and order negative numbers	Interpret negative numbers in context, count forwards and backwards with positive and negative whole numbers, including through zero	
Number – number and place value	15	Negative numbers	4	Find the difference	Interpret negative numbers in context, count forwards and backwards with positive and negative whole numbers, including through zero	
Measurement	16	Measure – converting units	1	Kilograms and kilometres	Convert between different units of metric measure (for example, kilometre and metre; centimetre and metre; centimetre and millimetre; gram and kilogram; litre and millilitre)	
Measurement	16	Measure – converting units	2	Millimetres and millilitres	Convert between different units of metric measure (for example, kilometre and metre; centimetre and metre; centimetre and millimetre; gram and kilogram; litre and millilitre)	
Measurement	16	Measure – converting units	3	Convert units of length	Convert between different units of metric measure (for example, kilometre and metre; centimetre and metre; centimetre and millimetre; gram and kilogram; litre and millilitre)	
Measurement	16	Measure – converting units	4	Imperial units of length	Understand and use approximate equivalences between metric units and common imperial units such as inches, pounds and pints	
Measurement	16	Measure – converting units	5	Imperial units of mass	Understand and use approximate equivalences between metric units and common imperial units such as inches, pounds and pints	

Strand	Unit		Lesson number	Lesson title	NC Objective 1	NC Objective 2
Measurement	16	Measure – converting units	6	Imperial units of capacity	Understand and use approximate equivalences between metric units and common imperial units such as inches, pounds and pints	
Measurement	16	Measure – converting units	7	Convert units of time	Solve problems involving converting between units of time	
Measurement	16	Measure – converting units	8	Timetables – calculating	Solve problems involving converting between units of time	
Measurement	16	Measure – converting units	9	Problem solving – units of measure (1)	Use all four operations to solve problems involving measure [for example, length, mass, volume, money] using decimal notation, including scaling	
Measurement	16	Measure – converting units	10	Problem solving – units of measure (2)	Use all four operations to solve problems involving measure [for example, length, mass, volume, money] using decimal notation, including scaling	
Measurement	17	Measure – volume	1	Cubic centimetres	Estimate volume [for example, using 1 cm^3 blocks to build cuboids (including cubes)] and capacity [for example, using water]	
Measurement	17	Measure – volume	2	Compare volumes	Estimate volume [for example, using 1 cm^3 blocks to build cuboids (including cubes)] and capacity [for example, using water]	
Measurement	17	Measure – volume	3	Estimate volume	Estimate volume [for example, using 1 cm^3 blocks to build cuboids (including cubes)] and capacity [for example, using water]	

Mindset: an introduction

Global research and best practice deliver the same message: learning is greatly affected by what learners perceive they can or cannot do. What is more, it is also shaped by what their parents, carers and teachers perceive they can do. Mindset – the thinking that determines our beliefs and behaviours – therefore has a fundamental impact on teaching and learning.

Everyone can!

Power Maths and mastery methods focus on the distinction between 'fixed' and 'growth' mindsets (Dweck, 2007).[1] Those with a fixed mindset believe that their basic qualities (for example, intelligence, talent and ability to learn) are pre-wired or fixed: 'If you have a talent for maths, you will succeed at it. If not, too bad!' By contrast, those with a growth mindset believe that hard work, effort and commitment drive success and that 'smart' is not something you are or are not, but something you become. In short, everyone can do maths!

Key mindset strategies

A growth mindset needs to be actively nurtured and developed. *Power Maths* offers some key strategies for fostering healthy growth mindsets in your classroom.

It is okay to get it wrong

Mistakes are valuable opportunities to re-think and understand more deeply. Learning is richer when children and teachers alike focus on spotting and sharing mistakes as well as solutions.

Praise hard work

Praise is a great motivator, and by focusing on praising effort and learning rather than success, children will be more willing to try harder, take risks and persist for longer.

Mind your language!

The language we use around learners has a profound effect on their mindsets. Make a habit of using growth phrases, such as, 'Everyone can!', 'Mistakes can help you learn' and 'Just try for a little longer'. The king of them all is one little word, 'yet'… I can't solve this…yet!' Encourage parents and carers to use the right language too.

Build in opportunities for success

The step-by-small-step approach enables children to enjoy the experience of success. In addition, avoid ability grouping and encourage every child to answer questions and explain or demonstrate their methods to others.

[1] Dweck, C (2007) *The New Psychology of Success*, Ballantine Books: New York

The *Power Maths* characters

The *Power Maths* characters model the traits of growth mindset learners and encourage resilience by prompting and questioning children as they work. Appearing frequently in the Textbooks and Practice Books, they are your allies in teaching and discussion, helping to model methods, alternatives and misconceptions, and to pose questions. They encourage and support your children, too: they are all hardworking, enthusiastic and unafraid of making and talking about mistakes.

Meet the team!

Creative Flo is open-minded and sometimes indecisive. She likes to think differently and come up with a variety of methods or ideas.

Determined Dexter is resolute, resilient and systematic. He concentrates hard, always tries his best and he'll never give up – even though he doesn't always choose the most efficient methods!

'Let's try again.'
'Mistakes are cool!'
'Have I found all of the solutions?'

'Let's try it this way…'
'Can we do it differently?'
'I've got another way of doing this!'

'I'm going to try this!'
'I know how to do that!'
'Want to share my ideas?'

Curious Ash is eager, interested and inquisitive, and he loves solving puzzles and problems. Ash asks lots of questions but sometimes gets distracted.

'What if we tried this…?'
'I wonder…'
'Is there a pattern here?'

Miaow!

Sparks the Cat

Brave Astrid is confident, willing to take risks and unafraid of failure. She's never scared to jump straight into a problem or question, and although she often makes simple mistakes she's happy to talk them through with others.

Mathematical language

Traditionally, we in the UK have tended to try simplifying mathematical language to make it easier for young children to understand. By contrast, evidence and experience show that by diluting the correct language, we actually mask concepts and meanings for children. We then wonder why they are confused by new and different terminology later down the line! *Power Maths* is not afraid of 'hard' words and avoids placing any barriers between children and their understanding of mathematical concepts. As a result, we need to be deliberate, precise and thorough in building every child's understanding of the language of maths. Throughout the Teacher Guides you will find support and guidance on how to deliver this, as well as individual explanations throughout the pupil Textbooks.

Use the following key strategies to build children's mathematical vocabulary, understanding and confidence.

Precise and consistent

Everyone in the classroom should use the correct mathematical terms in full, every time. For example, refer to 'equal parts', not 'parts'. Used consistently, precise maths language will be a familiar and non-threatening part of children's everyday experience.

Full sentences

Teachers and children alike need to use full sentences to explain or respond. When children use complete sentences, it both reveals their understanding and embeds their knowledge.

Stem sentences

These important sentences help children express mathematical concepts accurately, and are used throughout the *Power Maths* books. Encourage children to repeat them frequently, whether working independently or with others. Examples of stem sentences are:

'4 is a part, 5 is a part, 9 is the whole.'

'There are …. groups. There are …. in each group.'

Key vocabulary

The unit starters highlight essential vocabulary for every lesson. In the pupil books, characters flag new terminology and the Teacher Guide lists important mathematical language for every unit and lesson. New terms are never introduced without a clear explanation.

Mathematical signs

Mathematical signs are used early on so that children quickly become familiar with them and their meaning. Often, the *Power Maths* characters will highlight the connection between language and particular signs.

The role of talk and discussion

When children learn to talk purposefully together about maths, barriers of fear and anxiety are broken down and they grow in confidence, skills and understanding. Building a healthy culture of 'maths talk' empowers their learning from day one.

Explanation and discussion are integral to the *Power Maths* structure, so by simply following the books your lessons will stimulate structured talk. The following key 'maths talk' strategies will help you strengthen that culture and ensure that every child is included.

Sentences, not words

Encourage children to use full sentences when reasoning, explaining or discussing maths. This helps both speaker and listeners to clarify their own understanding. It also reveals whether or not the speaker truly understands, enabling you to address misconceptions as they arise.

Working together

Working with others in pairs, groups or as a whole class is a great way to support maths talk and discussion. Use different group structures to add variety and challenge. For example, children could take timed turns for talking, work independently alongside a 'discussion buddy', or perhaps play different *Power Maths* character roles within their group.

Think first – then talk

Provide clear opportunities within each lesson for children to think and reflect, so that their talk is purposeful, relevant and focused.

Give every child a voice

Where the 'hands up' model allows only the more confident child to shine, *Power Maths* involves everyone. Make sure that no child dominates and that even the shyest child is encouraged to contribute – and praised when they do.

Assessment strategies

Teaching for mastery demands that you are confident about what each child knows and where their misconceptions lie; therefore, practical and effective assessment is vitally important.

Formative assessment within lessons

> The **Think together** section will often reveal any confusions or insecurities; try ironing these out by doing the first **Think together** question as a class. For children who continue to struggle, you or your Teaching Assistant should provide support and enable them to move on.

> Performance in practice can be very revealing: check Practice Books and listen out both during and after practice to identify misconceptions.

> The **Reflect** section is designed to check on the all-important depth of understanding. Be sure to review how the children performed in this final stage before you teach the next lesson.

End of unit check – Textbook

Each unit concludes with a summative check to help you assess quickly and clearly each child's understanding, fluency, reasoning and problem solving skills. Your Teacher Guide will suggest ideal ways of organising a given activity and offer advice and commentary on what children's responses mean. For example, 'What misconception does this reveal?'; 'How can you reinforce this particular concept?'

Assessment with young children should always be an enjoyable activity, so avoid one-to-one individual assessments, which they may find threatening or scary. If you prefer, the End of unit check can be carried out as a whole-class group using whiteboards and Practice Books.

End of unit check – Practice Book

The Practice Book contains further opportunities for assessment, and can be completed by children independently whilst you are carrying out diagnostic assessment with small groups. Your Teacher Guide will advise you on what to do if children struggle to articulate an explanation – or perhaps encourage you to write down something they have explained well. It will also offer insights into children's answers and their implications for next learning steps. It is split into three main sections, outlined below.

My journal is designed to allow children to show their depth of understanding of the unit. It can also serve as a way of checking that children have grasped key mathematical vocabulary. The question children should answer is first presented in the Textbook in the Think! section. This provides an opportunity for you to discuss the question first as a class to ensure children have understood their task. Children should have some time to think about how they want to answer the question, and you could ask them to talk to a partner about their ideas. Then children should write their answer in their Practice Book, using the word bank provided to help them with vocabulary.

The **Power check** allows pupils to self-assess their level of confidence on the topic by colouring in different smiley faces. You may want to introduce the faces as follows:

Each unit ends with either a Power play or a Power puzzle. This is an activity, puzzle or game that allows children to use their new knowledge in a fun, informal way.

Progress Tests

There are *Power Maths* Progress Tests for each half term and at the end of the year, including an Arithmetic test and Reasoning test in each case. You can enter results in the online markbook to track and analyse results and see the average for all schools' results. The tests use a 6-step scale to show results against age-related expectation.

How to ask diagnostic questions

The diagnostic questions provided in children's Practice Books are carefully structured to identify both understanding and misconceptions (if children answer in a particular way, you will know why). The simple procedure below may be helpful:

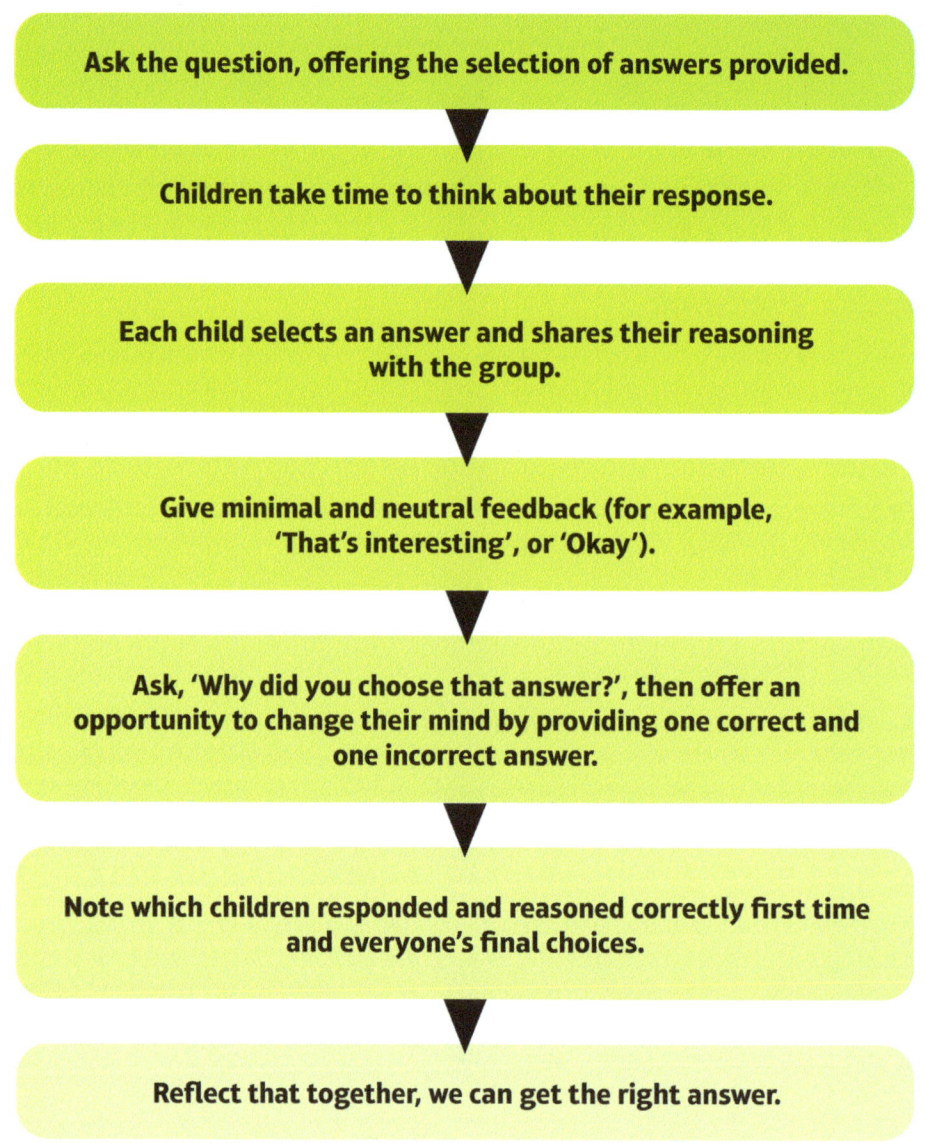

Keeping the class together

Traditionally, children who learn quickly have been accelerated through the curriculum. As a consequence, their learning may be superficial and will lack the many benefits of enabling children to learn with and from each other.

By contrast, *Power Maths'* mastery approach values real understanding and richer, deeper learning above speed. It sees all children learning the same concept in small, cumulative steps, each finding and mastering challenge at their own level. Remember that when you teach for mastery, EVERYONE can do maths! Those who grasp a concept easily have time to explore and understand that concept at a deeper level. The whole class therefore moves through the curriculum at broadly the same pace via individual learning journeys.

For some teachers, the idea that a whole class can move forward together is revolutionary and challenging. However, the evidence of global good practice clearly shows that this approach drives engagement, confidence, motivation and success for all learners, and not just the high flyers. The strategies below will help you keep your class together on their maths journey.

Mix it up

Do not stick to set groups at each table. Every child should be working on the same concept, and mixing up the groupings widens children's opportunities for exploring, discussing and sharing their understanding with others.

Recycling questions

Reuse the Textbook and Practice Book questions with concrete materials to allow children to explore concepts and relationships and deepen their understanding. This strategy is especially useful for reinforcing learning in same-day interventions.

Strengthen at every opportunity

The next lesson in a *Power Maths* sequence always revises and builds on the previous step to help embed learning. These activities provide golden opportunities for individual children to strengthen their learning with the support of Teaching Assistants.

Prepare to be surprised!

Children may grasp a concept quickly or more slowly. The 'fast graspers' won't always be the same individuals, nor does the speed at which a child understands a concept predict their success in maths. Are they struggling or just working more slowly?

Same-day intervention

Since maths competence depends on mastering concepts one by one in a logical progression, it is important that no gaps in understanding are ever left unfilled. Same-day interventions – either within or after a lesson – are a crucial safety net for any child who has not fully made the small step covered that day. In other words, intervention is always about keeping up, not catching up, so that every child has the skills and understanding they need to tackle the next lesson. That means presenting the same problems used in the lesson, with a variety of concrete materials to help children model their solutions.

We offer two intervention strategies below, but you should feel free to choose others if they work better for your class.

Within-lesson intervention

The **Think together** activity will reveal those who are struggling, so when it is time for practice, bring these children together to work with you on the first practice questions. Observe these children carefully, ask questions, encourage them to use concrete models and check that they reach and can demonstrate their understanding.

After-lesson intervention

You might like to use the **Think together** questions to recap the lesson with children who are working behind expectations during assembly time. Teaching Assistants could also work with these children at other convenient points in the school day. Some children may benefit from revisiting work from the same topic in the previous year group. Note also the suggestion for recycling questions from the Textbook and Practice Book with concrete materials on page 27.

The role of practice

Practice plays a pivotal role in the *Power Maths* approach. It takes place in class groups, smaller groups, pairs, and independently, so that children always have the opportunities for thinking as well as the models and support they need to practise meaningfully and with understanding.

Intelligent practice

In *Power Maths*, practice never equates to the simple repetition of a process. Instead we embrace the concept of intelligent practice, in which all children become fluent in maths through varied, frequent and thoughtful practice that deepens and embeds conceptual understanding in a logical, planned sequence. To see the difference, take a look at the following examples.

Traditional practice
- Repetition can be rote – no need for a child to think hard about what they are doing
- Praise may be misplaced
- Does this prove understanding?

Intelligent practice
- Varied methods – concrete, pictorial and abstract
- Equation expressed in different ways, requiring thought and understanding
- Constructive feedback

All practice questions are designed to move children on and reveal misconceptions.

Simple, logical steps build onto earlier learning.

C-P-A runs throughout – different ways of modelling and understanding the same concept.

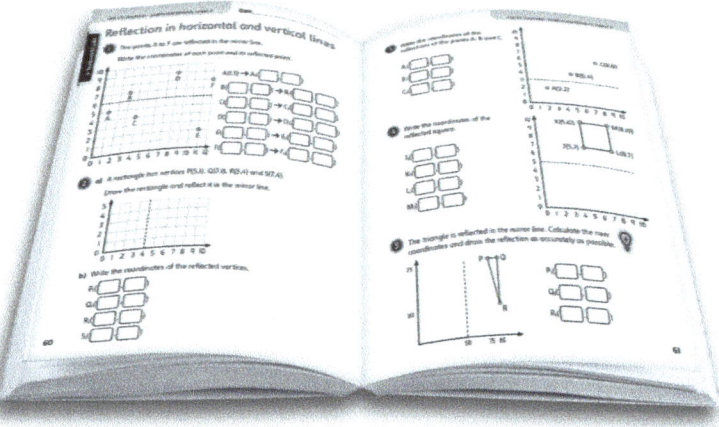

Conceptual variation – children work on different representations of the same maths concept.

Friendly characters offer support and encourage children to try different approaches.

A carefully designed progression

The Practice Books provide just the right amount of intelligent practice for children to complete independently in the final sections of each lesson. It is really important that all children are exposed to the practice questions, and that children are not directed to complete different sections. That is because each question is different and has been designed to challenge children to think about the maths they are doing. The questions become more challenging so children grasping concepts more quickly will start to slow down as they progress. Meanwhile, you have the chance to circulate and spot any misconceptions before they become barriers to further learning.

Homework and the role of parents and carers

While *Power Maths* does not prescribe any particular homework structure, we acknowledge the potential value of practice at home. For example, practising fluency in key facts, such as number bonds and times-tables, is an ideal homework task. You can share the Individual Practice Games for homework (see page 6), or parents and carers could work through uncompleted Practice Book questions with children at either primary stage.

However, it is important to recognise that many parents and carers may themselves lack confidence in maths, and few, if any, will be familiar with mastery methods. A Parents' and Carers' evening that helps them understand the basics of mindsets, mastery and mathematical language is a great way to ensure that children benefit from their homework. It could be a fun opportunity for children to teach their families that everyone can do maths!

Structures and representations

Unlike most other subjects, maths comprises a wide array of abstract concepts – and that is why children and adults so often find it difficult. By taking a concrete-pictorial-abstract (C-P-A) approach, *Power Maths* allows children to tackle concepts in a tangible and more comfortable way.

Non-linear stages

Concrete

Replacing the traditional approach of a teacher working through a problem in front of the class, the concrete stage introduces real objects that children can use to 'do' the maths – any familiar object that a child can manipulate and move to help bring the maths to life. It is important to appreciate, however, that children must always understand the link between models and the objects they represent. For example, children need to first understand that three cakes could be represented by three pretend cakes, and then by three counters or bricks. Frequent practice helps consolidate this essential insight. Although they can be used at any time, good concrete models are an essential first step in understanding.

Pictorial

This stage uses pictorial representations of objects to let children 'see' what particular maths problems look like. It helps them make connections between the concrete and pictorial representations and the abstract maths concept. Children can also create or view a pictorial representation together, enabling discussion and comparisons. The *Power Maths* teaching tools are fantastic for this learning stage, and bar modelling is invaluable for problem solving throughout the primary curriculum.

Abstract

Our ultimate goal is for children to understand abstract mathematical concepts, symbols and notation and of course, some children will reach this stage far more quickly than others. To work with abstract concepts, a child must be comfortable with the meaning of and relationships between concrete, pictorial and abstract models and representations. The C-P-A approach is not linear, and children may need different types of models at different times. However, when a child demonstrates with concrete models and pictorial representations that they have grasped a concept, we can be confident that they are ready to explore or model it with abstract symbols such as numbers and notation.

Use at any time and with any age to support understanding

Variation helps visualisation

Children find it much easier to visualise and grasp concepts if they see them presented in a number of ways, so be prepared to offer and encourage many different representations.

For example, the number six could be represented in various ways:

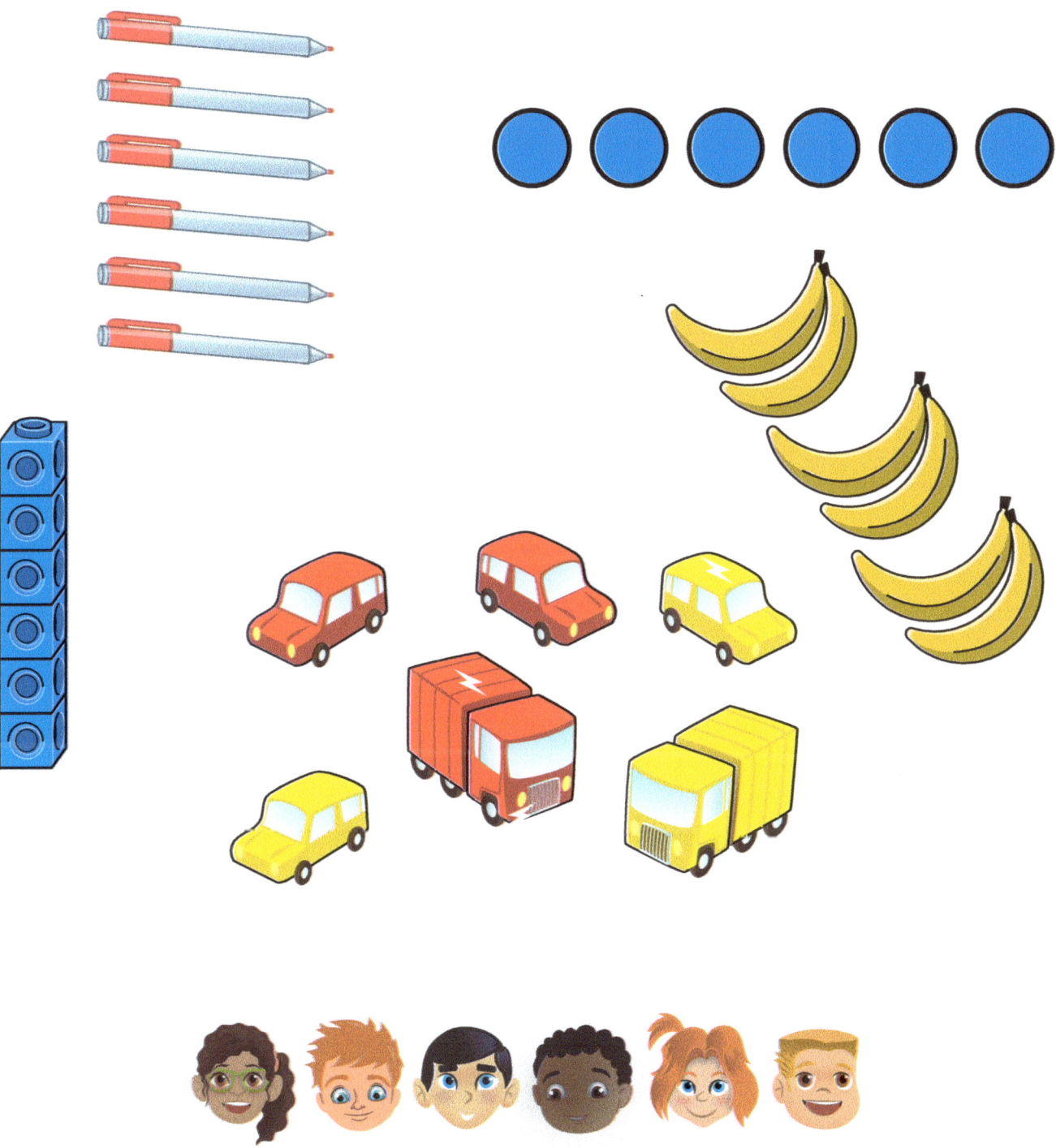

Practical aspects of *Power Maths*

One of the key underlying elements of *Power Maths* is its practical approach, allowing you to make maths real and relevant to your children, no matter their age.

Manipulatives are essential resources for both key stages and *Power Maths* encourages teachers to use these at every opportunity, and to continue the Concrete-Pictorial-Abstract approach right through to Year 6.

The Textbooks and Teacher Guides include lots of opportunities for teaching in a practical way to show children what maths means in real life.

Discover and Share

The **Discover** and **Share** sections of the Textbook give you scope to turn a real-life scenario into a practical and hands-on section of the lesson. Use these sections as inspiration to get active in the classroom. Where appropriate, use the **Discover** contexts as a springboard for your own examples that have particular resonance for your children – and allow them to get their hands dirty trying out the mathematics for themselves.

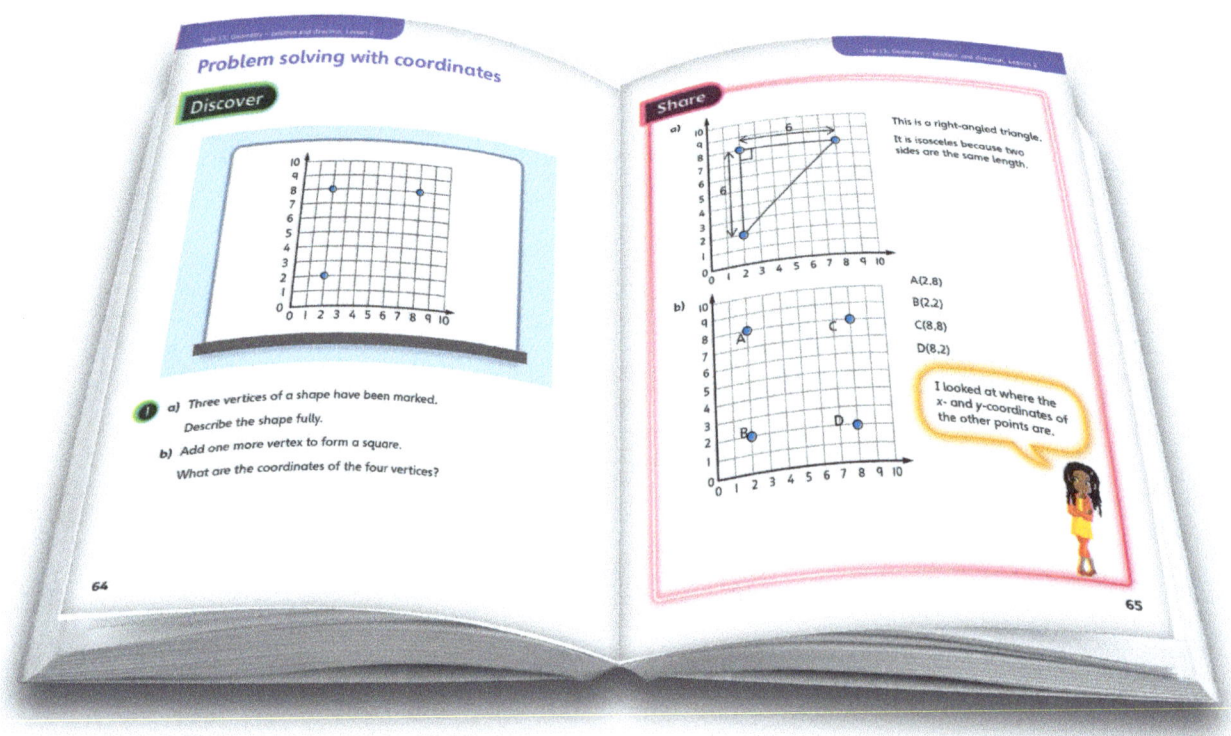

Unit videos

Every term has one unit video which incorporates real-life classroom sequences.

These videos show you how the reasoning behind mathematics can be carried out in a practical manner by showing real children using various concrete and pictorial methods to come to the solution. You can see how using these practical models, such as part-whole and bar models, helps them to find and articulate their answer.

Mastery tips

Mastery Experts give anecdotal advice on where they have used hands-on and real-life elements to inspire their children.

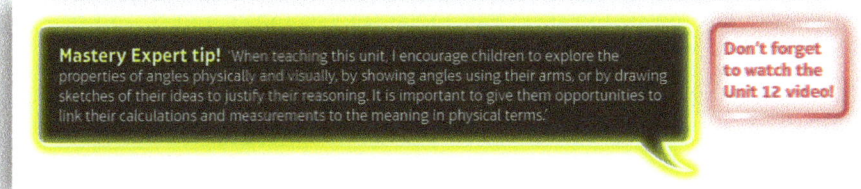

Concrete-Pictorial-Abstract (C-P-A) approach

Each **Share** section uses various methods to explain an answer, helping children to access abstract concepts by using concrete tools, such as counters. Remember, this isn't a linear process, so even children who appear confident using the more abstract method can deepen their knowledge by exploring the concrete representations. Encourage children to use all three methods to really solidify their understanding of a concept.

Pictorial representation – drawing the problem in a logical way that helps children visualise the maths

Concrete representation – using manipulatives to represent the problem. Encourage children to physically use resources to explore the maths.

Abstract representation – using words and calculations to represent the problem.

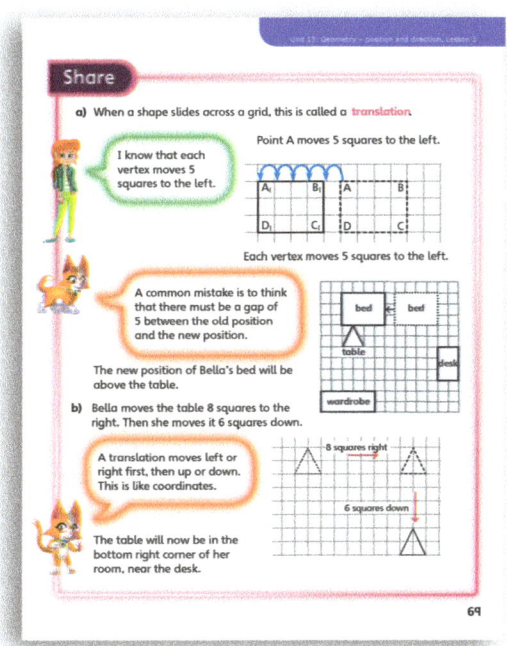

Practical tips

Every lesson suggests how to draw out the practical side of the **Discover** context.

You'll find these in the **Discover** section of the Teacher Guide for each lesson.

PRACTICAL TIPS Have children model this context using a 500 ml jug and 100 ml beakers. Tell children that each beaker is $\frac{1}{5}$ of the jug. Allow them to explore how many beakers fill the jug. Then ask them to explain this using a fraction strip. Encourage children to use pictorial support such as bar models to give another visual representation.

Resources

Every lesson lists the practical resources you will need or might want to use. There is also a summary of all of the resources used throughout the term on page 39 to help you be prepared.

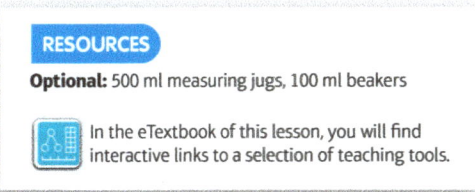

33

Working with children below age-related expectation

This section offers advice on using *Power Maths* with children who are significantly behind age-related expectation. Teacher judgement will be crucial in terms of where and why children are struggling, and in choosing the right approach. The suggestions can of course be adapted for children with special educational needs, depending on the specific details of those needs.

General approaches to support children who are struggling

Keeping the pace manageable
Remember, you have more teaching days than *Power Maths* lessons so you can cover a lesson over more than one day, and revisit key learning, to ensure all children are ready to move on. You can use the + and – buttons to adjust the time for each unit in the online planning. The NCETM's Ready-to-Progress criteria can be used to help determine what should be highest priority.

Same-day intervention
You could go over the Textbook pages or revisit the previous year's work if necessary (see Addressing gaps). Remember that same-day intervention can be within the lesson, as well as afterwards (see page 28). As children start their independent practice, you can work with those who found the first part of the lesson difficult, checking understanding using manipulatives.

Fluency sessions
Fit in as much practice as you can for number bonds and times-tables, etc., at other times of the day. If you can, plan a short 'maths meeting' for this in the afternoon. You might choose to use a Power Up you haven't used already.

Addressing gaps
Use material from the same topic in the previous year to consolidate or address gaps in learning, e.g. Textbook pages and Strengthen activities. The End of unit check will help gauge children's understanding.

Pre-teaching
Find a 5- to 10-minute slot before the lesson to work with the children you feel would benefit. The afternoon before the lesson can work well, because it gives children time to think in between. Recap previous work on the topic (addressing any gaps you're aware of) and do some fluency practice, targeting number facts etc. that will help children access the learning.

Focusing on the key concepts
If children are a long way behind, it can be helpful to take a step back and think about the key concepts for children to engage with, not just the fine detail of the objective for that year group (e.g. addition with a specific number of columns). Bearing that in mind, how could children advance their understanding of the topic?

Providing extra support within the lesson

Support in the Teacher Guide
First of all, use the Strengthen support in the Teacher Guide for guided and independent work in each lesson, and share this with Teaching Assistants, where relevant. As you read through the lesson content and corresponding Teacher Guide pages before the lesson, ask yourself what key idea or nugget of understanding is at the heart of the lesson. If children are struggling, this should help you decide what's essential for all children before they move on.

Annotating pages
You can annotate questions to provide extra scaffolding or hints if you need to, but aim to build up children's ability to access questions independently wherever you can. Children tend to get used to the style of the *Power Maths* questions over time.

Quick recap as lesson starter
The Quick recap for each lesson in the Teacher Guide is an alternative starter activity to the Power Up. You might choose to use this with some or all children if you feel they will need support accessing the main lesson.

Consolidation questions
If you think some children would benefit from additional questions at the same level before moving on, write one or two similar questions on the board. (This shouldn't be at the expense of reasoning and problem-solving opportunities: take longer over the lesson if you need to.)

Hard copy Textbooks
The Textbooks help children focus in more easily on the mathematical representations, read the text more comfortably, and revisit work from a previous lesson that you are building on, as well as giving children ownership of their learning journey. In main lessons, it can work well to use the e-Textbook for **Discover** and give out the books when discussing the methods in the **Share** section.

Reading support
It's important that all children are exposed to problem solving and reasoning questions, which often involve reading. For whole-class work you can read questions together. For independent practice you could consider annotating pages to help children see what the question is asking, and stem sentences to help structure their answer. A general focus on specific mathematical language and vocabulary will help children access the questions. You could consider pairing weaker readers with stronger readers, or read questions as a group if those who need support are on the same table.

Providing extra depth and challenge with *Power Maths*

Just as prescribed in the National Curriculum, the goal of *Power Maths* is never to accelerate through a topic but rather to gain a clear, deep and broad understanding. Here are some suggestions to help ensure all children are appropriately challenged as you work with the resources.

Overall approaches

First of all, remember that the materials are designed to help you keep the class together, allowing all children to master a concept while those who grasp it quickly have time to explore it in more depth. Use the Deepen support in the Teacher Guide (see below) to challenge children who work through the questions quickly. Here are some questions and ideas to encourage breadth and depth during specific parts of the lesson, or at any time (where no part of the lesson sequence is specified):

- **Discover**: 'Can you demonstrate your solution another way?'
- **Share**: Make sure every child is encouraged to give answers and engage with the discussion, not just the most confident.
- **Think together**: 'Can you model your answers using concrete materials? Can you explain your solution to a partner?'
- Practice: Allow all children to work through the full set of questions, so that they benefit from the logical sequence.
- **Reflect**: 'Is there another way of working out the answer? And another way?'
  'Have you found all the solutions?'
  'Is that always true?'
  'What's different between this question and that question? And what's the same?'

Note that the **Challenge** questions are designed so that all children can access and attempt them, if they have worked through the steps leading up to them. There may be some children in a given lesson who don't manage to do the **Challenge**, but it is not supposed to be a distinct task for a subset of the class. When you look through the lesson materials before teaching, think about what each question is specifically asking, and compare this with the key learning point for the lesson. This will help you decide which questions you feel it's essential for all children to answer, before moving on. You can at least aim for all children to try the **Challenge**!

Deepen activities and support

The Teacher Guide provides valuable support for each stage of the lesson. This includes Deepen tips for the guided and independent practice sections, which will help you provide extra stretch and challenge within your lesson, without having to organise additional tasks. If you have a Teaching Assistant, they can also make use of this advice. There are also suggestions for the lesson as a whole in the 'Going Deeper' section on the first page of the Teacher Guide section for that lesson. Every class is different, so you can always go a bit further in the direction indicated, if appropriate, and build on the suggestions given.

There is a Deepen activity for each unit. These are designed to follow on from the End of unit check, stretching children who have a firm understanding of the key learning from the unit. Children can work on them independently, which makes it easier for the teacher to facilitate the Strengthen activity for children who need extra support. Deepen activities could also be introduced earlier in the unit if the necessary work has been covered. The Deepen activities are on *ActiveLearn* on the Planning page for each unit, and also on the Resources page).

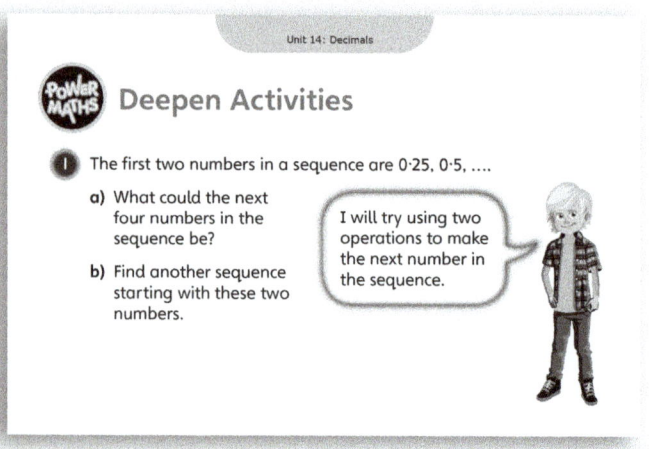

Using the questions flexibly to provide extra challenge

Sometimes you may want to write an extra question on the board or provide this on paper. You can usually do this by tweaking the lesson materials. The questions are designed to form a carefully structured sequence that builds understanding step by step, but, with careful thought about the purpose of each question, you can use the materials flexibly where you need to. Sometimes you might feel that children would benefit from another similar question for consolidation before moving on to the next one, or you might feel that they would benefit from a harder example in the same style. It should be quick and easy to generate 'more of the same' type questions where this is the case.

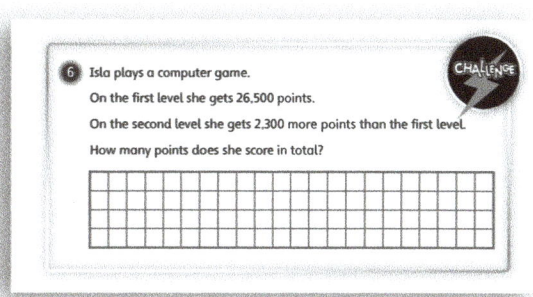

When you see a question like this one (from Unit 3, Lesson 3), it's easy to make harder examples to do afterwards if you need them. Any two numbers will generate a new multi-step problem, and you could make the numbers fiddly rather than round.

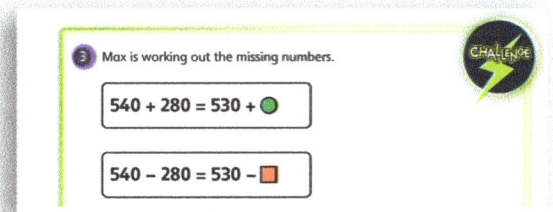

For this example (from Unit 3, Lesson 12), you could ask children to make up their own question(s) for a partner to solve. They could vary the number of digits and the operations. (In fact, for any of these examples you could ask early finishers to create their own question for a partner.)

Here's an example (from Unit 3, Lesson 4) where some of the journeys in the table feature as questions in the lesson, but others don't. Clearly there are any number of extra questions you could ask using the same table (the lesson includes multi-step journeys). Children could calculate journeys for their own itineraries, for example they could look for a multi-step journey that covers between 16,000 and 17,000 km, or they could even look for the shortest itinerary that visits all the destinations.

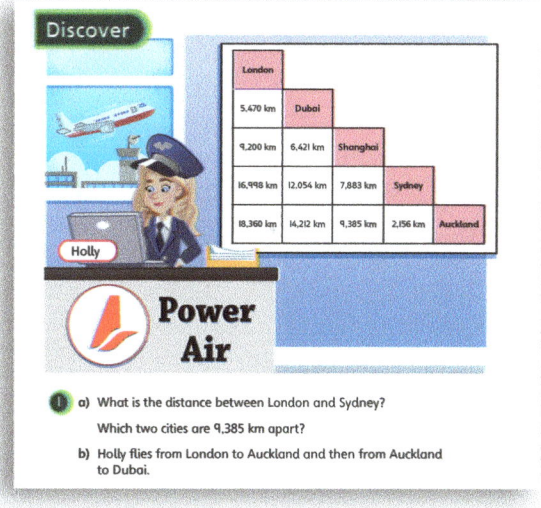

Besides creating additional questions, you should be able to find a question in the lesson that you can adapt into a game or open-ended investigation, if this helps to keep everyone engaged. It could simply be that, instead of answering 5 × 5 etc on the page, they could build a robot with 5 lots of 5 cubes.

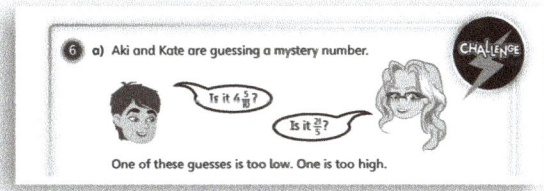

With a question like this (from Unit 5, Lesson 8), children could play a game where they have to guess their partner's mystery number, finding out each time if the guess is too high or too low.

See the bullets above for some general ideas that will help with 'opening out' questions in the books, e.g. 'Can you find all the solutions?' type questions.

Other suggestions

Another way of stretching children is through mixed ability pairs, or via other opportunities for children to explain their understanding in their own way. This is a good way of encouraging children to go deeper into the learning, rather than, for instance, tackling questions that are computationally more challenging but conceptually equivalent in level.

Using *Power Maths* with mixed age classes

Overall approaches

There are many variables between schools that would make it inadvisable to recommend a one-size-fits-all approach to mixed age teaching with *Power Maths*. These include how year groups are merged, availability of Teaching Assistants, experience and preference of teaching staff, range in pupil attainment across years, classroom space and layout, level of flexibility around timetables, and overall organisational structure (whether the school is part of a trust).

Some schools will find it best to timetable separate maths lessons for the different year groups. Others will aim to teach the class together as much as possible using the mixed age planning support on *ActiveLearn* (see the lesson exemplars for ways of organising lessons with strong/medium/weak correlation between year groups). There will also be ways of adapting these general approaches. For example, offset lessons where Year A start their lesson with the teacher, while Year B work independently on the practice from the previous lesson, and then start the next lesson with the teacher while Year A work independently; or teachers may choose to base their provision around the lesson from one year group and tweak the content up/down for the other group.

Key strategies for mixed age teaching

The mixed age teaching webinar on *ActiveLearn* provides advice on all aspects of mixed age teaching, including more detail on the ideas below.

Developing independence over time
Investing time in building up children's independence will pay off in the medium term.

Clear rationale
If someone asked, 'Why did you teach both Unit 3 and 4 in the same lesson/separate lessons?', what would your answer be?

Designing a lesson
1. Identify the core learning for each group
2. Identify any number skills necessary to access the core
3. Consider the flow of concepts and how one core leads to the other

Challenging all children
The questions are designed to build understanding step by step, but with careful thought about the purpose of each question you can tweak them to increase the challenge.

Multiple years combined
With more than two years together, teachers will inevitably need to use the resources flexibly if delivering a single lesson.

Enjoy the positives!

Comparison deepens understanding and there will be lots of opportunities for children, as well as misconceptions to explore. There is also in-built pre-teaching and the chance to build up a concept from its foundations. For teachers there is double the material to draw on! Mixed age teachers require a strong understanding of the progression of ideas across year groups, which is highly valuable for all teachers. Also, it is necessary to engage deeply with the lesson to see how to use the materials flexibly – this is recommended for all teachers and will help you bring your lesson to life!

List of practical resources

Year 5C Mandatory resources

Resource	Lesson
Base 10 equipment	**Unit 14** Lesson 13
Comparison bar models (blank)	**Unit 14** Lesson 10
Counters	**Unit 14** Lessons 12, 14
Cubes (cm³)	**Unit 17** Lessons 1, 2, 3
Measuring jug	**Unit 14** Lesson 8
Mirrors	**Unit 13** Lesson 5
Paper (isometric)	**Unit 13** Lesson 5 **Unit 17** Lesson 1
Paper (square dotted)	**Unit 13** Lesson 5
Paper (squared)	**Unit 13** Lesson 5
Paper circles (or card)	**Unit 12** Lesson 5
Place value counters	**Unit 14** Lessons 1, 2, 3, 5, 6, 8, 13, 15
Place value grids	**Unit 14** Lessons 12, 14
Protractors	**Unit 12** Lessons 2, 3, 4, 6, 8, 11
Rulers	**Unit 12** Lessons 2, 4, 7, 8, 9, 11 **Unit 13** Lesson 5
Weighing scales	**Unit 14** Lessons 8, 10

Year 5C Optional resources

Resource	Lesson	Resource	Lesson
2D shapes (cut out of squared paper)	**Unit 13** Lesson 3	Flashcards (for imperial and metric units of mass)	**Unit 16** Lesson 5
3D shapes (models of)	**Unit 12** Lesson 12	Flashcards (two-sided, for conversions)	**Unit 16** Lesson 5
Angle measurer (paper)	**Unit 12** Lesson 2		
Arrow spinner	**Unit 12** Lesson 1	Gattegno charts	**Unit 14** Lessons 12, 14
Bar models (blank)	**Unit 16** Lesson 9	Geoboards	**Unit 12** Lessons 8, 10, 11
Base 10 equipment	**Unit 14** Lessons 12, 14	Geostrips	**Unit 12** Lessons 8, 10
Bottles (with different capacities in ml)	**Unit 16** Lesson 2	Glue	**Unit 14** Lesson 1
Bottles and containers (different-sized)	**Unit 17** Lesson 1	Grids (first quadrant, blank)	**Unit 13** Lesson 1
		Masses (examples, such as 1 oz and 1 lb)	**Unit 16** Lesson 5
Boxes (1 large and 10 small)	**Unit 14** Lesson 15	Measuring items (various)	**Unit 17** Lesson 1
Building blocks	**Unit 14** Lesson 13	Measuring jugs	**Unit 16** Lesson 2
Calendars	**Unit 16** Lesson 7	Measuring jugs (imperial and metric)	**Unit 16** Lesson 6
Card strips	**Unit 12** Lesson 11		
Chalk	**Unit 13** Lesson 3	Measuring tape (cm)	**Unit 16** Lesson 10
Clock faces (analogue, with movable hands)	**Unit 16** Lessons 8, 10	Measuring wheels	**Unit 16** Lesson 1
		Metre rulers (marked in mm)	**Unit 16** Lesson 2
Clocks (analogue and digital)	**Unit 16** Lesson 7	Metre sticks	**Unit 14** Lessons 4, 7
Construction materials (such as strips of card)	**Unit 12** Lesson 8	Milk cartons	**Unit 16** Lesson 6
		Mini-figures	**Unit 12** Lesson 1
Containers (for liquid, such as water bottles)	**Unit 16** Lesson 6	Mini-whiteboards	**Unit 16** Lesson 8
		Mirrors	**Unit 13** Lesson 6
Coordinate grids (printed)	**Unit 13** Lesson 2	Number cards	**Unit 15** Lesson 4
Cubes	**Unit 14** Lesson 13	Number cards (0 to 9)	**Unit 14** Lesson 7
Cubes (interlinking)	**Unit 16** Lesson 7	Number cards (decimal)	**Unit 14** Lesson 11
Dice (two different colours)	**Unit 13** Lesson 4	Number lines	**Unit 14** Lesson 11 **Unit 16** Lesson 9
Digit cards	**Unit 16** Lessons 1, 3, 9		
Flashcards	**Unit 16** Lessons 9, 10		

Year 5C Optional resources – *continued*

Resource	Lesson
Paper	**Unit 14** Lesson 1
Paper (isometric)	**Unit 17** Lesson 2
Paper (large sheets)	**Unit 16** Lesson 2
Paper (squared dotted)	**Unit 12** Lesson 11 **Unit 13** Lesson 3 **Unit 17** Lesson 2
Paper (squared)	**Unit 12** Lesson 8 **Unit 13** Lessons 3, 4, 6
Paper money	**Unit 14** Lesson 6
Paper money (or plastic)	**Unit 14** Lesson 5
Paper rectangles	**Unit 16** Lesson 6
Paper semicircles	**Unit 12** Lesson 6
Paper squares and rectangles	**Unit 12** Lesson 7
Paper strips	**Unit 12** Lesson 11 **Unit 16** Lessons 8, 10
Picture (of items for sale in the shop)	**Unit 14** Lesson 5
Place value counters	**Unit 14** Lessons 7, 11
Place value grids	**Unit 14** Lesson 14 Unit 16 Lessons 3, 9
Place value grids (printed)	**Unit 14** Lessons 13, 15
Plant growth chart	**Unit 14** Lesson 11
Protractors	**Unit 12** Lessons 5, 10
Ramp	**Unit 12** Lesson 2
Reflections (images of)	**Unit 13** Lesson 5
Rice and pasta (dried)	**Unit 16** Lesson 10
Rulers	**Unit 12** Lesson 5 **Unit 16** Lesson 10 **Unit 17** Lesson 2
Rulers (selection, showing different metric units)	**Unit 16** Lesson 3
Rulers (some marked in cm, some in mm)	**Unit 16** Lesson 2
Rulers and tape measures (showing imperial and metric units)	**Unit 16** Lesson 4
Scales	**Unit 16** Lesson 1
Scales (g)	**Unit 16** Lesson 10
Scissors (safety)	**Unit 12** Lesson 6
Split pins	**Unit 12** Lessons 8, 11
Sticks	**Unit 12** Lesson 9
Sticky labels	**Unit 16** Lessons 4, 5
String	**Unit 16** Lesson 10
Suitcase (with a selection of items to fill it)	**Unit 16** Lesson 1
Thermometers	**Unit 15** Lessons 1, 2, 3
Thermometers (laminated paper)	**Unit 15** Lesson 4
Timetables (different kinds)	**Unit 16** Lesson 8
Torch	**Unit 12** Lesson 12
Toy items (fruit, to buy)	**Unit 14** Lesson 6
Toy train tracks	**Unit 14** Lesson 1
Weighing scales	**Unit 14** Lessons 4, 9 **Unit 16** Lesson 5
Year planners	**Unit 16** Lesson 7

Getting started with *Power Maths*

As you prepare to put *Power Maths* into action, you might find the tips and advice below helpful.

STEP 1: Train up!

A practical, up-front full day professional development course will give you and your team a brilliant head-start as you begin your *Power Maths* journey. You will learn more about the ethos, how it works and why.

STEP 2: Check out the progression

Take a look at the yearly and termly overviews. Next take a look at the unit overview for the unit you are about to teach in your Teacher Guide, remembering that you can match your lessons and pacing to match your class.

STEP 3: Explore the context

Take a little time to look at the context for this unit: what are the implications for the unit ahead? (Think about key language, common misunderstandings and intervention strategies, for example.) If you have the online subscription, don't forget to watch the corresponding unit video.

STEP 4: Prepare for your first lesson

Familiarise yourself with the objectives, essential questions to ask and the resources you will need. The Teacher Guide offers tips, ideas and guidance on individual lessons to help you anticipate children's misconceptions and challenge those who are ready to think more deeply.

STEP 5: Teach and reflect

Deliver your lesson — and enjoy!

Afterwards, reflect on how it went… Did you cover all five stages? Does the lesson need more time? How could you improve it? What percentage of your class do you think mastered the concept? How can you help those that didn't?

Unit 12
Geometry – properties of shapes

Mastery Expert tip! 'When teaching this unit, I encourage children to explore the properties of angles physically and visually, by showing angles using their arms, or by drawing sketches of their ideas to justify their reasoning. It is important to give them opportunities to link their calculations and measurements to the meaning in physical terms.'

Don't forget to watch the Unit 12 video!

WHY THIS UNIT IS IMPORTANT

This unit is important because it develops geometric reasoning alongside key measurement skills. Many children find the protractor a difficult tool to master, so it is important that they practise in a meaningful context. The skill of measurement is developed alongside reasoning and calculating to allow children to make predictions, to check their calculations and to discuss the properties of shapes and angles that they are going to explore.

WHERE THIS UNIT FITS

→ Unit 11: Graphs and tables
→ **Unit 12: Geometry – properties of shapes**
→ Unit 13: Geometry – position and direction

This unit builds on children's work from Year 4 where they identified properties of angles. They will also be given the grounding to work on the following unit where they will be asked to reason about the lengths and angles of quadrilaterals.

Before they start this unit, it is expected that children:
- know the difference between clockwise and anticlockwise turns
- understand the concept of an angle as a measure of turn
- can structure calculations with more than one step
- are familiar with right angles.

ASSESSING MASTERY

Children who have mastered this unit will be able to measure angles accurately in degrees, create given angles and calculate missing angles. They will also be able to justify logical reasoning about missing angles and lengths based on the properties of shapes and angles that they have learnt.

COMMON MISCONCEPTIONS	STRENGTHENING UNDERSTANDING	GOING DEEPER
Children may find the protractor difficult to use if the angle is not presented with a horizontal 'base'.	Encourage children to rotate the page to help them position the protractor. This encourages them to think of how to manipulate a problem in a way that supports understanding.	Explore the possibilities of drawing shapes and designs with greater accuracy by using the protractor to measure to the nearest degree. Children may create their own designs.
Once children master the skill of measurement, they may revert to using it at times when it could be more efficient and accurate to calculate based on known properties of the shape.	Encourage children to look for what is known and what is unknown before deciding how best to tackle the problem.	Children can investigate general statements about the angles around a point and on a straight line.

Unit 12: Geometry – properties of shapes

UNIT STARTER PAGES

Use these pages to introduce the focus of the unit to children. You can use the characters to explore different ways of working too.

STRUCTURES AND REPRESENTATIONS

Angle diagrams: Use these to help children justify reasoning based on the fractions of a turn.

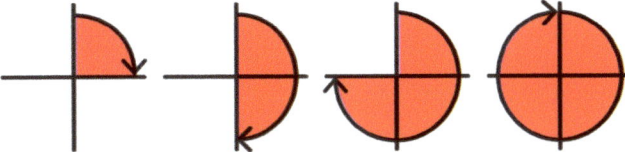

Right angles: The properties of right angles will recur and be important as the unit progresses.

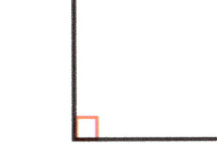

Protractor: Children will spend much of the unit developing their understanding of angles through the use of a protractor to measure and draw acute and obtuse angles.

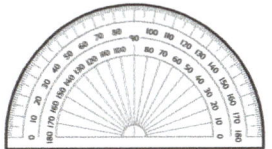

KEY LANGUAGE

There is some key language that children will need to know as part of the learning in this unit.

- angle, turn
- whole turn, half turn, quarter turn
- acute angle, right angle, obtuse angle, reflex angle
- degrees (°)
- 90 degrees
- 180 degrees, 360 degrees
- interior angle
- protractor
- clockwise, anticlockwise
- perpendicular, parallel, regular, irregular
- top view, side view, plan view

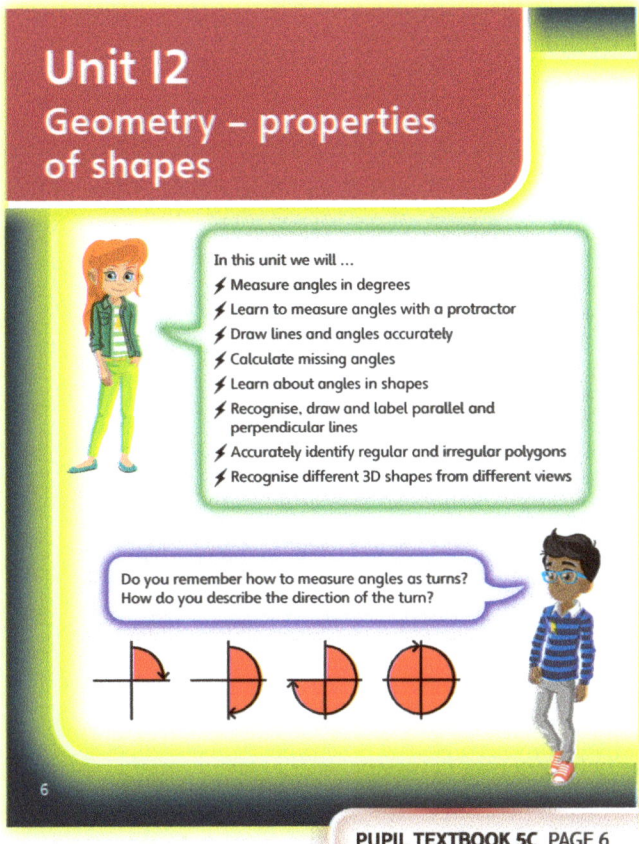

PUPIL TEXTBOOK 5C PAGE 6

PUPIL TEXTBOOK 5C PAGE 7

Unit 12: Geometry – properties of shapes, Lesson 1

Understand and use degrees

Learning focus
In this lesson, children will learn to use degrees as a unit for measure of turn, focusing on common angles of 90°, 180°, 270° and 360° and linking with their understanding of whole turns, half turns and quarter turns. They will also encounter turns in multiples of 45°.

Before you teach
- Can children define 90-degree angles from their work in Years 3 and 4?
- Are children confident in identifying angles?

NATIONAL CURRICULUM LINKS

Year 5 Geometry – properties of shapes

Know angles are measured in degrees: estimate and compare acute, obtuse and reflex angles.

Identify:
- angles at a point and one whole turn (total 360°)
- angles at a point on a straight line and $\frac{1}{2}$ a turn (total 180°)
- other multiples of 90°.

ASSESSING MASTERY

Children can describe and follow turns in multiples of 90°, and also in multiples of 45°. They can explain their reasoning based on fractions of a whole turn.

COMMON MISCONCEPTIONS

Children may be confused by the fact that degrees are also used as a unit of measure for temperature. Ask:
- Have you heard 'degrees' used before?

STRENGTHENING UNDERSTANDING

Ask children to act out the turns themselves, by standing on a point with an arm stretched out in front. They should then perform different turns by rotating on the spot, noticing the sweep of their arm as they rotate.

GOING DEEPER

The lesson challenges children to recognise and explore angles in an eight-point grid, which requires a deep understanding of angles with different start and end positions, and in different orientations. Children can extend this activity by adding new turns to the table or by making up their own similar problem for a partner to solve.

KEY LANGUAGE

In lesson: degrees (°), clockwise, anticlockwise, half turn, quarter turn, whole turn, right angle, acute angle, obtuse angle, reflex angle

STRUCTURES AND REPRESENTATIONS

Angle diagrams

RESOURCES

Optional: arrow spinner, mini-figures

 In the eTextbook of this lesson, you will find interactive links to a selection of teaching tools.

Quick recap
Ask children to stand and face a particular direction. Ask them to make different turns, for example: a half turn, a quarter turn right, a quarter turn left, clockwise, anticlockwise, a three-quarter turn. Ask them to look at each other and notice which way they are facing.

Unit 12: Geometry – properties of shapes, Lesson 1

Discover

WAYS OF WORKING Pair work

ASK

- Question 1 a): *Who is Lexi facing to begin with?*
- Question 1 a): *Have you heard of a 180-degree turn before? What do you think 180° means? What about a 360-degree turn?*
- Question 1 b): *What direction could Lexi turn?*

IN FOCUS Questions 1 a) and b) are the first time that children encounter the word 'degrees' as a unit for the measure of turn. They may very well have heard phrases such as 'do a 180' or 'turn 360' or '90-degree angle' in other contexts such as sport, design and technology, gymnastics, skateboarding or video games. These questions cement this vocabulary within geometry. Make sure that children understand that Lexi will be facing the opposite direction to the one shown in the picture when she makes a 180° turn. In question 1 b), Lexi could turn either clockwise or anticlockwise so could be facing the bench or the flower bed. See **Share** to help with this.

PRACTICAL TIPS A practical version of this problem could be acted out by children very simply, perhaps in the school hall, the playground or a cleared space in the classroom.

All children could practise turning by different fractions of a turn, especially to revise the difference between clockwise and anticlockwise turns. It may also help to use a mini-figure to act out the turns or to use an arrow spinner.

ANSWERS

Question 1 a): Lexi will be facing Lee.

Question 1 b): She could be facing the flower bed or the bench after a 90° turn.

PUPIL TEXTBOOK 5C PAGE 8

Share

WAYS OF WORKING Whole class teacher led

ASK

- Question 1 a): *What do you notice about the numbers 180 and 360?*
- Question 1 a): *Does it matter whether you turn 180° clockwise or anticlockwise? Why?*
- Question 1 b): *What is another name for a 90-degree turn?*

IN FOCUS In this lesson, children are learning how to use degrees as an accurate measure of turn. In question 1 a), the key angles 180° and 90° are described. The focus is on understanding the measure of degrees for common angles in terms of the fractions of a whole turn. Children may ask why 360° represents a whole turn instead of perhaps 100 or 1? One reason is that 360 divides exactly into more fractions than 100. Some children might use division to explore this at a later stage.

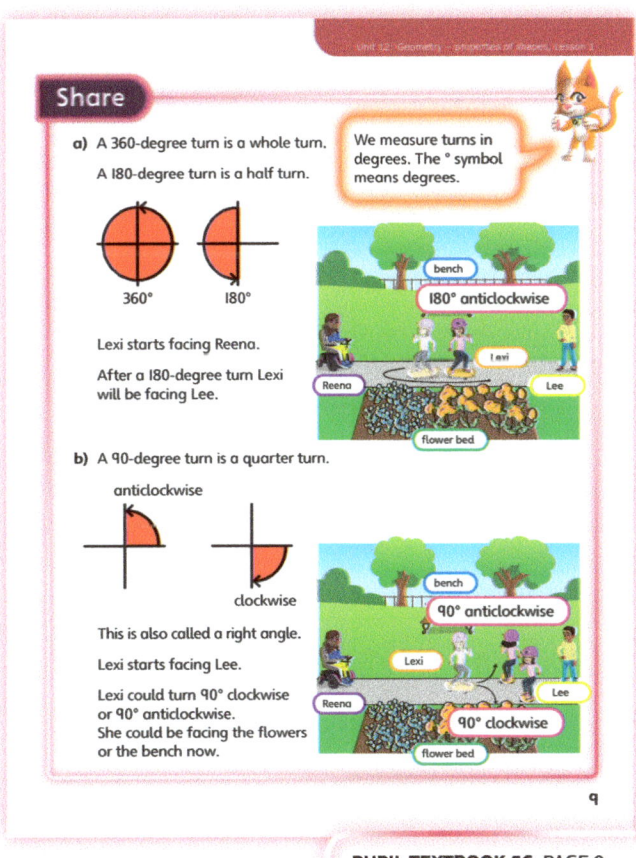

PUPIL TEXTBOOK 5C PAGE 9

45

Unit 12: Geometry – properties of shapes, Lesson 1

Think together

WAYS OF WORKING Whole class teacher led (I do, We do, You do)

ASK

- Question ②: *What is the starting position? What is the turn? What is the end position?*
- Question ③: *Show me a 90-degree turn. What is half of the 90-degree turn?*

IN FOCUS Question ③ introduces an eight-point grid, so that children explore the turns in different orientations, and also includes an understanding of turns that are 45° and 135°. This challenges them to solve problems where different pieces of information are missing, such as start direction, turn in degrees or end direction. This requires a deep understanding rather than simple procedural understanding.

STRENGTHEN Encourage children to use a pencil, a mini-figure or an arrow to act out the turn on the page, as a way of being able to envisage the turn from the perspective of the problem.

DEEPEN Children could make up their own problems involving turns and then swap with a partner or share with the class.

ASSESSMENT CHECKPOINT Answers to question ③ will indicate whether or not children are using the terms clockwise and anticlockwise correctly.

ANSWERS

Question ①: Lexi will turn 360°, a whole turn, so will be back facing Lee again.

Question ②: Amelia has turned 270° anticlockwise.

Question ③ a):

Start	Turn	Finish
facing B	180°	facing F
facing A	90° anticlockwise	facing G
facing E	90° anticlockwise	facing C
facing G	90° clockwise	facing A
facing G	45° clockwise	facing H
facing a	45° clockwise	facing B

Question ③ b): 135° clockwise or 215° anticlockwise.

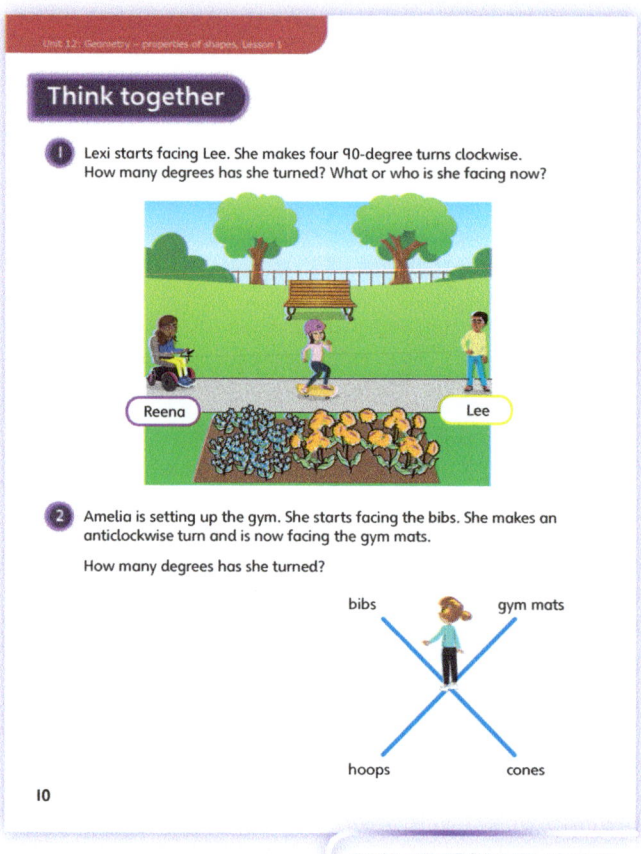

PUPIL TEXTBOOK 5C PAGE 10

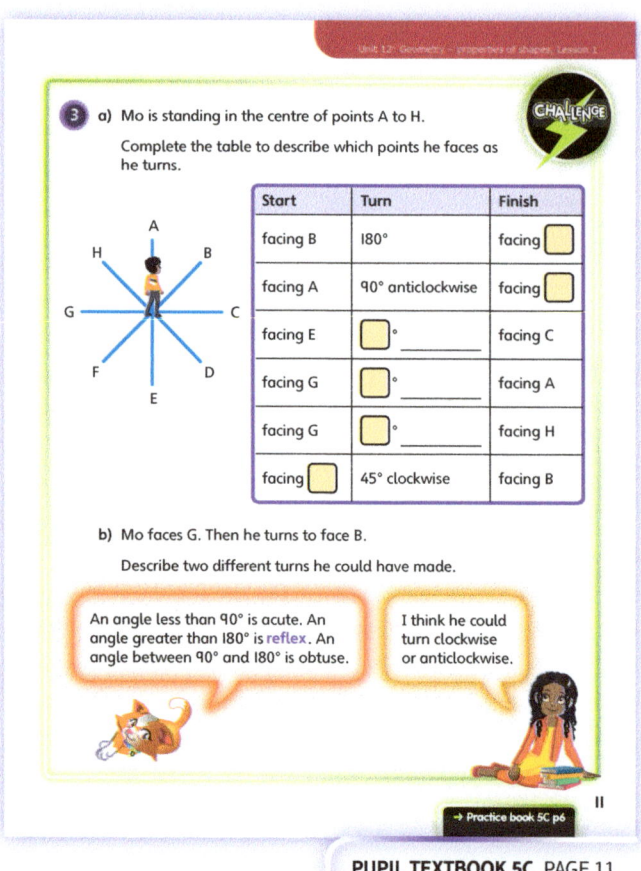

PUPIL TEXTBOOK 5C PAGE 11

46

Unit 12: Geometry – properties of shapes, Lesson 1

Practice

WAYS OF WORKING Independent thinking

IN FOCUS Question ❶ gives practice for recognition of common turns measured in degrees and requires children to recognise the turns in different orientations.

Question ❷ deepens understanding by asking for different aspects of the information, including start position, turn and end position.

Question ❸ looks at describing turns on an eight-point grid, including 45-degree turns or multiples of 45°.

STRENGTHEN Children could use mini-figures or even act out the turns using role play on the playground or in the school hall.

DEEPEN Extend question ❹ by asking children to find new combinations using the buttons if the robot moves from two different points on the grid.

ASSESSMENT CHECKPOINT Accurate answers to questions ❷ and ❸ show good understanding of the key skills. Look at incorrect answers to check whether children have a secure knowledge of clockwise and anticlockwise. Question ❸ asks children to calculate turns from different starting positions – wrong answers here might indicate that children have not fully grasped this visualisation.

ANSWERS Answers for the **Practice** part of the lesson can be found in the *Power Maths* online subscription.

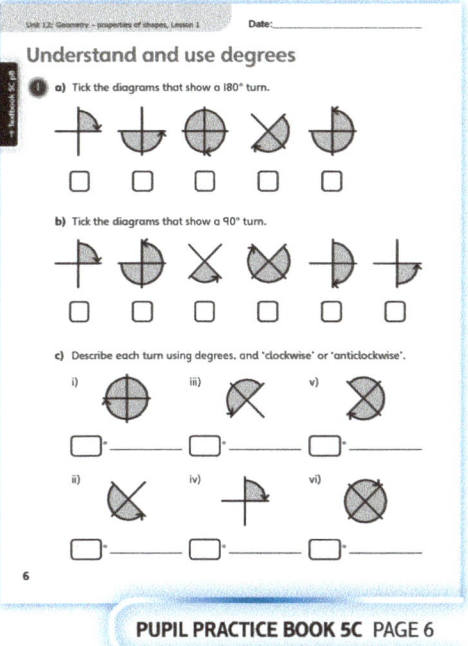

PUPIL PRACTICE BOOK 5C PAGE 6

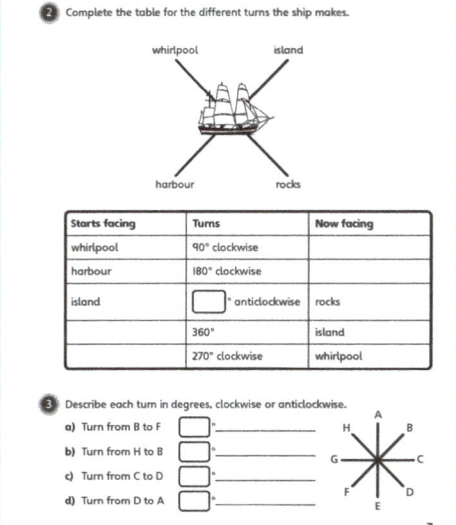

PUPIL PRACTICE BOOK 5C PAGE 7

Reflect

WAYS OF WORKING Independent thinking

IN FOCUS This **Reflect** question allows children to draw their own diagrams of the key angles covered in the lesson. This ensures they understand how each angle looks and how they correspond to each other in terms of size.

ASSESSMENT CHECKPOINT Children's diagrams should show understanding of the relationship between the different angles.

ANSWERS Answers for the **Reflect** part of the lesson can be found in the *Power Maths* online subscription.

After the lesson ⏸

- Could children follow turns from different starting positions, rather than always from a north position?
- Were children confident explaining how the fractions of a turn relate to 45-, 90- and 180-degree turns?

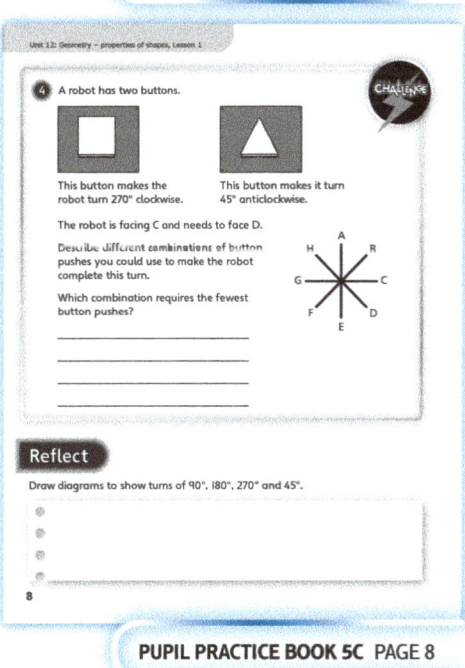

PUPIL PRACTICE BOOK 5C PAGE 8

Unit 12: Geometry – properties of shapes, Lesson 2

Measure acute angles

Learning focus
In this lesson, children will be introduced to the protractor and how to use it to measure acute angles.

Before you teach
- Are children confident with the terminology of acute and obtuse angles?
- How confidently do children measure angles? Do they measure accurately?

NATIONAL CURRICULUM LINKS

Year 5 Geometry – properties of shapes

Know angles are measured in degrees: estimate and compare acute, obtuse and reflex angles.

ASSESSING MASTERY

Children can measure acute angles in different orientations and are able to explain which scale to use for a given angle.

COMMON MISCONCEPTIONS

Children may need support to measure angles where there is not a horizontal 'base' line to orient the protractor. Ask:
- *What would you do first to measure this angle?*

Children may find it difficult to know how to place the protractor accurately, with the cross-hairs precisely on the turn and the base lined up with one of the angles. Ask:
- *How would you place the protractor to measure this angle?*

STRENGTHENING UNDERSTANDING

Encourage children to decide whether an angle to be measured is acute or obtuse by making a visual judgement and then use this to support their choice of scale on the protractor.

GOING DEEPER

Challenge children to measure precisely and to explain their reasoning when selecting a measure between multiples of 10 on a protractor scale.

KEY LANGUAGE

In lesson: protractor, degrees (°), acute, angle, scale

Other language to be used by the teacher: obtuse

STRUCTURES AND REPRESENTATIONS

Angle diagrams

RESOURCES

Mandatory: protractor, ruler

Optional: ramp, paper angle measurer

 In the eTextbook of this lesson, you will find interactive links to a selection of teaching tools.

Quick recap

Repeat the activity from the previous lesson. Ask children to face the front of the class and then ask them to make 90-, 180-, 270- and 360-degree turns. Use words like clockwise and anticlockwise. You may also want to ask children what turns these are equivalent to – for example, quarter turn, half turn and so on).

Unit 12: Geometry – properties of shapes, Lesson 2

Discover

WAYS OF WORKING Pair work

ASK

- Question 1 a): *Does the angle look greater than or less than 90°? What about 45°?*
- Question 1 a): *Can you estimate the angle the ramp makes with the table?*
- Question 1 a): *What tool could you use to measure the angle? Would a ruler help or does it require a different tool?*

IN FOCUS Questions 1 a) and b) are important as this is the first introduction to the protractor as a tool. Many children find it very challenging to use a protractor to measure angles accurately and so may need plenty of practice over the course of this lesson. Encourage them to make an estimate, based on their understanding of 90° and 45°.

PRACTICAL TIPS This is a version of a simple experiment that children can try in class. They could lift a ramp to a different angle and then test at which point an object begins to slip. However, it is difficult to measure an angle exactly using a practical ramp. It may make sense to create an angle measure from a large sheet of paper, which can be held up against the ramp, rather than having to use a protractor on a practical experiment. The angle measurer could be marked in multiples of 10°, from 0° up to 90°.

Children could then be introduced to the protractor as a tool for measuring angles precisely on a drawing.

ANSWERS

Question 1 a): The ramp is now at an angle of 30°.

Question 1 b): Amal's mistake is reading the wrong scale.

PUPIL TEXTBOOK 5C PAGE 12

Share

WAYS OF WORKING Whole class teacher led

ASK

- Question 1 a): *Where should you place the protractor?*
- Question 1 a): *Where is the turning point of the angle?*
- Question 1 b): *Which scale on the protractor is correct for this angle?*

IN FOCUS The key point of question 1 a) is to learn the correct procedure for using the protractor accurately. Make sure that children know that they need to:
- line up the base line with one of the lines of the angle
- line up the cross hair with the point of the angle
- read the correct scale on the protractor

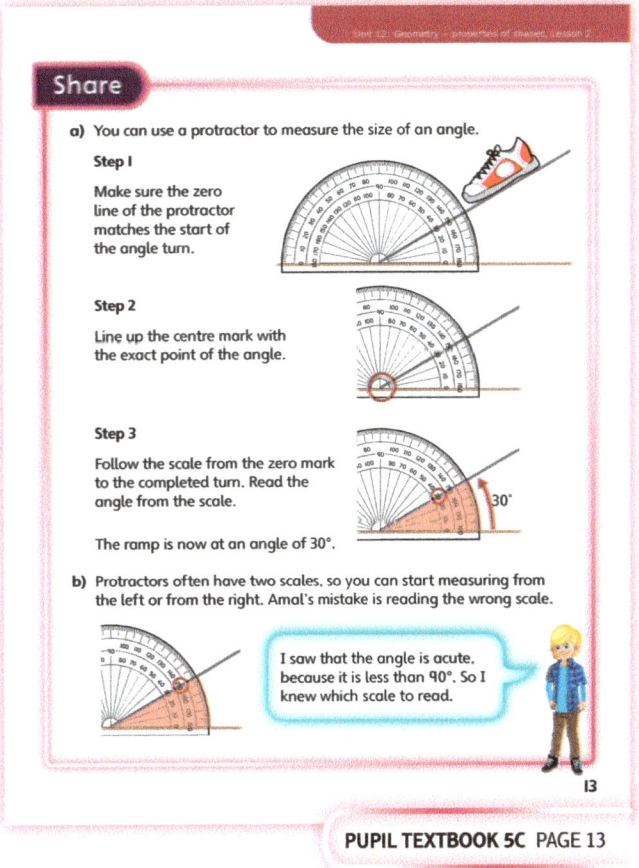

PUPIL TEXTBOOK 5C PAGE 13

49

Unit 12: Geometry – properties of shapes, Lesson 2

Think together

WAYS OF WORKING Whole class teacher led (I do, We do, You do)

ASK
- Questions ① and ②: *Show me how you would find the starting position for measuring this angle.*
- Question ②: *How would you decide which scale to use to measure these angles?*

IN FOCUS Question ② introduces angles in different orientations from a horizontal line and includes an angle that is not a multiple of 10°.

Question ③ includes angles in different orientations that are not multiples of 10 and are angles inside a shape. It also encourages children to question whether the orientation of an angle affects the size of the angle.

STRENGTHEN Encourage children to rotate the page when measuring angles in different orientations.

DEEPEN Ask children to measure to the nearest degree. This requires them to ensure that they are following the three steps for correct measuring to get a precise measurement.

ASSESSMENT CHECKPOINT If children are able to measure the angles in the triangles in question ③ accurately enough to discover that the triangles are similar, they have accurate measuring skills.

ANSWERS

Question ①: Between 30° and 70°.

Question ②: a = 60°
b = 50°
c = 45°

Question ③: Approximately two angles of 75° and one angle of 30° for both triangles.

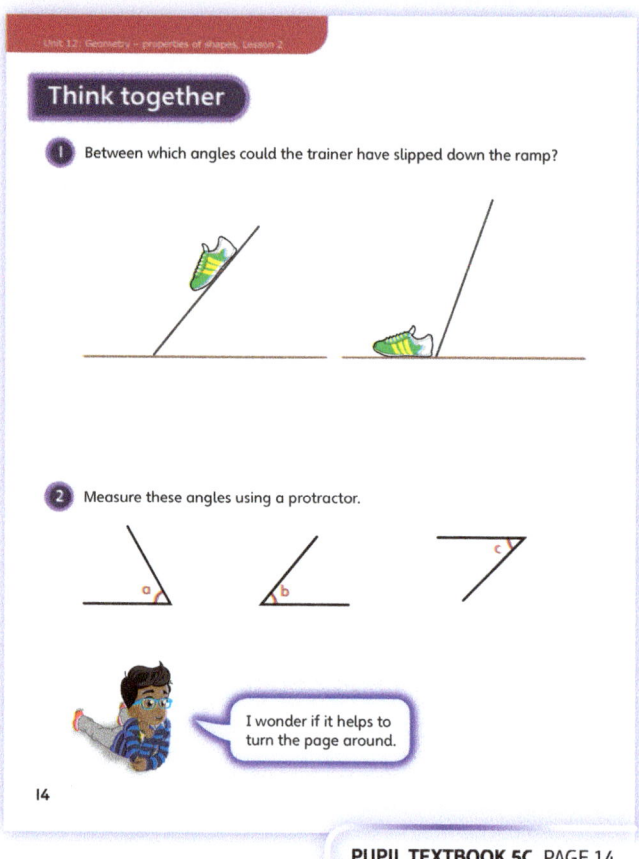

PUPIL TEXTBOOK 5C PAGE 14

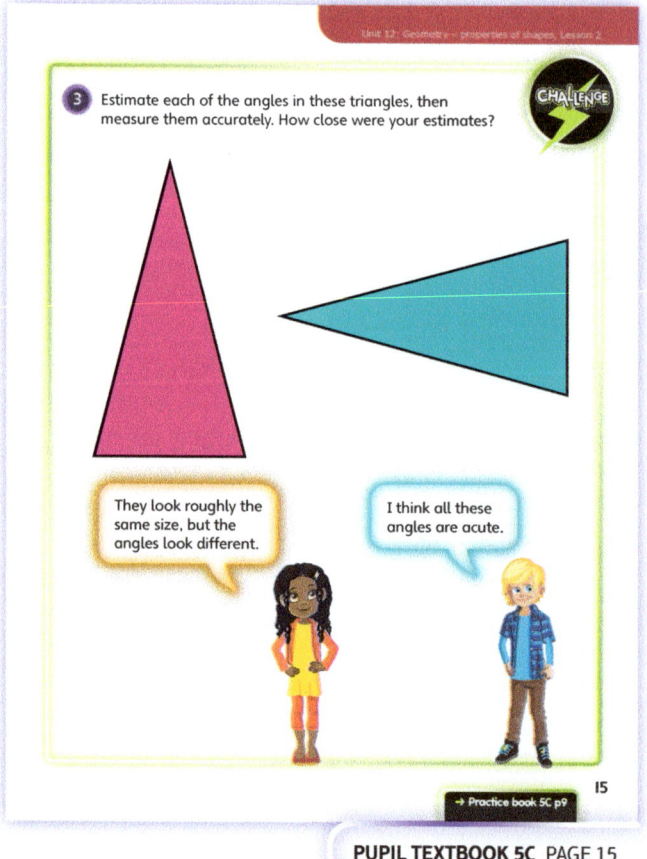

PUPIL TEXTBOOK 5C PAGE 15

Unit 12: Geometry – properties of shapes, Lesson 2

Practice

WAYS OF WORKING Independent thinking

IN FOCUS Question ❶ practises the skill of measuring angles where there is a horizontal base, and the first two examples are multiples of 10° and the protractor is already placed correctly. This pictorially reinforces how children should use a protractor before reaching question ❷ where they need to use a protractor themselves.

STRENGTHEN For question ❺, suggest that children can extend one line by drawing with a pencil and ruler to allow them to measure more accurately. Encourage children to do this very carefully so that the line they draw is directly straight with the short line ray.

DEEPEN Can children draw angles using a protractor for a partner to accurately measure?

THINK DIFFERENTLY Question ❹ shows two common errors when placing the protractor for children to spot and explain. Richard has read the wrong scale on the protractor; Emma has not lined either of the angle's lines with the baseline of the protractor. By explaining these misconceptions, children will become more aware of avoiding this when doing the measuring themselves.

ASSESSMENT CHECKPOINT Questions ❷ and ❸ will show that children can confidently use a protractor themselves to measure angles accurately. Responses to question ❹ will reveal any potential misconceptions children have when using protractors.

ANSWERS Answers for the **Practice** part of the lesson can be found in the *Power Maths* online subscription.

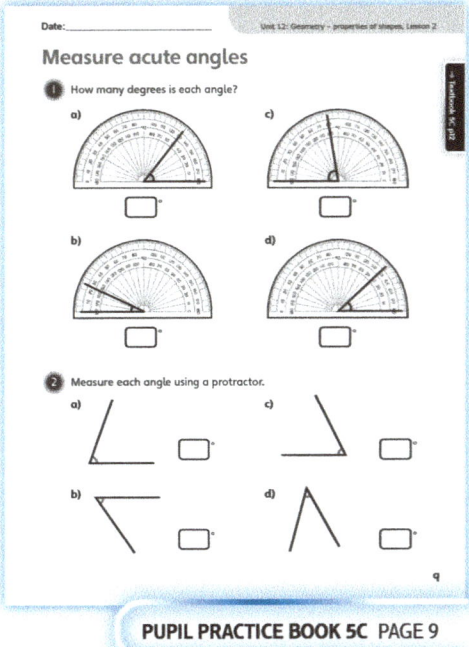

PUPIL PRACTICE BOOK 5C PAGE 9

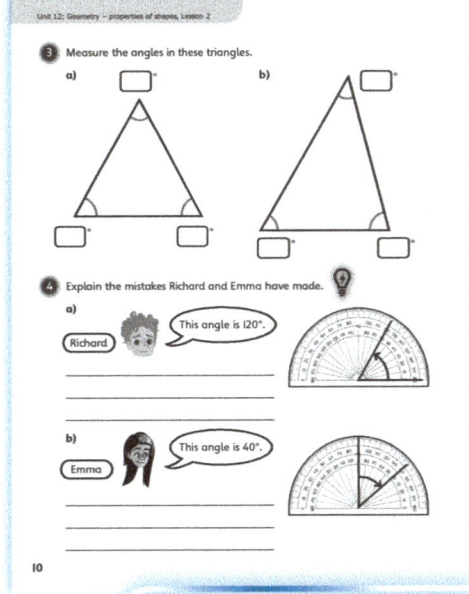

PUPIL PRACTICE BOOK 5C PAGE 10

Reflect

WAYS OF WORKING Independent thinking

IN FOCUS The practical skill of measuring is procedural in nature and this challenges children to explain the process in their own words.

ASSESSMENT CHECKPOINT Can children explain how each step ensures that they measure accurately?

ANSWERS Answers for the **Reflect** part of the lesson can be found in the *Power Maths* online subscription.

After the lesson ⏸
- Were children able to place the protractor accurately when angles were presented in different orientations?
- Can children explain which scale to use on the protractor to measure accurately, by considering the zero line and also in terms of acute and obtuse angles?

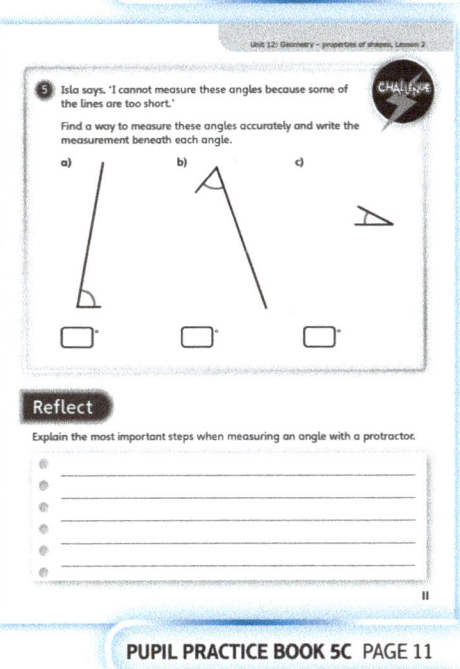

PUPIL PRACTICE BOOK 5C PAGE 11

Unit 12: Geometry – properties of shapes, Lesson 3

Measure angles up to 180°

Learning focus
In this lesson, children will continue to develop their protractor skills by measuring obtuse angles accurately.

Before you teach
- Are children familiar with the difference between acute and obtuse angles?
- Can children confidently use a protractor?

NATIONAL CURRICULUM LINKS

Year 5 Geometry – properties of shapes

Know angles are measured in degrees: estimate and compare acute, obtuse and reflex angles.

Draw given angles, and measure them in degrees (°).

ASSESSING MASTERY

Children can measure obtuse angles accurately and can select the correct scale to use by considering the size of the angle in relation to 90°.

COMMON MISCONCEPTIONS

Children may need support to see which scale to use when measuring a given angle. Ask:
- Would you use the inside or the outside scale for this angle? How do you know?

Children may need support to measure accurately when the angle is not a multiple of 10. Ask:
- What is this angle to the nearest 10°?

STRENGTHENING UNDERSTANDING

Encourage children to count from the zero mark on the protractor and follow the scale in multiples of 10 until they reach the completed angle.

GOING DEEPER

Challenge children to use their knowledge of acute and obtuse angles to make predictions on the angle size then justify or check answers based on their reasoning.

KEY LANGUAGE

In lesson: obtuse, angle, protractor, greatest, smallest, hexagon, scale

Other language to be used by the teacher: acute, interior angle, right angle, degrees (°)

STRUCTURES AND REPRESENTATIONS

Angle diagrams, 2D shapes

RESOURCES

Mandatory: protractors

 In the eTextbook of this lesson, you will find interactive links to a selection of teaching tools.

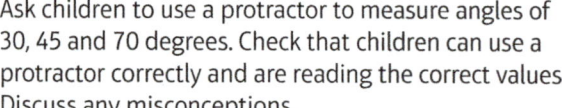

Quick recap
Ask children to use a protractor to measure angles of 30, 45 and 70 degrees. Check that children can use a protractor correctly and are reading the correct values. Discuss any misconceptions.

Unit 12: Geometry – properties of shapes, Lesson 3

Discover

WAYS OF WORKING Pair work

ASK

- Question 1 a): *Is the angle less than or greater than 90°?*
- Questions 1 a) and b): *What are the main things to remember when measuring angles with a protractor?*
- Question 1 b): *Can you make a reasoned estimate of that angle?*

IN FOCUS Questions 1 a) and b) prompt children to think about measuring angles inside a shape (interior angles) and to notice that the angles are greater than 90°. Children should make estimates of the angles and consider different justifications for those estimates. Encourage children to discuss how this relates to their learning in the previous lesson, especially with reference to any difficulties they encountered, such as knowing which scale to use or placing a protractor accurately in order to measure the angle.

PRACTICAL TIPS The context in the **Discover** image can be recreated in the classroom and could be modelled as a game of pass the parcel or passing a ball around the group. Children should make exaggerated turn movements, like robots passing boxes with arms straightened out in front. The idea is to sense that the turn they are making is the angle to be measured and to feel that it is greater than 90°.

ANSWERS

Question 1 a): Mo turns an angle of 120°.

Question 1 b): Emma turns an angle of 120°.

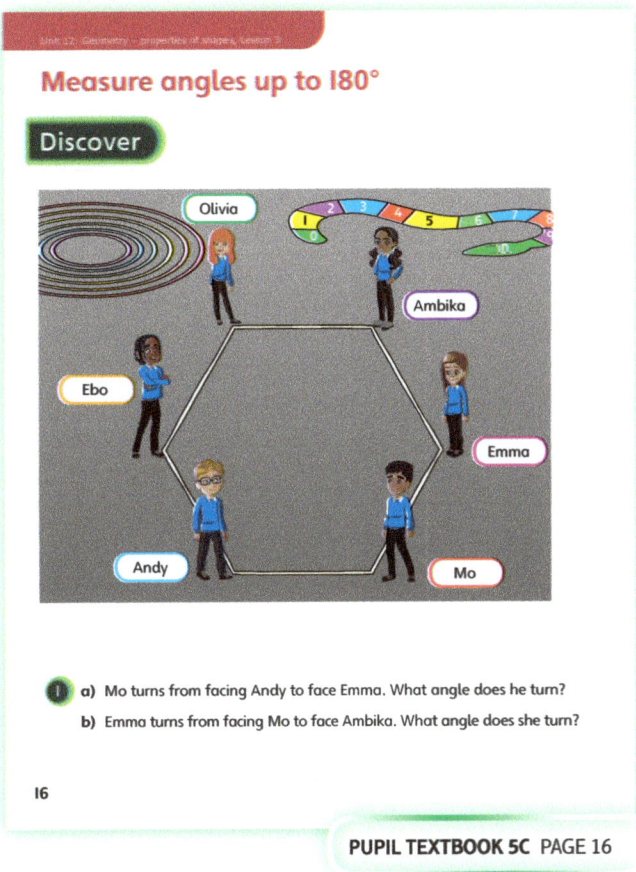

PUPIL TEXTBOOK 5C PAGE 16

Share

WAYS OF WORKING Whole class teacher led

ASK

- Question 1 a): *Where is the correct point to place the protractor?*
- Question 1 a): *Which scale would you use to measure this angle?*

IN FOCUS The main focus is to recognise that the angle is obtuse and use this to support a judgement about which scale to use when measuring with the protractor.

Children may explore whether all angles are equal in this shape and discuss whether this is true for all hexagons.

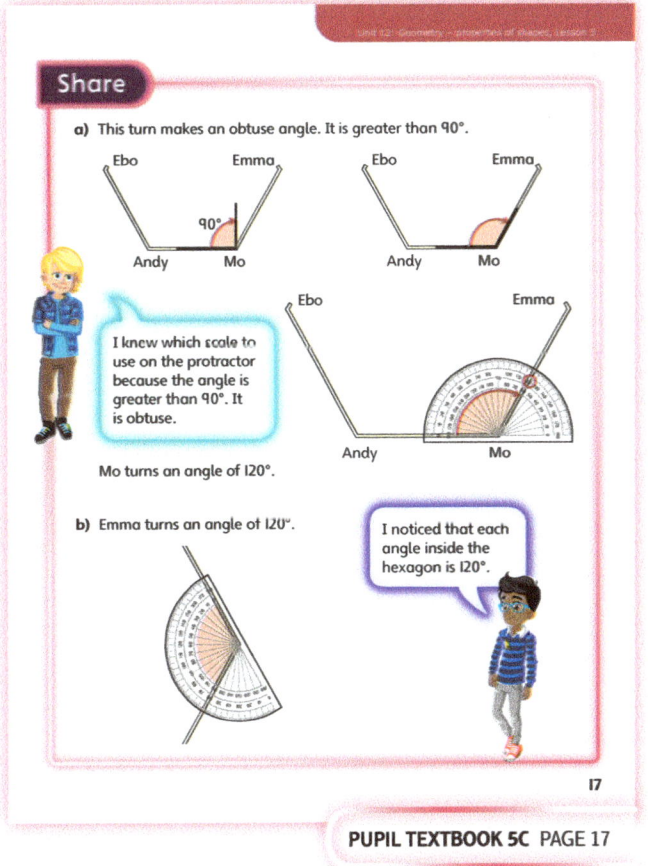

PUPIL TEXTBOOK 5C PAGE 17

53

Unit 12: Geometry – properties of shapes, Lesson 3

Think together

WAYS OF WORKING Whole class teacher led (I do, We do, You do)

ASK

- Question ❶: *Where would you place the protractor for this angle?*
- Question ❷: *Would it help to rotate the page?*
- Question ❷: *Can you make a reasoned estimate before measuring? How can you read the angle correctly if it is between two 10-degree marks?*

IN FOCUS Question ❶ is important as it provides basic practice at reading angles in different orientations. It refreshes understanding of working with protractors from the last lesson.

Question ❷ encourages children to estimate first. It can be partially solved by reasoning about the acute and right angles, then measuring to check the two obtuse angles.

Question ❸ challenges children to measure the acute angle from A to C. Can they find the obtuse angle that Emma would turn through if she turned in the opposite direction?

STRENGTHEN Encourage children to rotate the page when measuring angles in different orientations. Support children with placing the protractor by modelling how to find the 'baseline' first, then slide to line up the cross-hair.

DEEPEN With question ❸, challenge children to explain how the angle for a turn from C to A is the same as the turn from C to F, even though F is further away than A.

ASSESSMENT CHECKPOINT Question ❷ will show whether children are confident measuring angles at different orientations as well as showing their reasoning skills when judging angle size.

ANSWERS

Question ❶: a) 120°

Question ❶: b) 170°

Question ❷: d, a, b, c

Question ❸ a): Amelia turns 140–145° or 215–220°.

Question ❸ b): Amelia turns 140–145° or 215–220°. If Amelia turns in the same direction both times, the sum of her two turns is 360°. If she turns one way from A to C and then the opposite way from C to F, she will turn through the same angle both times.

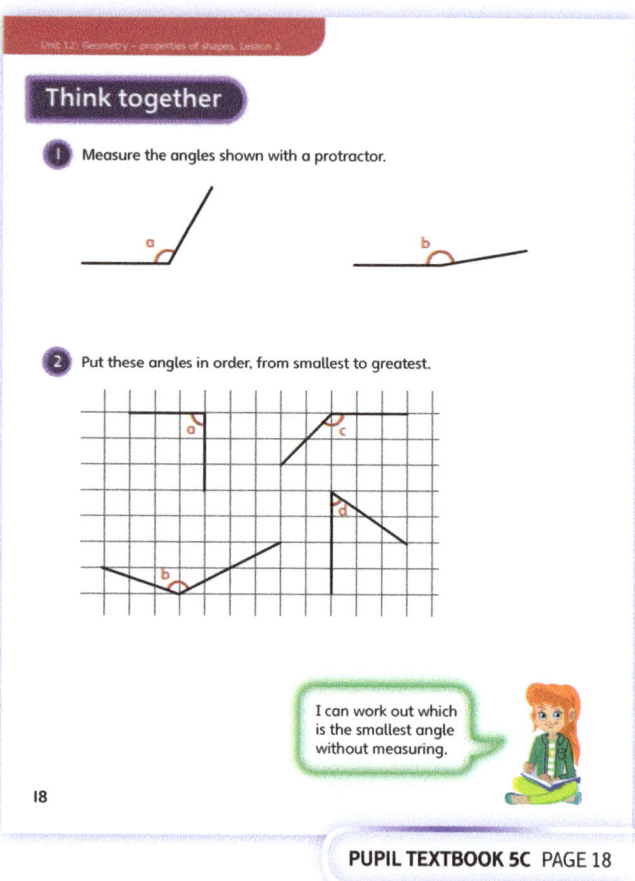

PUPIL TEXTBOOK 5C PAGE 18

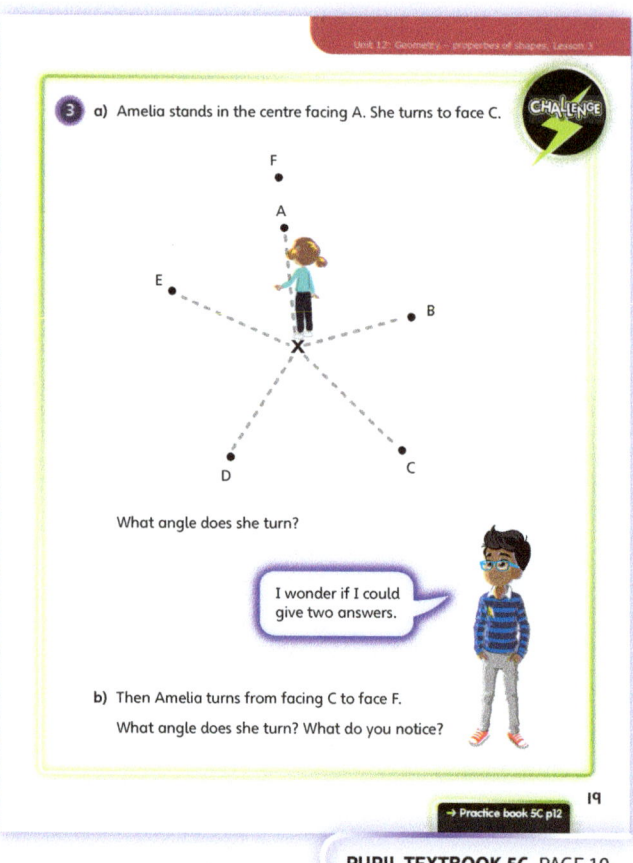

PUPIL TEXTBOOK 5C PAGE 19

Unit 12: Geometry – properties of shapes, Lesson 3

Practice

WAYS OF WORKING Independent thinking

IN FOCUS Questions ① and ② focus on the correct placement of the protractor, especially when measuring in different orientations. This reinforces the learning from this and the previous lesson.

Question ③ extends learning by introducing the concept of reasoning as well as measurement, because children should be able to identify the acute angle as the smallest, and the right angle as the next smallest, without having to measure. Children may then compare the two obtuse angles by measuring.

Question ④ asks children to measure angles inside a polygon, ensuring that they understand the concept of interior angles and are able to measure angles in all orientations.

STRENGTHEN Ask children to talk through the different steps of using a protractor with a partner. Fully explaining each step will help cement the process.

DEEPEN Extend question ⑤ by adding more turns to the table for children to measure and complete.

THINK DIFFERENTLY Children may spot that there are symmetries in the shapes in question ④ which can reduce the amount of measuring they need to do.

ASSESSMENT CHECKPOINT Questions ② and ④ test the core skill of measuring using a protractor. Wrong answers here indicate that children's learning of the steps for accurate measurement is not secure and needs reinforcing.

ANSWERS Answers for the **Practice** part of the lesson can be found in the *Power Maths* online subscription.

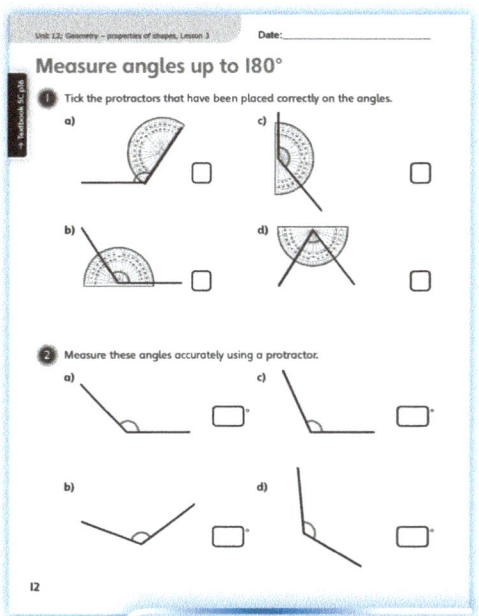

PUPIL PRACTICE BOOK 5C PAGE 12

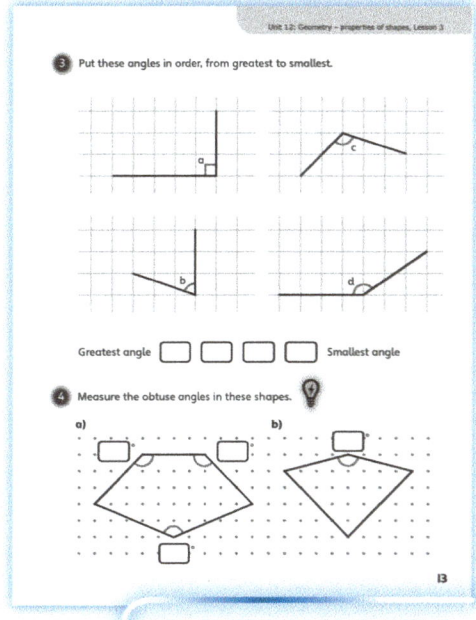

PUPIL PRACTICE BOOK 5C PAGE 13

Reflect

WAYS OF WORKING Independent thinking

IN FOCUS The **Reflect** question ensures children are confident with key terms such as 'obtuse angle' and relate this to their knowledge of how to accurately use a protractor.

ASSESSMENT CHECKPOINT Can children verbalise the link between obtuse and acute angles with the two scales on a protractor?

ANSWERS Answers for the **Reflect** part of the lesson can be found in the *Power Maths* online subscription.

After the lesson ⏸

- Could children recognise obtuse angles?
- Could children read angles that fall between multiples of 10°?
- How did children manipulate the page or the protractor when the angles were presented in different orientations?

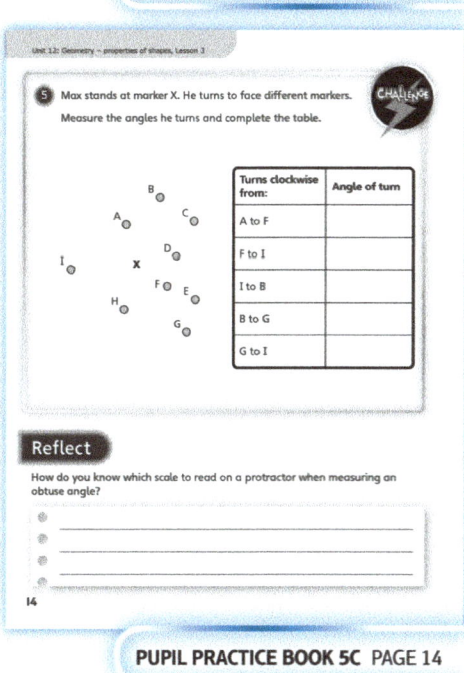

PUPIL PRACTICE BOOK 5C PAGE 14

Unit 12: Geometry – properties of shapes, Lesson 4

Draw lines and angles accurately

Learning focus

In this lesson, children will continue to use a protractor to draw angles accurately. They combine this with drawing lines accurately to the nearest millimetre.

Before you teach

- How will you ensure that children are confident and accurate when using a protractor?
- How will you refer back to previous learning to explore with children what is the same and what is different about, for example, 55 mm and 5.5 cm?

NATIONAL CURRICULUM LINKS

Year 5 Geometry – properties of shapes

Draw given angles, and measure them in degrees (°).

ASSESSING MASTERY

Children can draw angles accurately up to 180°. They can draw lines to the nearest millimetre, given measurements in units of cm and mm, and can follow instructions to create a design.

COMMON MISCONCEPTIONS

Children will need to have mastered protractor skills from the previous two lessons before tackling this. They might still need support with positioning the protractor accurately and choosing the appropriate scale to read. Ask:
- *Explain how to measure accurately using a protractor. What are the key points you need to think about?*

STRENGTHENING UNDERSTANDING

This lesson practises drawing angles in different orientations, both in isolation and by following a design, in order to give some purpose to the activity. Some children may need to practise more basic angle drawing before tackling the independent work. Give children a range of challenges for drawing angles in different orientations before moving on.

GOING DEEPER

Challenge children to explain how small errors at the start of a design can cause quite large differences in the finished results.

KEY LANGUAGE

In lesson: angle, protractor, millimetre (mm), centimetre (cm)

Other language to be used by the teacher: acute angle, degrees (°)

STRUCTURES AND REPRESENTATIONS

Angle diagrams

RESOURCES

Mandatory: protractor, ruler

 In the eTextbook of this lesson, you will find interactive links to a selection of teaching tools.

Quick recap

Ask children to measure a variety of different angles up to 180 degrees. Check that children can use a protractor accurately. This is a new skill they have learnt recently, and this recap aims to remind them how to use it.

Discover

WAYS OF WORKING Pair work

ASK

- Question ① a): *What tools will you need to use to draw an angle rather than just measure an angle?*
- Question ① a): *Which line would you draw first? Why? What do you then do with the protractor?*
- Question ① b): *How is this the same or different to the angle you have just drawn? How long are the lines? How can you make sure this is an exact copy?*

IN FOCUS In questions ① a) and b), children are building on skills from the previous two lessons and now add the skill of drawing a given angle. The difficulty level builds up over the course of the lesson, to allow children to build the skill set gradually. In question ① b), it is important that children are measuring line lengths too, to make their drawings as accurate as possible.

PRACTICAL TIPS Have protractors and rulers available for children. It may help some children to work in pairs to understand the four steps involved in drawing an angle, and to discuss the order in which they need to be done. Encourage discussion such as: 'We need a straight line to place the protractor on, so we need to draw a straight line first.'

ANSWERS

Question ① a):

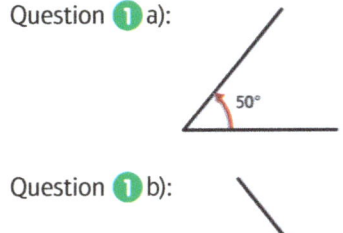

Question ① b):

Share

WAYS OF WORKING Whole class teacher led

ASK

- Question ① a): *Talk me through how you drew an angle of 50 degrees. What did you do first?*
- Question ① b): *What is the first step to copy the angle? What are the lengths of the lines we need to draw?*

IN FOCUS In question ① a), children focus on drawing an angle of 50 degrees with any line length. In question ① b), children now try to copy the exact measurements, including the lengths of the lines. Ask children to use a protractor to check each other's angles. The major new skill is marking the correct angle on the edge of the protractor, then using the ruler to join the point of the angle with the new mark and drawing a line of the exact length required. This is a difficult skill for children to master, so encourage them to have more than one go at it where necessary.

PUPIL TEXTBOOK 5C PAGE 20

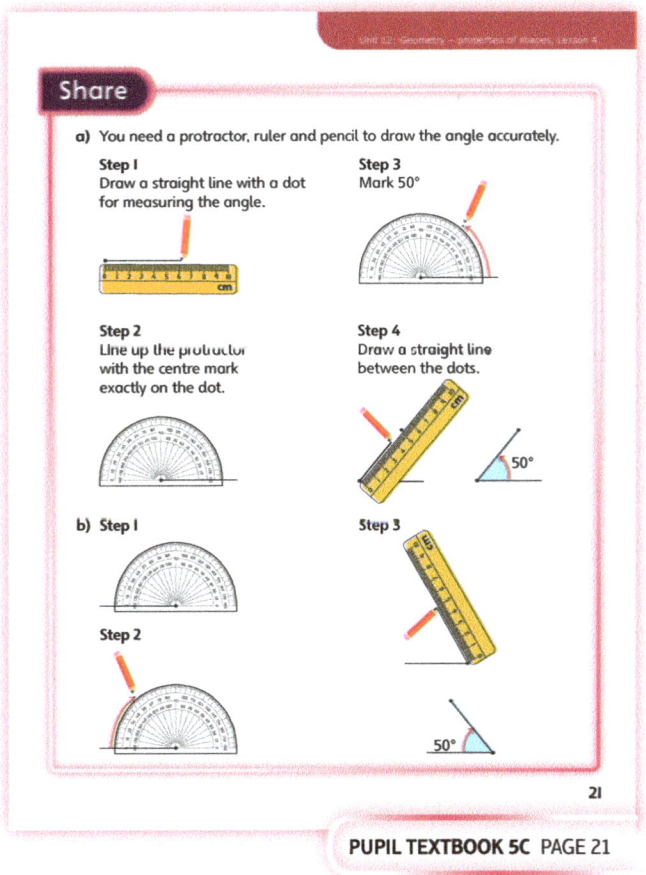

PUPIL TEXTBOOK 5C PAGE 21

Unit 12: Geometry – properties of shapes, Lesson 4

Think together

WAYS OF WORKING Whole class teacher led (I do, We do, You do)

ASK

- Question 1: *What do you need to do first? What is the next step? How can you make sure the angle is accurate?*
- Question 2: *Why are these angles more challenging than those in question 1? What can you do first? Do the angles have to be in the same orientation for you to be able to measure their size?*
- Question 3: *When you draw an angle of 55°, this angle will be between two 10-degree marks on your protractor. How should you use the protractor to do this? Would it be easier if you rotated the page?*

IN FOCUS In questions 1, 2 and 3, children practise drawing angles of increasing complexity. This is a pretty tricky skill so ensure that all children have sufficient time to practise. If needed, add more simple angles for children to draw. In question 2, children may find it easier to rotate the page. This means they can position their protractor as they did in question 1.

Question 3 shows a more complicated design involving multiple angles, so children will need to work out the correct placement of the protractor. They will also need to draw an angle of 55°, which will be between two 10-degree marks on their protractor. Ask: *Which two marks will 55° lie between?* (50° and 60°). Ensure children draw the angle half-way between the two marks.

STRENGTHEN Encourage children to rotate the page and to work in pairs to agree on the placement of the protractor before drawing. All children could work on scrap paper, as mistakes are very likely.

DEEPEN In question 2, you could challenge children not to rotate the page but instead to draw the angles in the same orientation. Once step 2 of the design in question 3 has been completed, ask children to label each angle as 'acute' or 'obtuse'.

ASSESSMENT CHECKPOINT Use questions 1 and 2 to check that children can accurately use a protractor to draw acute and obtuse angles.

ANSWERS

Question 1: Compare children's angles for both a) and b) with those on the page.

Question 2: Compare children's angles for both a) and b) with those on the page.

Question 3 c): The missing angles are 50° and 60°. The missing line measurement is 62 mm.

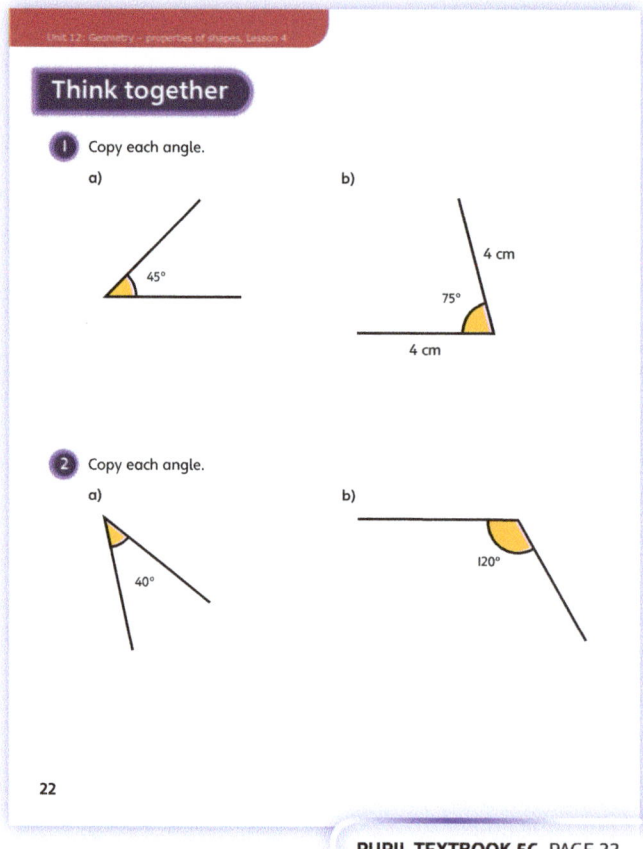

PUPIL TEXTBOOK 5C PAGE 22

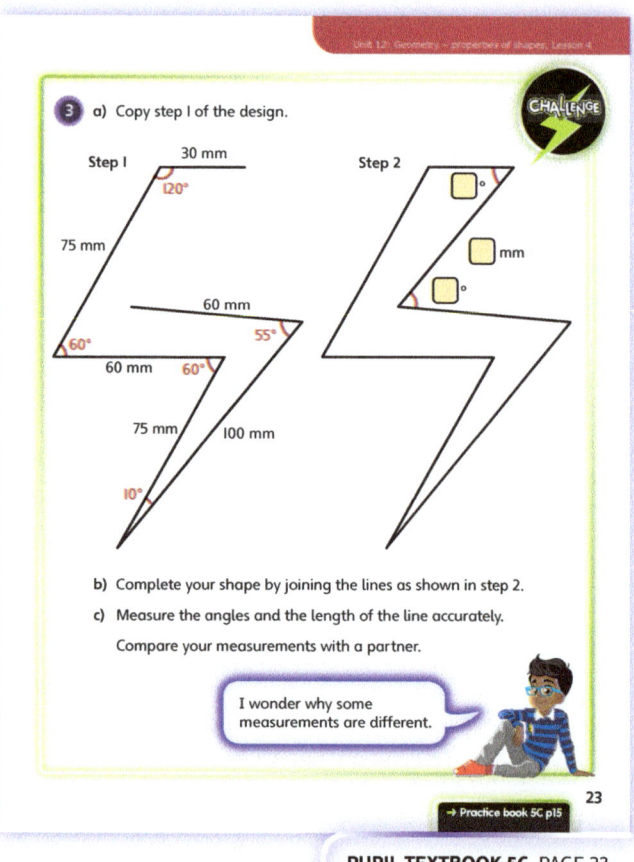

PUPIL TEXTBOOK 5C PAGE 23

Unit 12: Geometry – properties of shapes, Lesson 4

Practice

WAYS OF WORKING Independent thinking

IN FOCUS Question ❶ reinforces the learning with protractors by asking children to complete angles of varying sizes. Angles include acute and obtuse and children should be able to draw both, reading from the relevant scale. In question ❷, children are given more freedom to draw three angles of the same size. This emphasises the learning that there are multiple ways to position an angle, and that the orientation of the protractor helps them to draw angles at different orientations. Encourage children to draw angles in different orientations so that they do not all look the same.

Question ❸ challenges children to follow the steps of a design to ensure they accurately use a protractor and ruler together.

Question ❹ asks children to consider whether the sides of a triangle can all be the same length if the angles are different. Encourage children to use trial and error to investigate the question.

STRENGTHEN Encourage children to work on scrap paper to try out their designs, as they may need to redraft if errors creep in, which is a likely occurrence given the practical difficulty of this skill.

DEEPEN Children could experiment by changing the angle by a small amount and seeing what happens to the design. This would give additional practice in drawing accurately and also underline the importance of accurate measurement.

ASSESSMENT CHECKPOINT Correctly following the steps of question ❸ requires very good protractor skills. Errors here could be caused by inaccurate measurements or by children forgetting that 'not to scale' means they can't find the missing angles by measuring.

ANSWERS Answers for the **Practice** part of the lesson can be found in the *Power Maths* online subscription.

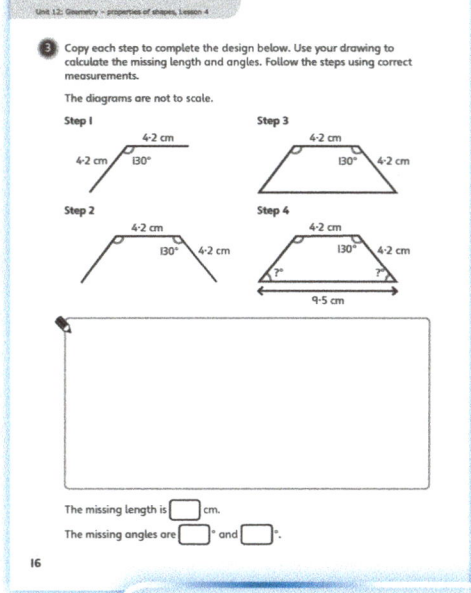

Reflect

WAYS OF WORKING Independent thinking

IN FOCUS Discuss with children how confident they feel with using a protractor. Ask: *What steps do you find easier? Which angles do you find easier to draw? Which mistakes can be easy to make when using a protractor?* Discuss if there are ways children can check the accuracy of their measurements.

ASSESSMENT CHECKPOINT Ask children to draw an angle of 70 degrees and then check one another's drawings by measuring with a protractor. This will show whether children are drawing and measuring accurately.

ANSWERS Answers for the **Reflect** part of the lesson can be found in the *Power Maths* online subscription.

After the lesson ⏸

- Can children draw angles in different orientations?
- Can children explain what error has been made if the design 'goes wrong'?
- Are children confident following steps of a design independently?

Unit 12: Geometry – properties of shapes, Lesson 5

Calculate angles around a point

Learning focus
In this lesson, children will learn to calculate missing angles around a point, by reasoning about the 360° in a whole turn.

Before you teach
- Are there any resources or visual aids that would help with this lesson? Making an angle-maker in advance would enable you to demonstrate this.
- How secure are children with their vocabulary (acute, right angle, obtuse)? Consider doing a quick refresher on this as a lesson starter.

NATIONAL CURRICULUM LINKS

Year 5 Geometry – properties of shapes

Identify:
- angles at a point and one whole turn (total 360°)
- angles at a point on a straight line and $\frac{1}{2}$ a turn (total 180°)
- other multiples of 90°.

ASSESSING MASTERY
Children can use reasoning to calculate missing angles around a point.

COMMON MISCONCEPTIONS
Some children may need support to structure the calculations where there are more than two angles around a point. Ask:
- *What is the whole and what are the parts? What calculations do you need to do to work out the missing part?*

STRENGTHENING UNDERSTANDING
Children can create and use the angle-makers to explore the way two parts of 360° vary as one increases and one decreases.

GOING DEEPER
Ask children to draw reflex angles. Can they identify the most efficient way to draw these when their protractor only goes as far as 180°?

KEY LANGUAGE
In lesson: angles around a point, reflex angle, degrees (°), quarter turn, whole turn, obtuse angle

Other language to be used by the teacher: half turn, acute angle, right angle

STRUCTURES AND REPRESENTATIONS
Angle diagrams, 2D shapes

RESOURCES
Mandatory: paper or card circles

Optional: protractor, ruler

 In the eTextbook of this lesson, you will find interactive links to a selection of teaching tools.

Quick recap
Ask children to make a quarter turn clockwise. Ask: *What angle have you turned through?* Ask them to make a half turn clockwise. Ask: *What angle have you turned through?* Keep repeating this for different turns, including full turns.

Unit 12: Geometry – properties of shapes, Lesson 5

Discover

WAYS OF WORKING Pair work

ASK

- Question 1 a): *What do we call a 90-degree angle?*
- Question 1 b): *Why is one angle greater than 180°?*
- Question 1 b): *What calculation should you use to find the other angle?*

IN FOCUS The focus of questions 1 a) and b) is to explore the relationship between two angles that form a whole of 360°. By creating the angle-maker themselves, children gradually find it easier to visualise how the angles fit together in a whole turn.

PRACTICAL TIPS Children could make angle-makers like those shown in the image. They should cut two paper circles of different colours and cut a line to the centre of each circle, then slide the two circles together. This then enables them to rotate the circles and show different angles.

ANSWERS

Question 1 a): A whole turn is 360°. A 90° angle is a quarter turn.
The angle-maker can be used in any of these ways to show a 90° angle:

Question 1 b): One right angle is a quarter turn, so there must be 3 quarter turns remaining.
90° + 90° + 90° = 270°
270° is a three-quarter turn.

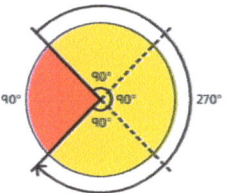

Share

WAYS OF WORKING Whole class teacher led

ASK

- Question 1 a): *How could you show 90° in different orientations?*
- Question 1 b): *What is the same and what is different about the two different ways to calculate the 270° angle?*

IN FOCUS The main focus of questions 1 a) and b) is for children to recognise that the whole turn is 360°, and how this can be used to calculate the missing angle around a point. It is important that they recognise how a subtraction can be a very useful calculation for questions like these and are able to explain why.

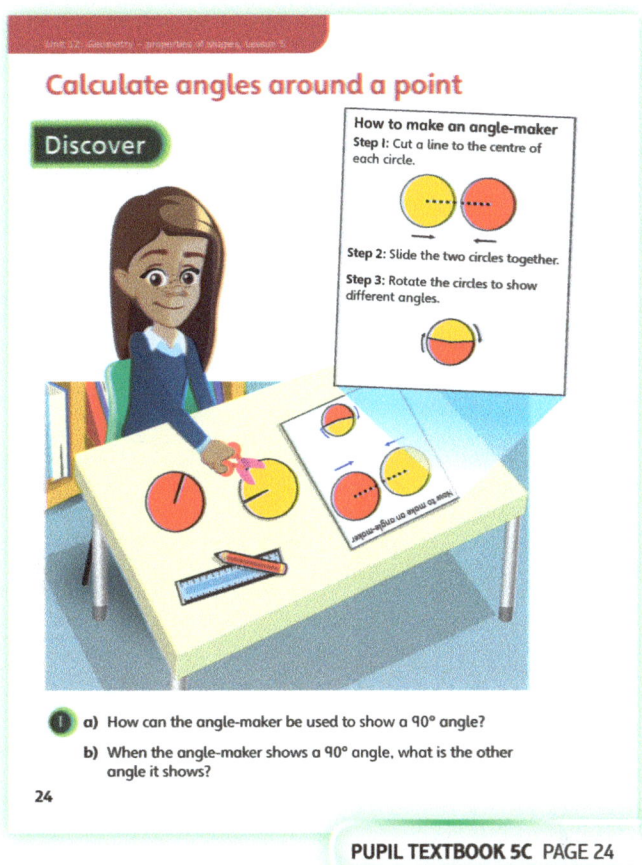

PUPIL TEXTBOOK 5C PAGE 24

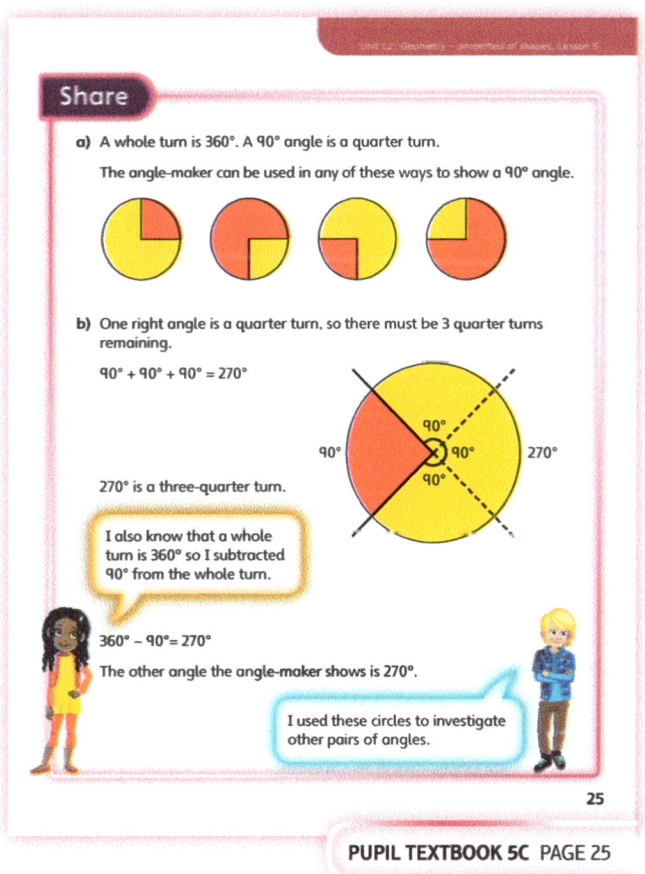

PUPIL TEXTBOOK 5C PAGE 25

Unit 12: Geometry – properties of shapes, Lesson 5

Think together

WAYS OF WORKING Whole class teacher led (I do, We do, You do)

ASK

- Question ❶: *What is the total of the parts you know? What is the whole?*
- Question ❶: *What calculation do you need to do to find the missing angle?*
- Question ❸: *How can you apply this to drawing an angle greater than 180°?*

IN FOCUS Question ❷ requires children to perform a multi-step calculation as there are three angles around a point. This reinforces that the angles around a point total 360°.

Question ❸ allows children to explore that one of the angles is a reflex angle. They may need to discuss that 'reflex' means greater than 180°.

STRENGTHEN Ask children to sketch the angles or make them with angle-makers. Allow them to explore the relationship between the parts and the whole using the angle-makers to create given angles, then calculate the other part to total 360°.

DEEPEN Question ❸ includes new language from this unit in the form of reflex angles. Provide children with a selection of angles around a point which they must calculate, using the knowledge from this lesson, and also label as acute, right angle, obtuse and reflex.

ASSESSMENT CHECKPOINT Question ❷ requires good understanding of the concept of using calculation rather than measuring to find missing angles. Have children grasped that they need to use the knowledge that the angles around a point total 360°? Do they recognise the right angle and that it is 90°?

ANSWERS

Question ❶: Approximately 48° and 312°.

Question ❷: a = 230°   b = 120°   c = 182°

Question ❸ a): Draw an angle of 130°. The remaining angle around the point will be 230°.

Question ❸ b): Children should accurately draw angles of 130° and 48°, labelling the reflex angles 230° and 312° respectively.

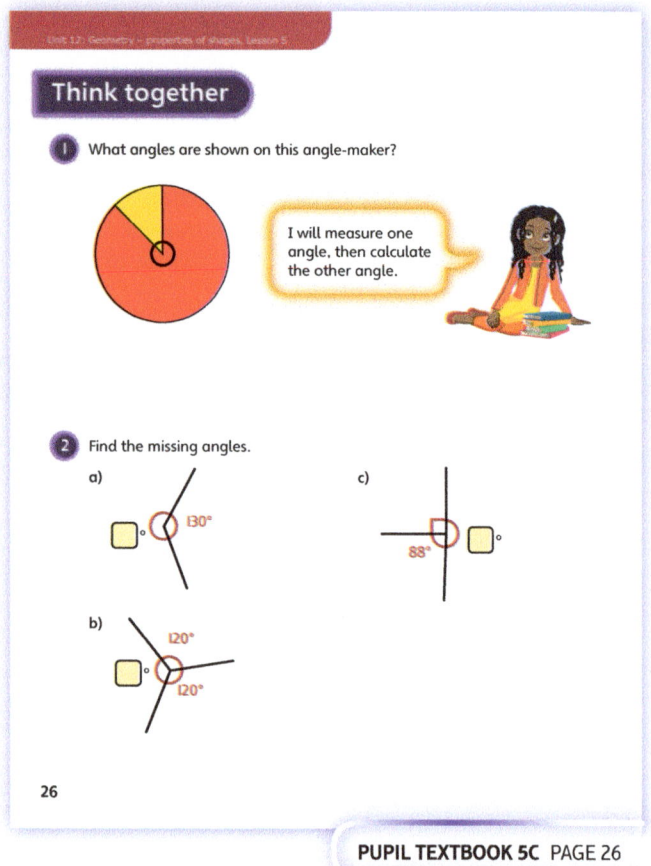

PUPIL TEXTBOOK 5C PAGE 26

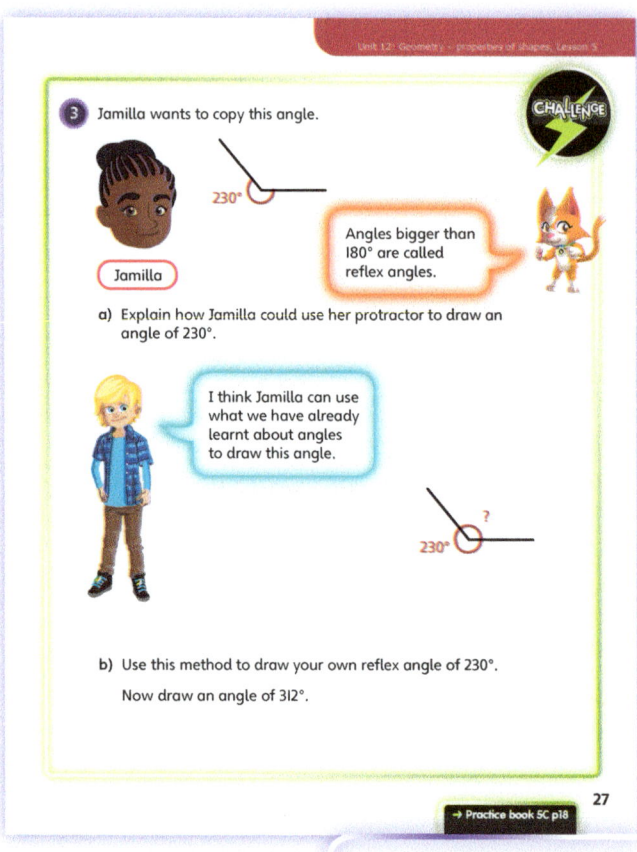

PUPIL TEXTBOOK 5C PAGE 27

Unit 12: Geometry – properties of shapes, Lesson 5

Practice

WAYS OF WORKING Independent thinking

IN FOCUS Question ❶ is basic practice at finding angles that add up to 360°, with the subtraction scaffolded to reinforce the efficient method of working out the missing angles.

Question ❷ requires children to calculate the missing angle where there are more than two parts. This shows that they need to add the given angles and then subtract the total from 360° to find the missing angle.

STRENGTHEN Some children may need to build their calculations step-by-step when working with more than two angles around a point.

DEEPEN Give children some examples of incorrect calculations and challenge them to find the mistakes. For example, the given angles could be added up wrongly or the total could be more or less than 360°. Children could then set similar problems for a partner to solve.

THINK DIFFERENTLY Question ❺ requires children to remember the definition of an obtuse angle and use their reasoning skills to describe why Reena is unable to make four obtuse angles within the circle.

ASSESSMENT CHECKPOINT Question ❷ shows deep understanding of the key skill. Incorrect answers here might indicate that children have not grasped that there are 360° around a point or they may be relying on measuring rather than using calculation.

ANSWERS Answers for the **Practice** part of the lesson can be found in the *Power Maths* online subscription.

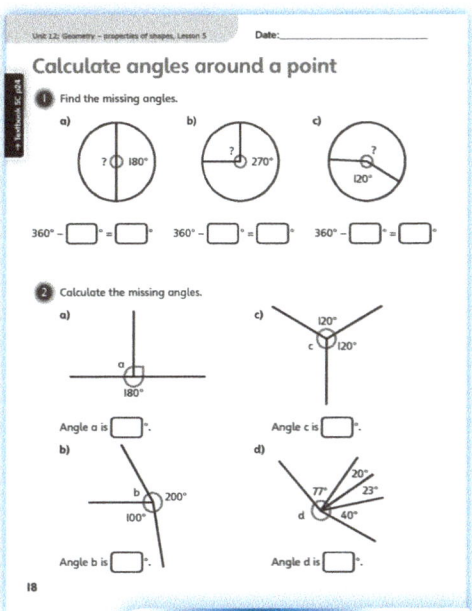

PUPIL PRACTICE BOOK 5C PAGE 18

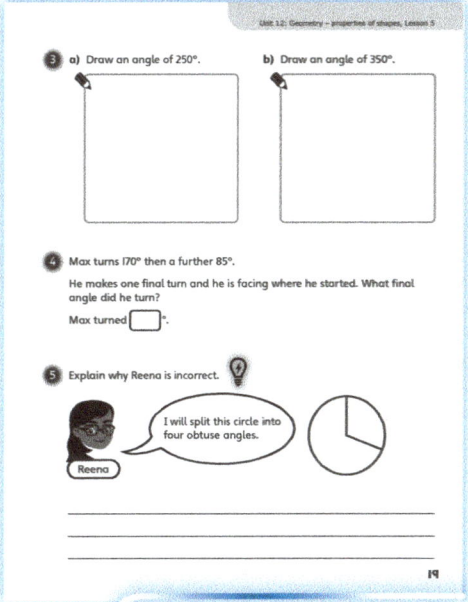

PUPIL PRACTICE BOOK 5C PAGE 19

Reflect

WAYS OF WORKING Independent thinking

IN FOCUS Children should be able to find multiple solutions and explain the relationship between the two missing angles and the whole.

ASSESSMENT CHECKPOINT Can children explain the sum of the two missing angles and that there are various possibilities? Have they fully explained which number they are subtracting from to find these possibilities?

ANSWERS Answers for the **Reflect** part of the lesson can be found in the *Power Maths* online subscription.

After the lesson ⏸

- Can children explain the reasoning based on the total angle of a whole turn?
- Are children able to follow a chain of reasoning where there are more than two angles around a point?
- Are children confident with the vocabulary of acute, right, obtuse and reflex angles?

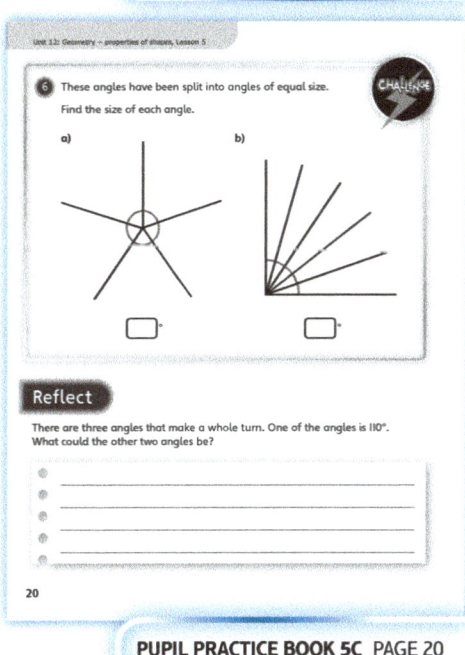

PUPIL PRACTICE BOOK 5C PAGE 20

Unit 12: Geometry – properties of shapes, Lesson 6

Calculate angles on a straight line

Learning focus

In this lesson, children will begin to understand that they can calculate missing angles on a straight line, based on their knowledge of 180° as a half turn.

Before you teach

- In this lesson, children will move from measuring angles to calculating them from known facts. Would children benefit from a lesson starter on refreshing calculation skills?
- Are there any resources that would be useful to underpin learning? For example, a visual reminder of the number of degrees in a whole, half or quarter turn.

NATIONAL CURRICULUM LINKS

Year 5 Geometry – properties of shapes

Identify:
- angles at a point and one whole turn (total 360°)
- angles at a point on a straight line and $\frac{1}{2}$ a turn (total 180°)
- other multiples of 90°.

ASSESSING MASTERY

Children can explain how to calculate the missing angles on a straight line by reasoning about their knowledge of a half turn.

COMMON MISCONCEPTIONS

It is a common error to find the complementary angles to 200° rather than 180°, as children are used to finding number bonds to 100 in their calculations. Ask:
- What should the angles total on a straight line?

STRENGTHENING UNDERSTANDING

Children could practise the skill by creating a 'windscreen wiper'. As the line rotates to create angles of 10°, 20°, 30° and so on, children can find the number bonds to 180 to calculate the missing angle.

GOING DEEPER

Ask questions that challenge children to reason based on their knowledge of equal and unequal parts.

KEY LANGUAGE

In lesson: angles, half turn, degrees (°), right angle, quarter turn, **interior angles**

Other language to be used by the teacher: total

STRUCTURES AND REPRESENTATIONS

Angle diagrams

RESOURCES

Mandatory: protractor

Optional: paper semicircles, safety scissors

 In the eTextbook of this lesson, you will find interactive links to a selection of teaching tools.

Quick recap

Play a game called 'Make 180'. Give children a number between 0 and 180 and ask them to work out how many they need to add on to make 180 degrees. Start with multiples of 10 and then move on to other values. Discuss different methods that children could use to find the answer.

Discover

WAYS OF WORKING Pair work

ASK

- Question 1 a): *Does angle a look acute or obtuse? How can you be sure?*
- Question 1 a): *What are the angles in a half turn?*
- Question 1 a): *Can you check your estimate by measuring? What should you use to do this?*

IN FOCUS The main focus of question 1 a) is for children to connect their understanding of angles in a half turn with the problem of finding the missing angle. Some children may revert to measurement, especially considering the effort they have put into learning that skill over the previous three lessons. Encourage children to compare the different approaches and to discuss how calculations may or may not be more efficient.

PRACTICAL TIPS This would be an excellent task for children to try out using lightweight card. Provide children with semicircles of paper and then ask them to measure and cut different angles from a point marked on the straight line. They could then calculate the remaining angle and measure to check their predictions. This could be extended into a game where all children measure an angle and cut it away and then, as a class, they try to match everyone's pairs.

ANSWERS

Question 1 a): Angle a is 100°.

Question 1 b): Max cuts two 90° angles.

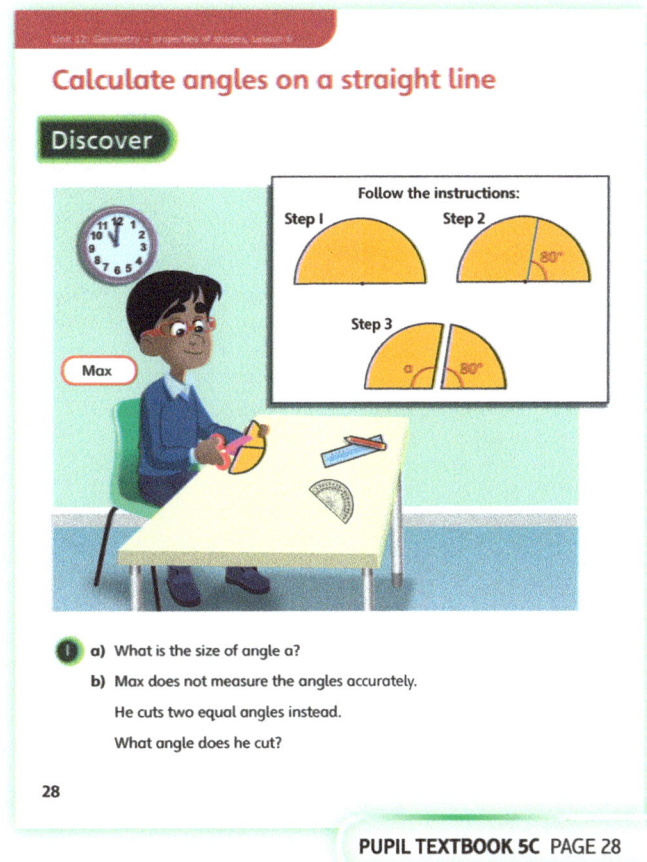

PUPIL TEXTBOOK 5C PAGE 28

Share

WAYS OF WORKING Whole class teacher led

ASK

- Question 1 a): *Which method is more efficient? Why?*
- Question 1 a): *Can you explain why the subtraction method works?*
- Question 1 a): *Would the subtraction method work for other angles?*

IN FOCUS The main focus of question 1 a) is for children to recognise that it is often more efficient and accurate to use their calculation skills rather than measurement to find the missing angles, based on their understanding of the degrees in half a turn.

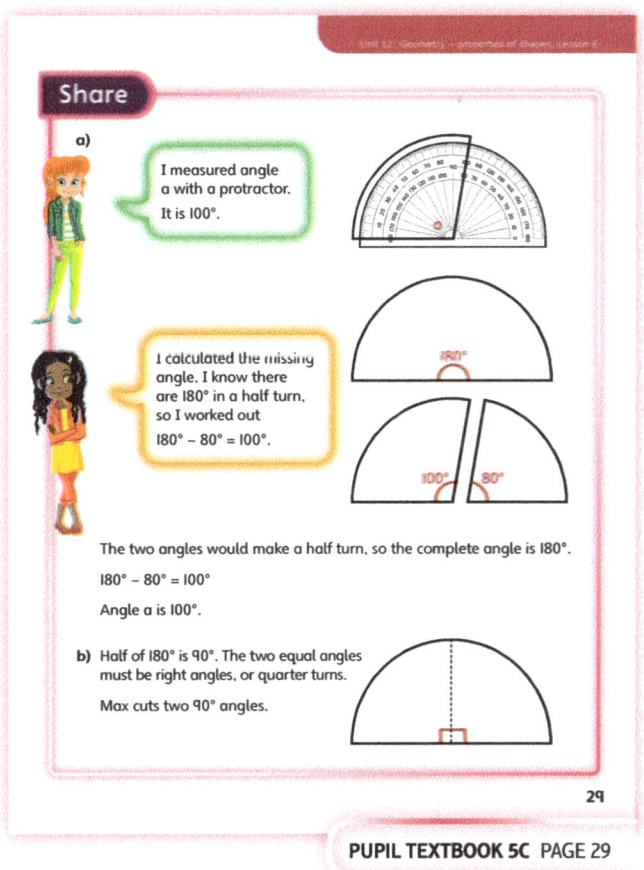

PUPIL TEXTBOOK 5C PAGE 29

Unit 12: Geometry – properties of shapes, Lesson 6

Think together

WAYS OF WORKING Whole class teacher led (I do, We do, You do)

ASK
- Question ①: *What is the total number of degrees in a half turn? What number do you need to subtract from?*
- Question ③: *Which operation can you use if the 180° are split into five equal parts?*

IN FOCUS Question ② requires children to measure the angles then find pairs that add to 180 to form a straight line, which reinforces that there are 180° along a straight line, as well as the concept of calculation that was introduced in the **Share** activity.

Question ② looks at an angle split into equal parts and so requires a division, demonstrating that (depending on the calculation) different operations may be necessary.

STRENGTHEN Encourage children to explore the relationship between angles on a line by looking at the two scales on a protractor. What do they notice?

DEEPEN Challenge children to continue the exercise in question ③ with different straight-sided 2D shapes. They could draw these or use 2D shape manipulatives.

ASSESSMENT CHECKPOINT Question ② helps to assess children's understanding that angles on a straight line must add up to 180° and that this can be in many different combinations. Question ④ assesses children's understanding of both the angles within a square and the angles on a straight line.

ANSWERS

Question ①: a = 140°   b = 105°   c = 10°

Question ②: a = 120°   c = 35°   e = 90°
b = 155°   d = 60°   f = 25°
a and d fit together to make a straight line.
b and f fit together to make a straight line.

Question ③: a = 36°

Question ④ a): Both angles will equal 45° (half of the remaining 90°).

Question ④ b): 30° and 60°

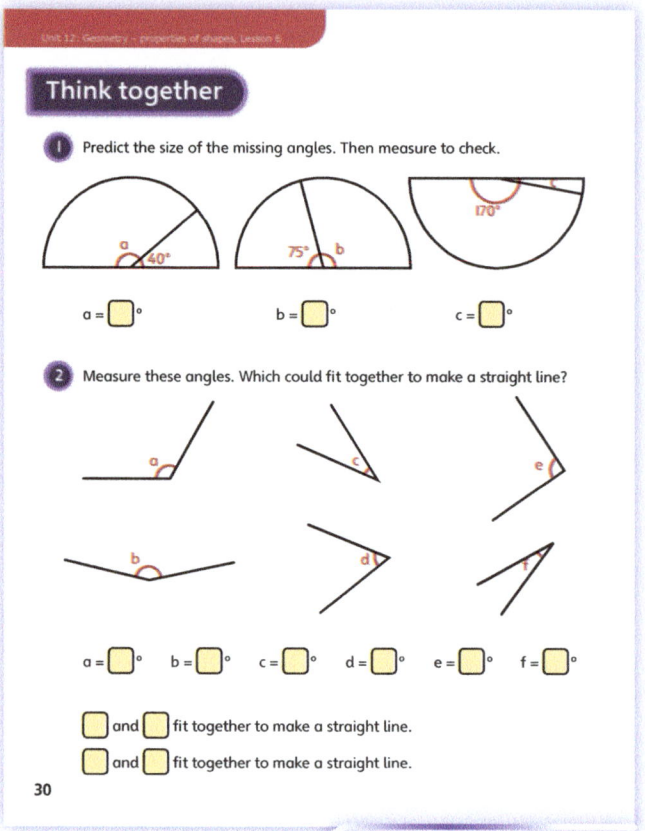

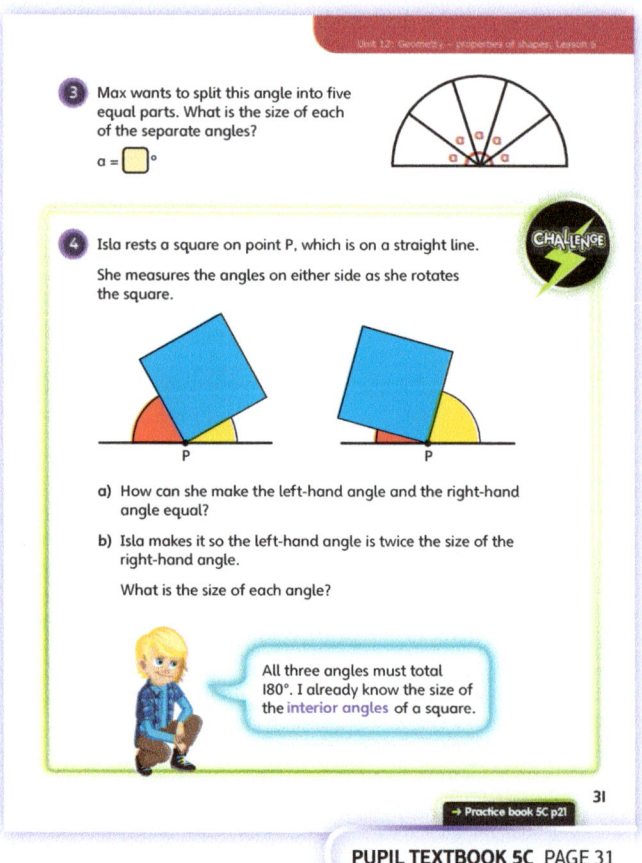

Unit 12: Geometry – properties of shapes, Lesson 6

Practice

WAYS OF WORKING Independent thinking

IN FOCUS Questions ❶ and ❷ reinforce the process of using the subtraction method to find missing angles. It is important to cement this understanding of the angles on a line making 180° so that children can complete the rest of the questions.

Question ❸ extends this learning by asking children to use both measurement and calculation. They must measure given angles then use calculation to find pairs and a triplet that sum to 180 degrees.

STRENGTHEN Encourage children to explore the bonds to 180° by totalling to 100, then adding on 80.

DEEPEN Extend learning in question ❹ by asking children to place a square along a straight line and measure one angle with a protractor. They can then swap with a partner to find the remaining angle.

ASSESSMENT CHECKPOINT Do children's answers to questions ❷ and ❹ show good understanding of the parts that sum to 180?

ANSWERS Answers for the **Practice** part of the lesson can be found in the *Power Maths* online subscription.

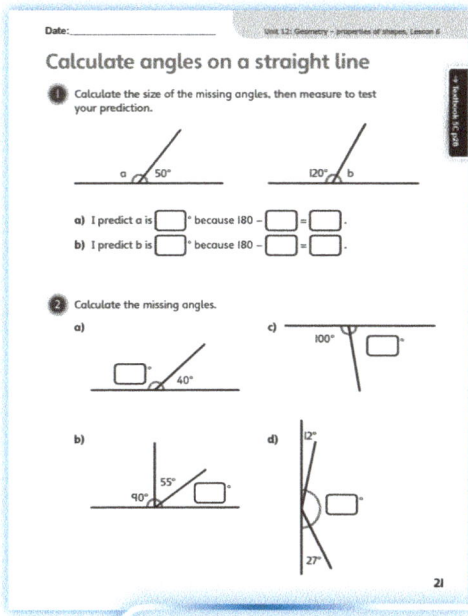

PUPIL PRACTICE BOOK 5C PAGE 21

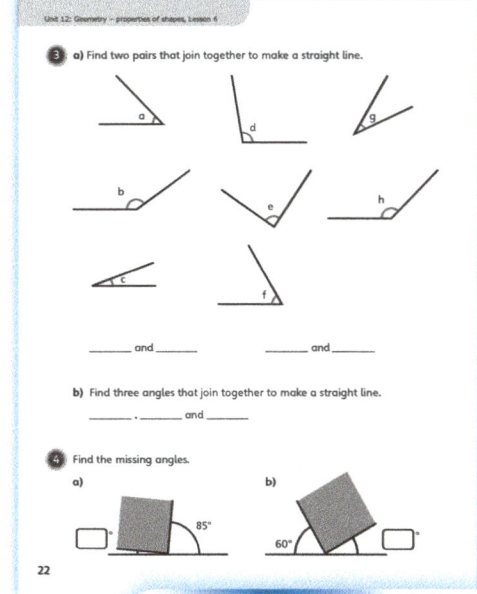

PUPIL PRACTICE BOOK 5C PAGE 22

Reflect

WAYS OF WORKING Independent thinking

IN FOCUS The **Reflect** question creates an opportunity for children to explain their understanding of angles along a straight line and also prove their calculation skills to 180.

ASSESSMENT CHECKPOINT Can children explain that Aki has summed to 190° instead of 180°? Have they explained what the angle must actually be? Are children able to say which part Aki has said correctly (the measurement of a right angle) and where his error appeared?

ANSWERS Answers for the **Reflect** part of the lesson can be found in the *Power Maths* online subscription.

After the lesson

- Were children able to follow the chains of reasoning required when there are more than two angles on a straight line?
- Could children explain why the angles on a straight line must sum to 180°, based on their knowledge of whole and half turns?

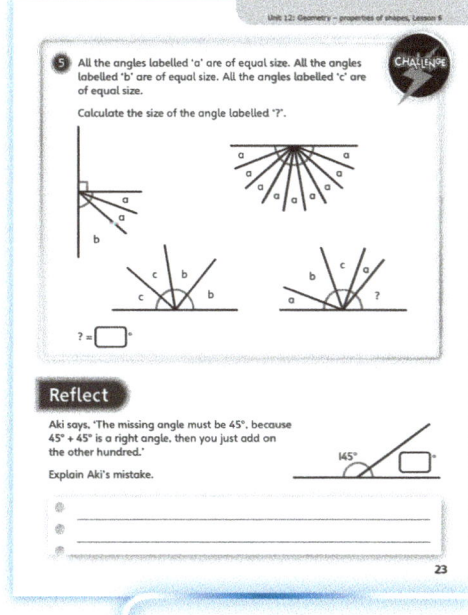

PUPIL PRACTICE BOOK 5C PAGE 23

Unit 12: Geometry – properties of shapes, Lesson 7

Lengths and angles in shapes

Learning focus
In this lesson, children will use reasoning about shapes to calculate missing angles and lengths. They use reasoning based on the properties of known lines, angles and shapes.

Before you teach
- You might like to provide some 2D and 3D shapes for children to look at and find 90-degree angles.
- Are children confident in knowing that right angles are always 90°?

NATIONAL CURRICULUM LINKS

Year 5 Geometry – properties of shapes

Use the properties of rectangles to deduce related facts and find missing lengths and angles.

ASSESSING MASTERY

Children can explain their reasoning based on known properties, lengths and composites of shapes or parts of shapes.

COMMON MISCONCEPTIONS

Children may try to spot lengths or angles that look the same rather than using reasoning based on properties. Ask:
- *How do you know that these lengths must be the same?*

STRENGTHENING UNDERSTANDING

Children can use and transform paper versions of the shapes to be used in the problems, manipulating them by cutting and rearranging.

GOING DEEPER

Challenge children to make general statements about the angles in the diagonals of squares and rectangles.

KEY LANGUAGE

In lesson: angle, length, interior angle, parallelogram, degrees (°)

Other language to be used by the teacher: right angle, measure, calculate, obtuse angle, acute angle

STRUCTURES AND REPRESENTATIONS

Angle diagrams, 2D shapes

RESOURCES

Mandatory: ruler

Optional: paper squares and rectangles

 In the eTextbook of this lesson, you will find interactive links to a selection of teaching tools.

Quick recap

Recap different 2D shapes by naming a shape and asking children to draw them. Include different types of quadrilaterals that children should know. Pick out two answers from children and ask: *What is the same? What is different?*

Unit 12: Geometry – properties of shapes, Lesson 7

Discover

WAYS OF WORKING Pair work

ASK

- Question 1 a): *What size are the angles in the original square?*
- Question 1 a): *How would you calculate the angles in the triangles that are formed?*
- Question 1 b): *How are the lengths combined?*

IN FOCUS In question 1 a), the key is for children to recognise that the angles and side lengths of the triangles can be deduced based on the properties of the original squares. They need to recall known facts about squares.

Question 1 b) asks children to consider the length of a diagonal. They may want to explore how the diagonal of a square or rectangle is always longer than the side length. This could be an extra challenge to deepen learning.

PRACTICAL TIPS Children should create the parallelogram by cutting their own squares of paper in the way the instructions show. They can work in pairs with one square each and put them together to form two parallelograms. This will enable them to see exactly how the squares become the parallelograms.

ANSWERS

Question 1 a): The interior angles of the parallelogram are 45°, 135°, 45° and 135°.

Question 1 b): Length A is 20 cm but length B is longer than 10 cm.
Lee is correct about length A but incorrect about length B.

Share

WAYS OF WORKING Whole class teacher led

ASK

- Question 1 a): *How are the 90-degree angles halved in the triangles?*
- Question 1 a): *Why is there one 90-degree angle in each triangle?*
- Question 1 b): *Why is the diagonal length longer than 10 cm?*

IN FOCUS The key to this activity is to reason about the angles and lengths, based on the properties of the original squares. Children will notice that the diagonal of a square is always longer than its sides.

STRENGTHEN Although children should not need to measure again or check with a protractor or ruler, some may find it increases their confidence in their calculations to do so.

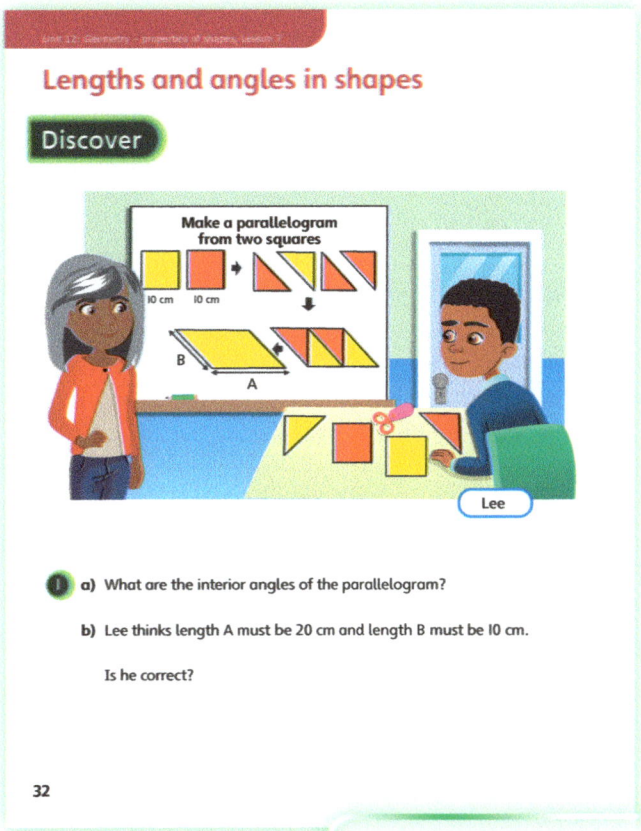

PUPIL TEXTBOOK 5C PAGE 32

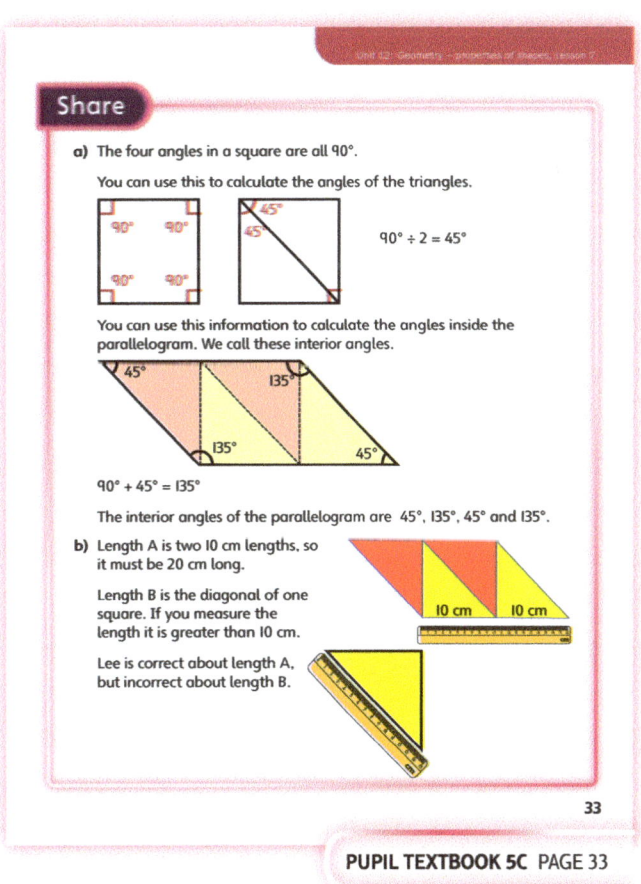

PUPIL TEXTBOOK 5C PAGE 33

Unit 12: Geometry – properties of shapes, Lesson 7

Think together

WAYS OF WORKING Whole class teacher led (I do, We do, You do)

ASK

- Question ❶: *How can you use your knowledge that a right angle is 90° to help you?*
- Question ❷: *Can you see how the shapes from Shape A have been rearranged to make Shape B?*
- Question ❷: *Which of the measurements on the first shape should you use to help you?*

IN FOCUS Question ❶ focuses on finding the composite angles and applying prior knowledge of squares and triangles, as well as angles around a point, to work out angles without using a protractor.

Question ❷ challenges children to focus on deducing side lengths by calculating, based on the given dimensions.

Question ❸ explores the potential assumption that splitting any rectangle in half creates 45°, as the 90-degree angles may be seen to be halved by the process. Children should use or create different rectangles in different proportions to check this conjecture.

STRENGTHEN It may help children to use paper versions of all the shapes to support their reasoning, as these can be manipulated physically to support and prompt, but not replace, reasoning.

DEEPEN Children can be given paper squares and rectangles and experiment with ways of cutting them in half. What different angles do they make? Can they work out the answers without measuring?

ASSESSMENT CHECKPOINT Can children justify their reasoning for each stage, based on the properties of the shapes given?

ANSWERS

Question ❶: p = 45°   q = 135°   r = 90°
              s = 135°   t = 45°   u = 270°

Question ❷: length = 200 mm   width = 45 mm

Question ❸: Max is incorrect. The diagonals of a rectangle do not split the 90° angles at the vertices in half, unless the rectangle is also a square.

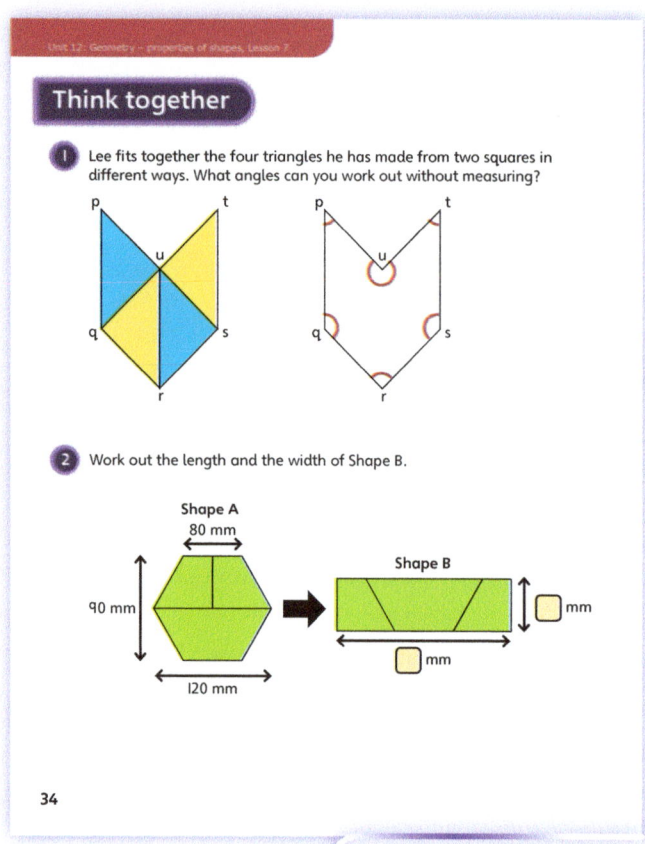

PUPIL TEXTBOOK 5C PAGE 34

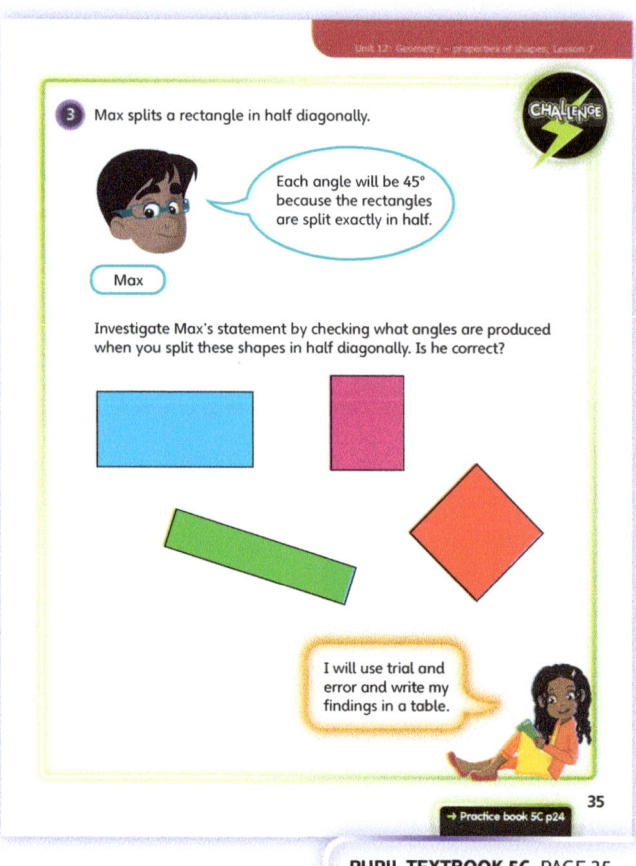

PUPIL TEXTBOOK 5C PAGE 35

Unit 12: Geometry – properties of shapes, Lesson 7

Practice

WAYS OF WORKING Independent thinking

IN FOCUS Question ❶ has been included as it allows children to practise the skill of deducing angles based on the properties of an original design.

Question ❷ requires children to recognise the composite angles and lengths shown.

Question ❸ asks children to apply the reasoning from their understanding of angles on a straight line and at a right angle. This takes the knowledge from the previous lessons, on angles on a line and around a point, and begins to apply it to properties of shapes.

STRENGTHEN Support children by creating paper or card versions of the shapes to be manipulated in the ways shown in the problems. They could explore the effects of halving different squares.

DEEPEN Extend learning on question ❶ by asking children to cut up a square into smaller squares and rectangles and to then calculate the interior angles and side lengths of their new shapes.

ASSESSMENT CHECKPOINT Question ❸ requires a deep understanding of how to make deductions about missing angles. Are children applying their knowledge of right angles and the sum of angles on straight lines and around points to calculate the answers?

ANSWERS Answers for the **Practice** part of the lesson can be found in the *Power Maths* online subscription.

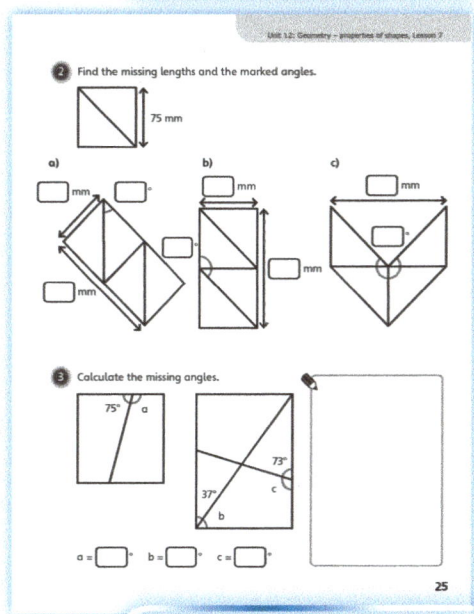

PUPIL PRACTICE BOOK 5C PAGE 24

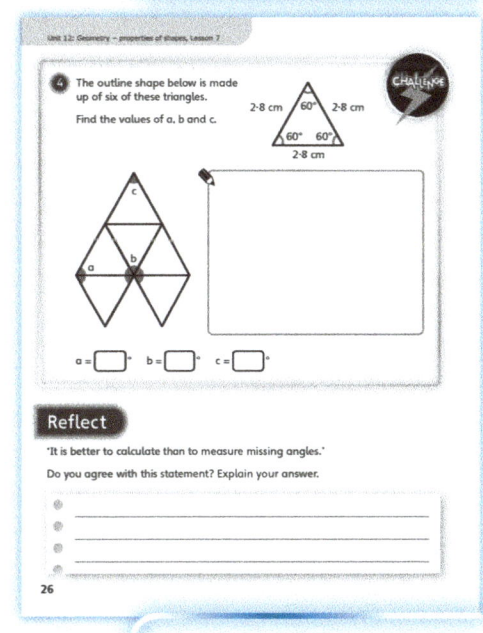

PUPIL PRACTICE BOOK 5C PAGE 25

Reflect

WAYS OF WORKING Independent thinking

IN FOCUS The **Reflect** section allows children to think back over the shapes they have worked on in this unit to decide if there are times when calculating is more efficient than measuring. It will provide them with an opportunity to fully digest the unit as a whole and reflect on how they will tackle geometry lessons in the future.

ASSESSMENT CHECKPOINT Can children justify when to measure and when to calculate, based on their knowledge of the properties of angles and shapes? Children should discuss how there are times when calculating is a more efficient or accurate method, but that sometimes it may be more practical to measure because the calculations may not be clear, such as finding the length of a diagonal.

ANSWERS Answers for the **Reflect** part of the lesson can be found in the *Power Maths* online subscription.

After the lesson ⏸

- Were children able to explain their reasoning?
- Could children apply their knowledge of angles around a point or on a straight line?

PUPIL PRACTICE BOOK 5C PAGE 26

Unit 12: Geometry – properties of shapes, Lesson 8

Regular and irregular polygons

Learning focus
In this lesson, children will deepen their understanding of the concepts of regular and irregular polygons by considering them in terms of their knowledge of angles and lengths.

Before you teach
- What do children remember about regular and irregular shapes?
- Can children define 'quadrilateral'?
- Do children know how to measure the angles in a given shape?

NATIONAL CURRICULUM LINKS

Year 5 Geometry – properties of shapes

Distinguish between regular and irregular polygons based on reasoning about equal sides and angles.

ASSESSING MASTERY

Children can recognise common regular polygons and can justify why a given polygon does or does not meet the criteria necessary to be regular, based on angle size or lengths of sides.

COMMON MISCONCEPTIONS

Children may rely on visual intuition which will lead to misidentification in certain cases. Ask:
- *How can you tell for sure whether this shape is regular or irregular? What equipment can you use to check?*

STRENGTHENING UNDERSTANDING

Children should explore different shapes by manipulating them on geoboards or using construction materials to form models of 2D shapes. Encourage children to explore whether sides are the same length and to investigate the size of the angles in the shapes.

GOING DEEPER

Challenge children to consider how the properties of different grids restrict the formation of certain regular polygons. Do the properties of grids prevent children from making certain shapes with all the same length sides and the same sized angles?

Challenge children to make and justify general statements about regular polygons.

KEY LANGUAGE

In lesson: regular, irregular, polygon, rhombus, rectangle, octagon, pentagon, sides, angles

Other language to be used by the teacher: polygon, quadrilateral, vertices

STRUCTURES AND REPRESENTATIONS

2D shapes

RESOURCES

Mandatory: protractors, rulers

Optional: geoboards, squared paper, construction materials such as strips of card, split pins, geostrips

 In the eTextbook of this lesson, you will find interactive links to a selection of teaching tools.

Quick recap

Ask children what polygon names they know. Ask: *What is a 3-sided polygon called? What about a 4-sided polygon? Can you have a 2-sided polygon?* Work through the polygons up to a 10-sided polygon.

Unit 12: Geometry – properties of shapes, Lesson 8

Discover

WAYS OF WORKING Pair work

ASK

- Question 1 a): *What does 'polygon' mean?*
- Question 1 a): *How could you find the size of the interior angles?*
- Question 1 a): *Could the grid help identify the size of each angle?*
- Question 1 b): *What are the properties of a regular shape?*

IN FOCUS Question 1 a) encourages children to consider what is meant by the interior angles of a shape and discuss different ways to work out each angle. For question 1 b), children should discuss what is meant by the phrase 'regular polygon' in general and 'regular octagon' in particular.

PRACTICAL TIPS This could be turned into a practical task by giving children geoboards to model shapes on. Ask children to copy the octagon from the **Discover** task. Alternatively, children could draw the octagon onto squared paper, making sure they carefully match what they see in the picture.

ANSWERS

Question 1 a): Every interior angle is 135°. Isla is correct.

Question 1 b): The sides are not all the same length, so Richard is not correct.

PUPIL TEXTBOOK 5C PAGE 36

Share

WAYS OF WORKING Whole class teacher led

ASK

- Question 1 a): *Which is the most accurate or efficient method for finding the size of these angles?*
- Question 1 a): *What do you notice when you compare each angle?*
- Question 1 a): *Can you always use the grid to find the size of an angle?*
- Question 1 b): *Can you prove that the diagonals are longer than the other lines? Are diagonal lines always longer?*

IN FOCUS The main point of question 1 is to recap the core concepts of regular and irregular polygons: that regular polygons have angles that are all the same and have sides that are all the same length.

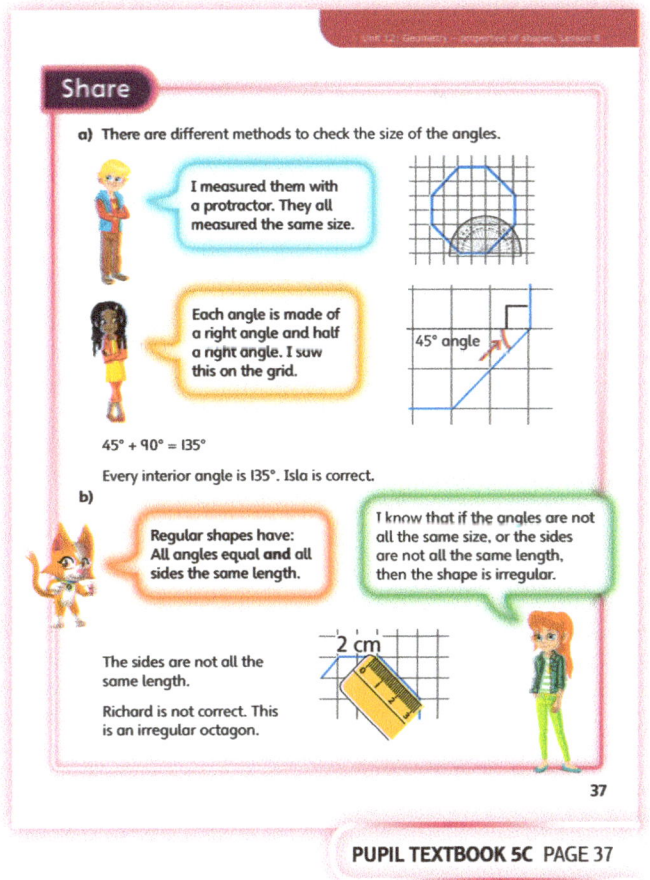

PUPIL TEXTBOOK 5C PAGE 37

Unit 12: Geometry – properties of shapes, Lesson 8

Think together

WAYS OF WORKING Whole class teacher led (I do, We do, You do)

ASK

- Question ❶: *Is a shape definitely regular if its sides are all the same length?*
- Question ❶: *What two aspects do you need to check to decide if a shape is regular?*
- Question ❷: *Can you convince me that the angles are, or are not, all the same size?*
- Question ❸: *How many different regular quadrilaterals are there?*

IN FOCUS Question ❶ focuses on the variations that cause irregularity: same lengths but different angles, same angles but different lengths, different lengths and different angles. This reinforces how important it is for children to check everything before making a decision on whether a shape is regular or irregular.

Question ❸ challenges children to consider how different geoboards and grids support different types of regular shapes.

STRENGTHEN It would be beneficial if children had access to geoboards to explore the construction of different regular and irregular shapes on different grids. Alternatively, children could make shapes with strips of card and split pins.

DEEPEN Challenge children to make and test various general statements about regular and irregular shapes, such as 'A regular shape has right angles'.

ASSESSMENT CHECKPOINT Question ❷ requires a sound understanding of the core concept about regular and irregular polygons. Look out for clear understanding of the two different aspects required for a regular polygon: to have both equal sides and equal angles. The question also allows assessment of accurate measuring skills with a ruler and protractor.

ANSWERS

Question ❶: From left to right: the angles are not all equal; the sides are not all equal; neither the sides nor the angles are all equal.

Question ❷: Shape A has measurements of 21 mm (top lines), 20·5 mm (side lines) and 25 mm (bottom line). Shape B is 28 mm all sides. Shape C has measurements of (clockwise from top line): 26 mm, 26 mm, 33 mm, 23 mm and 35 mm. Shape B is regular as all the sides are equal and all the angles are equal.

Question ❸ a): Max should use the left board to make a square, the only regular quadrilateral. Ambika should use the right board to make a hexagon with side length 1 or 2.

Question ❸ b): Board 1: Square or equilateral triangle. Board 2: Equilateral triangle.

Question ❸ c): Any other polygons, apart from those mentioned in b). For example, a pentagon, an octagon or a decagon.

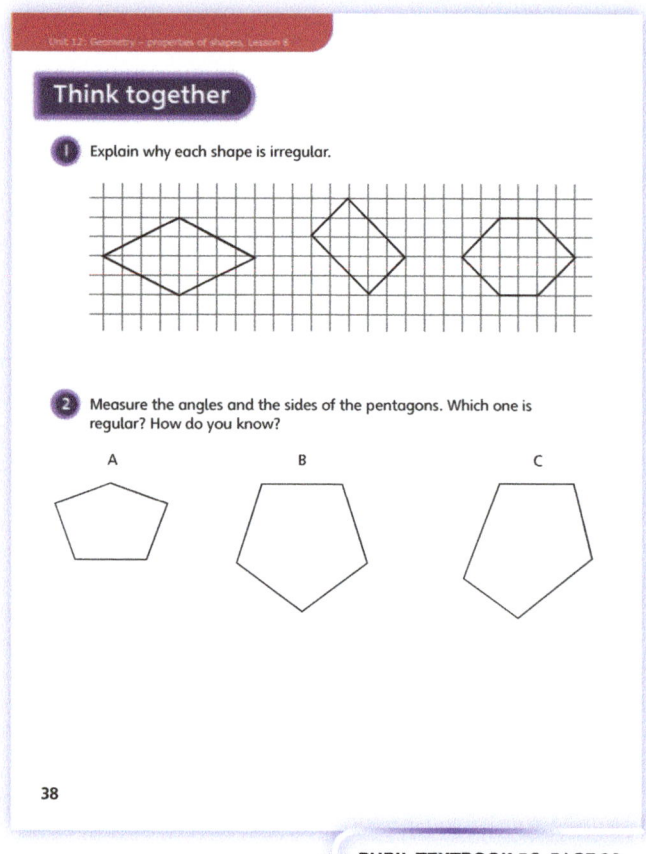

PUPIL TEXTBOOK 5C PAGE 38

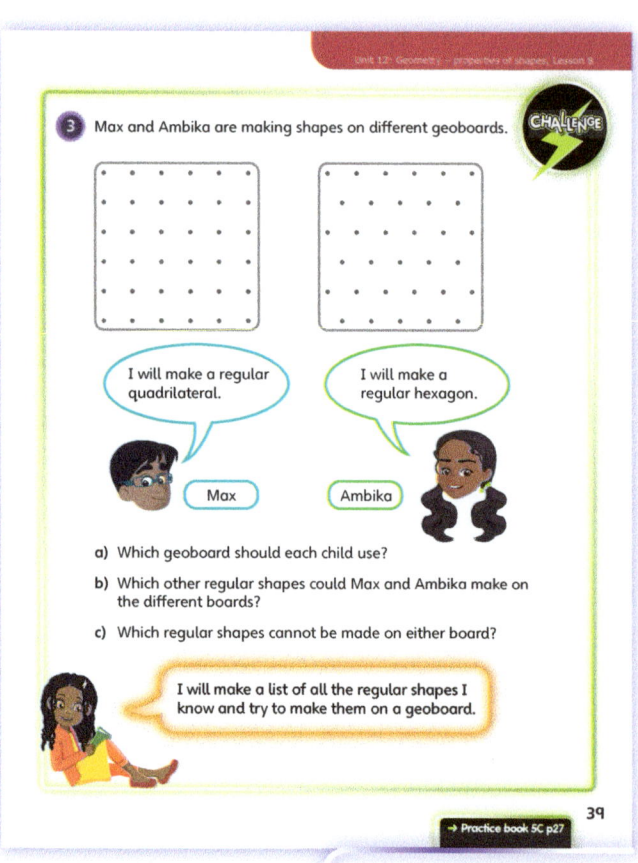

PUPIL TEXTBOOK 5C PAGE 39

Unit 12: Geometry – properties of shapes, Lesson 8

Practice

WAYS OF WORKING Independent thinking

IN FOCUS Question ❶ requires children to reason about whether or not different polygons meet both the criteria for being regular. This cements the thinking that children need to look at both the angles and the lengths of the sides in order to determine whether a shape is regular.

Question ❷ prompts consideration of the case of regular triangles (equilateral triangles).

STRENGTHEN Encourage children to draw copies of the shapes in questions ❹ and ❺ onto squared or isometric paper, then cut them out and move them around to help visualise how the shapes can fit together.

DEEPEN Deepen learning for question ❹ by providing children with new shapes and asking them to predict which shapes will join together to make a regular shape.

THINK DIFFERENTLY Question ❸ requires children to explain the mistaken assumption that having equal angles is sufficient to meet the requirements for a shape being regular.

ASSESSMENT CHECKPOINT Questions ❶ and ❷ cover the fundamental concepts. In question ❷, children may mistake a right-angled triangle for a regular shape, as it looks even. Use children's responses to check that they understand that both sides and angles have to be equal for a shape to be regular.

ANSWERS Answers for the **Practice** part of the lesson can be found in the *Power Maths* online subscription.

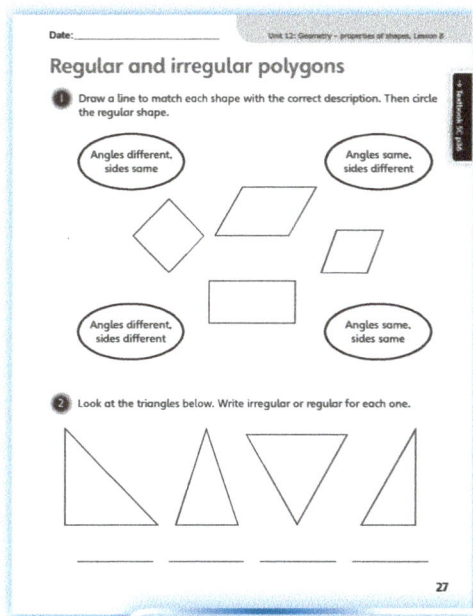

PUPIL PRACTICE BOOK 5C PAGE 27

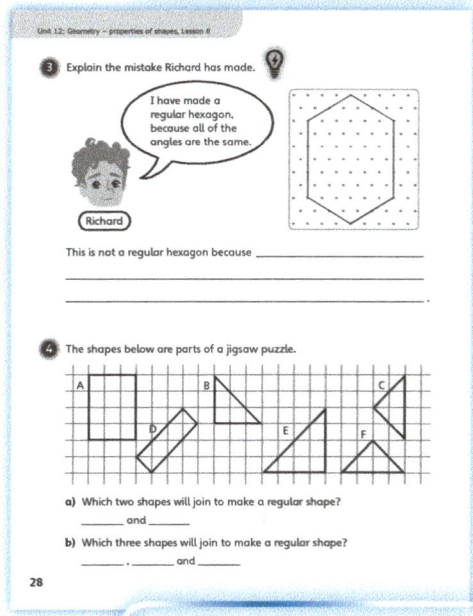

PUPIL PRACTICE BOOK 5C PAGE 28

Reflect

WAYS OF WORKING Pair work

IN FOCUS Children consider the points they have learnt throughout the lesson. Encourage them to explain to a partner the different ways of checking whether a shape is regular or irregular.

ASSESSMENT CHECKPOINT Do children include both angle size and side length as requirements for a regular shape?

ANSWERS Answers for the **Reflect** part of the lesson can be found in the *Power Maths* online subscription.

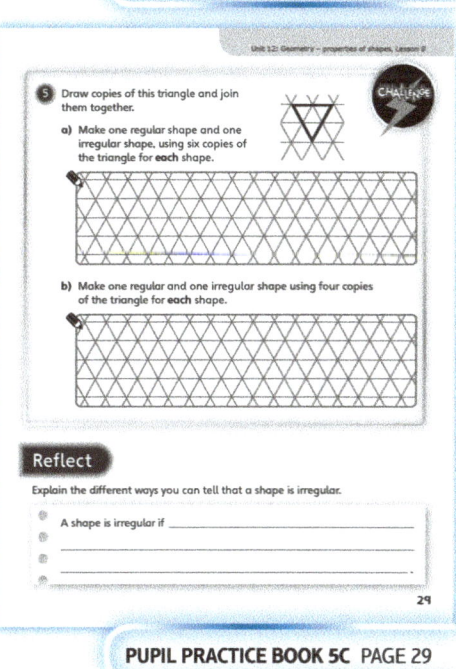

PUPIL PRACTICE BOOK 5C PAGE 29

After the lesson ⏸

- Can children explain confidently that a shape must meet both criteria to be regular?
- Are children able to justify their reasoning based on the properties of the shapes or the grid on which the shapes are constructed?

Unit 12: Geometry – properties of shapes, Lesson 9

Parallel lines

Learning focus
In this lesson, children will develop their understanding of parallel lines, including the use of arrow notation to distinguish sets of parallel lines. Children will recognise and draw parallel lines in different orientations.

Before you teach
- What do children remember about the word 'parallel'?
- Can children give examples of where they can see parallel lines?
- Do children know what 'horizontal' and 'vertical' mean?

NATIONAL CURRICULUM LINKS

Year 3 Geometry – properties of shapes

Identify horizontal and vertical lines and pairs of perpendicular and parallel lines.

ASSESSING MASTERY

Children can use arrow notation to indicate sets of lines that are parallel and can use the properties of a grid to reason about parallel and non-parallel lines.

COMMON MISCONCEPTIONS

Children may think that parallel lines must be exactly the same length. Point to two parallel lines of different lengths. Ask:
- *Can these two lines be called parallel? Do you look at the length to find that out?*

Children may fail to recognise that there can be more than two lines parallel to one another. Ask:
- *Are all parallel lines in pairs?*

STRENGTHENING UNDERSTANDING

Encourage children to discuss different ways to describe parallel lines, recapping previous work with them. Parallel lines are a constant width apart; they would never meet, even if they continued forever, and they go in exactly the same direction.

GOING DEEPER

Challenge children to use the properties of a grid to create lines that are parallel. Encourage them to draw a line on a grid for a partner to draw a parallel line, based on the grid.

KEY LANGUAGE

In lesson: parallel, horizontal, vertical, arrow marking, diagonal

Other language to be used by the teacher: length, constant, width

RESOURCES

Mandatory: rulers

Optional: sticks

 In the eTextbook of this lesson, you will find interactive links to a selection of teaching tools.

Quick recap

Play a game of geometry bingo. On the board write 12 words related to geometry and ask children to pick 6 to put in a bingo grid. Then read out a definition of each word and children cross the word out if they have it. Include the words horizontal and vertical.

Discover

WAYS OF WORKING Pair work

ASK

- Question 1 a): *How can you tell if two lines are parallel?*
- Question 1 b): *How far apart are these lines? Does the distance apart change?*
- Question 1 b): *Would these lines cross if they continued further?*

IN FOCUS Questions 1 a) and b) contain examples of parallel lines that are horizontal and vertical. They also contain clear examples of non-parallel lines, discussion of which should deepen understanding of the concept.

Children should have a good basic understanding of parallel lines from previous years' learning, but it will be important to recap the key ideas that parallel lines are a constant width apart and would never touch or cross, even if extended indefinitely.

PRACTICAL TIPS The gates in the image show parallel lines in the horizontal and vertical orientations very clearly. There will be many examples of parallel lines in the school environment. The class could go on a parallel line treasure hunt around the school, perhaps capturing many different examples on a digital camera, to be displayed or printed for inspection by the class.

ANSWERS

Question 1 a): All the horizontal and all the vertical lines on the gates are parallel.

Question 1 b): The diagonal lines on the gates are not parallel because if they continued they would cross over.

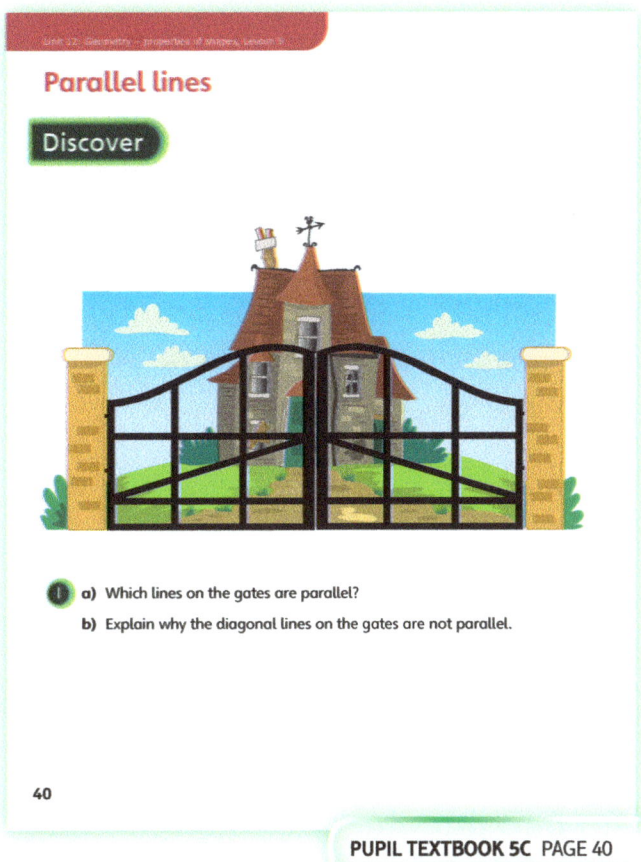

PUPIL TEXTBOOK 5C PAGE 40

Share

WAYS OF WORKING Whole class teacher led

ASK

- Question 1 a): *Why do some lines have one arrow mark and others have two or three?*
- Question 1 b): *What is the proof that the diagonal lines are not parallel?*
- Question 1 b): *Can parallel lines ever be diagonal?*

IN FOCUS The focus in questions 1 a) and b) is on recapping the main definitions of parallel lines and learning how to use arrow notation to indicate different sets of parallel lines.

Children should discuss how all the vertical lines in this picture are parallel to each other (as are all the horizontal lines). Parallel lines do not have to come solely in pairs.

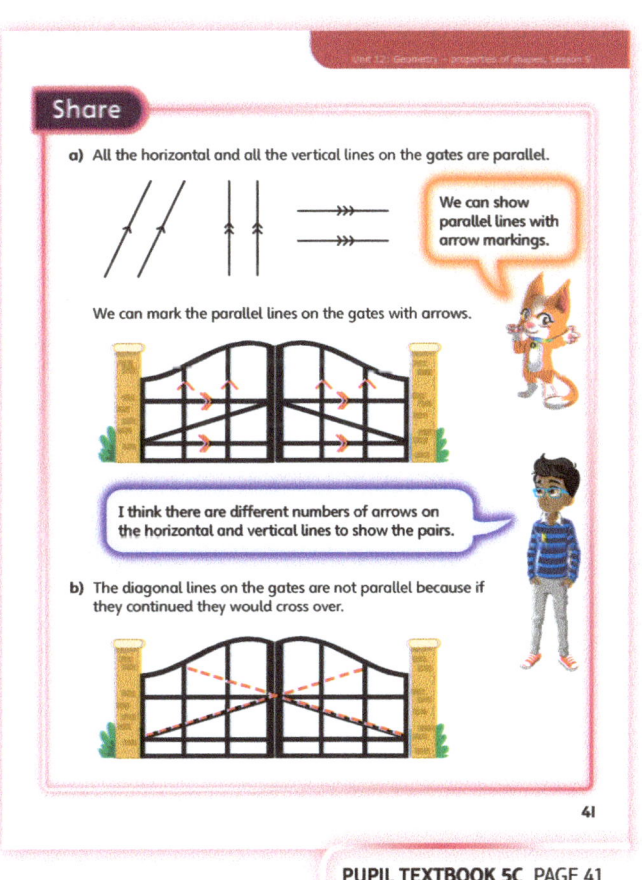

PUPIL TEXTBOOK 5C PAGE 41

Unit 12: Geometry – properties of shapes, Lesson 9

Think together

WAYS OF WORKING Whole class teacher led (I do, We do, You do)

ASK

- Question ❶: *The gate has fallen, so the vertical lines are no longer vertical. Does this change whether the lines are parallel? Are the horizontal lines still parallel on the fallen gate?*
- Question ❷: *How can you make sure you copy the shapes accurately?*
- Question ❸ a): *What letters do you use to describe a line?*
- Question ❸ c): *How will you make sure your lines are parallel?*

IN FOCUS Question ❶ has been included as it covers sets of parallel lines that are not oriented horizontally or vertically. Question ❷ introduces the idea of sets of parallel lines within shapes. Question ❸ introduces the notation for a line by naming the vertices, such as 'line BC'. Questions ❷ and ❸ require children to use the properties of the grid to identify, form and justify sets of parallel lines.

STRENGTHEN Encourage children to place long, straight sticks or rulers over lines, to support their understanding of whether parallel lines cross if extended.

DEEPEN Challenge children to recognise that the properties of a grid can support their reasoning about parallel lines. Ask: *How can you use the grid to prove your lines are parallel?*

ASSESSMENT CHECKPOINT Question ❸ a) requires a deep understanding of parallel lines and the vocabulary introduced in this lesson. Are children using the grid lines to help make their identification of parallel lines more efficient?

ANSWERS

Question ❶: The same pairs of lines are still parallel.

Question ❷: Shape b) has no parallel lines.

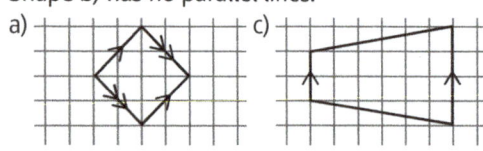

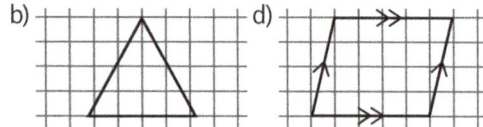

Question ❸ a): Children should point to lines BC and EF.

Question ❸ b): AB and DE are parallel. The distance between the lines stays the same.

Question ❸ c):

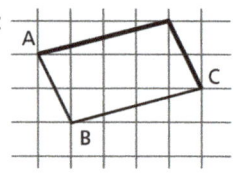

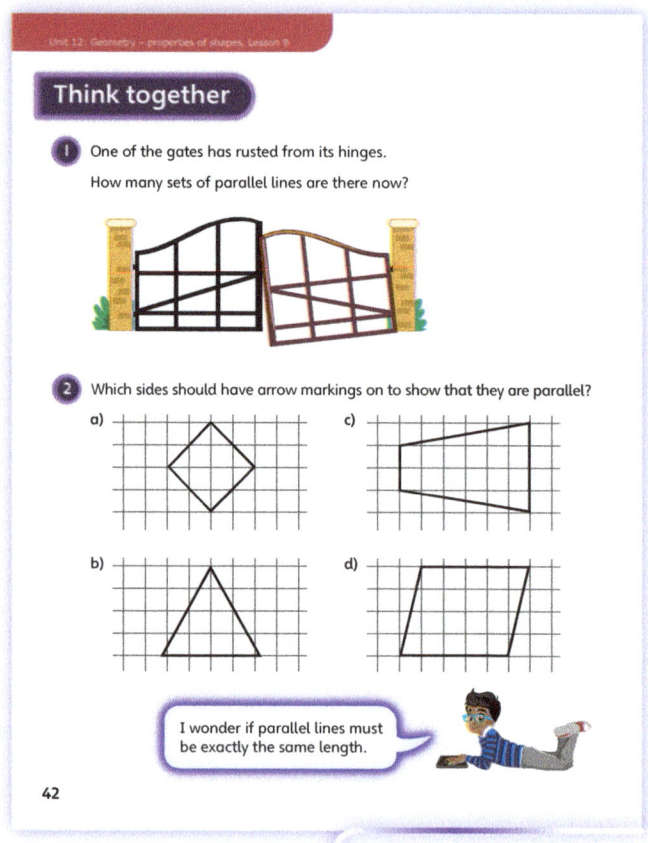

PUPIL TEXTBOOK 5C PAGE 42

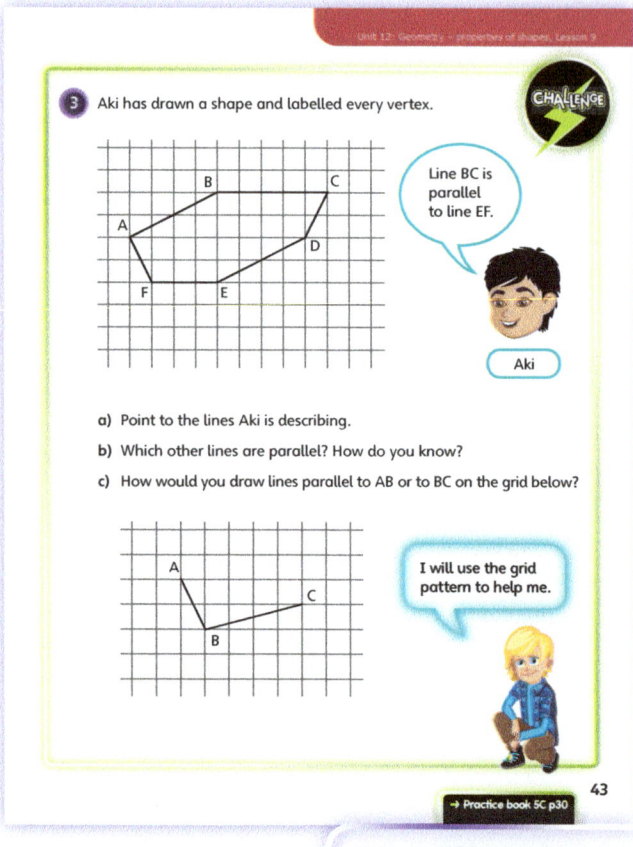

PUPIL TEXTBOOK 5C PAGE 43

Unit 12: Geometry – properties of shapes, Lesson 9

Practice

WAYS OF WORKING Independent thinking

IN FOCUS Question ❶ is included here as it reinforces the skill of using arrow notation to indicate parallel lines.

Question ❷ requires children to recognise parallel lines in shapes and to remember to use the grid lines to aid them in this task.

Question ❸ focuses on the idea that parallel lines need not be the same length.

STRENGTHEN Encourage children to practise forming parallel lines by arranging pairs of pencils or rulers in different orientations.

DEEPEN Question ❺ challenges children to consider examples of parallel and non-parallel lines, by using the properties of an isometric grid to deepen their understanding. Are children able to draw shapes of their own on isometric paper for their partner to find the parallel lines?

THINK DIFFERENTLY Question ❹ challenges children to consider hypothetical lines, which are not yet drawn, but which are indicated by vertex label notation.

ASSESSMENT CHECKPOINT Identifying and marking the parallel lines in different orientations in questions ❶ and ❷ requires a good understanding of the basic concept, allowing you to assess children's grasp of it through their answers.

ANSWERS Answers for the **Practice** part of the lesson can be found in the *Power Maths* online subscription.

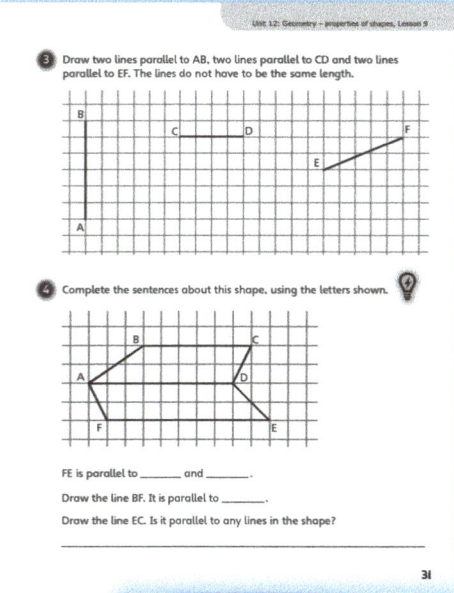

PUPIL PRACTICE BOOK 5C PAGE 30

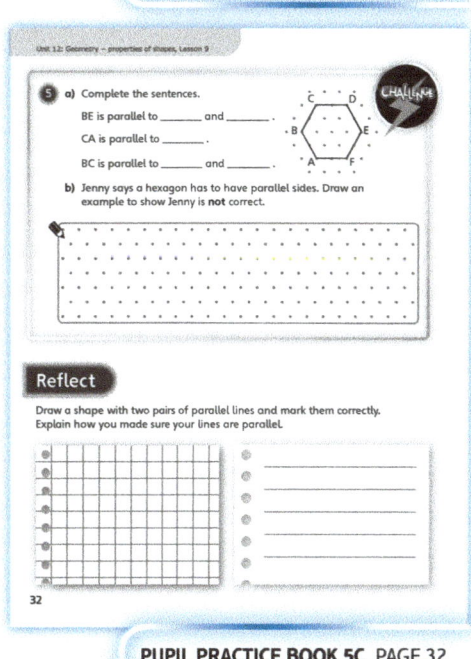

PUPIL PRACTICE BOOK 5C PAGE 31

Reflect

WAYS OF WORKING Independent thinking

IN FOCUS Children could sketch different drafts on a whiteboard or on scrap paper, then share their ideas with a partner, before deciding on the best way to meet the task.

ASSESSMENT CHECKPOINT Can children justify why there are two pairs of parallel lines based on the properties of the grid?

ANSWERS Answers for the **Reflect** part of the lesson can be found in the *Power Maths* online subscription.

After the lesson ⏸

- Do children recognise parallel lines in different orientations?
- Are all children confident in identifying parallel lines that are different lengths?
- Can children use the properties of a grid to draw and identify parallel lines, and justify when lines are not parallel?

PUPIL PRACTICE BOOK 5C PAGE 32

Unit 12: Geometry – properties of shapes, Lesson 10

Perpendicular lines

Learning focus
In this lesson, children will build on their knowledge of angles to deepen their understanding of perpendicular lines, including recognising, labelling and drawing lines that are perpendicular to one another.

Before you teach
- Can children recognise a 90° angle?
- Can children show how a horizontal and a vertical line would meet if they continued?
- Do children remember the word 'perpendicular' from previous learning?

NATIONAL CURRICULUM LINKS

Year 3 Geometry – properties of shapes

Identify horizontal and vertical lines and pairs of perpendicular and parallel lines.

ASSESSING MASTERY

Children can justify a statement that two lines are or are not perpendicular based on their understanding of right angles and the properties of a grid.

COMMON MISCONCEPTIONS

Children may not recognise that two lines can be perpendicular even if they do not touch or cross. Ask:
- *At what angle would these lines meet if they were both continued?*

Children may assume vertical lines or horizontal lines are always perpendicular to another line, even if the line is oriented differently/diagonally in relation to them. Ask:
- *Are all vertical lines perpendicular to other lines?*

STRENGTHENING UNDERSTANDING

Children should develop their understanding of right angles by exploring lines that cross at different angles, perhaps working as a group using lengths of string. They can vary the angle at which the lines cross, and call 'Stop!' when they are perpendicular.

GOING DEEPER

Challenge children to complete rectangles in different orientations by forming perpendicular lines at the vertices and by using the properties of a grid.

KEY LANGUAGE

In lesson: perpendicular, degrees (°), right angle, vertical, horizontal, box marking, protractor

Other language to be used by the teacher: right angle notation

RESOURCES

Optional: protractor, geostrips, geoboards

 In the eTextbook of this lesson, you will find interactive links to a selection of teaching tools.

Quick recap
On the board display a geoboard. Ask one child to come up and draw one side of a square and then ask another child to come up and draw a second side. Keep going until all sides are drawn. Repeat this collaborative activity with different shapes.

Unit 12: Geometry – properties of shapes, Lesson 10

Discover

WAYS OF WORKING Pair work

ASK

- Question 1 a): *What does the word 'perpendicular' mean?*
- Question 1 a): *What are you looking for to determine if the streetlamps are perpendicular?*
- Question 1 b): *How is the word 'perpendicular' different from the word 'vertical'?*

IN FOCUS Questions 1 a) and b) prompt children to discuss the meaning of the word perpendicular, which they originally met in Year 3. They may notice that one set of streetlamps looks precarious. The important aspect to focus on is the difference in meaning between vertical and perpendicular, and how these meanings intersect.

Some children may discuss the relationship between vertical structures and engineering. Ask questions to encourage this, such as: *Why do you not build leaning structures? Why are some houses built in a way that is not perpendicular to the street?*

PRACTICAL TIPS This is presented as an engineering problem, and it could be recreated practically by pushing straws into modelling clay formed into the different gradients of the street. This could also be shown using gymnastics equipment raised to form different inclines.

ANSWERS

Question 1 a): The streetlamps on the top of the hill and on the right are perpendicular to the road because they make right angles.

Question 1 b): The streetlamps on the top of the hill are both vertical and perpendicular to the road, as they make right angles.

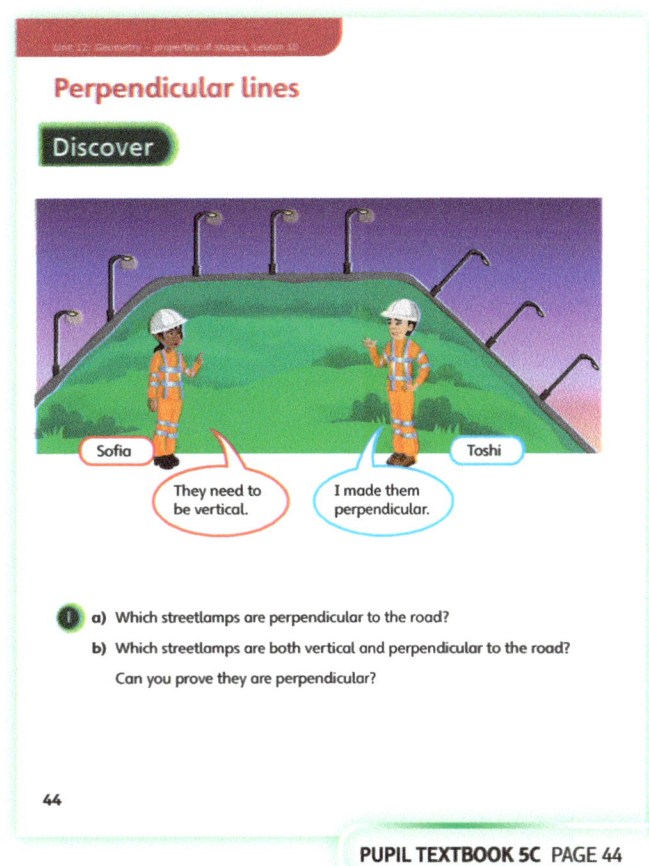

PUPIL TEXTBOOK 5C PAGE 44

Share

WAYS OF WORKING Whole class teacher led

ASK

- Question 1 a): *Look at the little square. Have you seen this before? What does it tell us?*
- Question 1 a): *How can you check if an angle is a right angle?*
- Question 1 b): *Is something that is horizontal always perpendicular to something that is vertical?*

IN FOCUS In question 1 a), children recap that 90° angles are also called right angles, and then learn to use right-angle notation to indicate lines that are perpendicular. They should discuss how perpendicular lines need not be oriented horizontally and vertically. Children should become more confident recognising and using the box marking to indicate a right angle in different orientations.

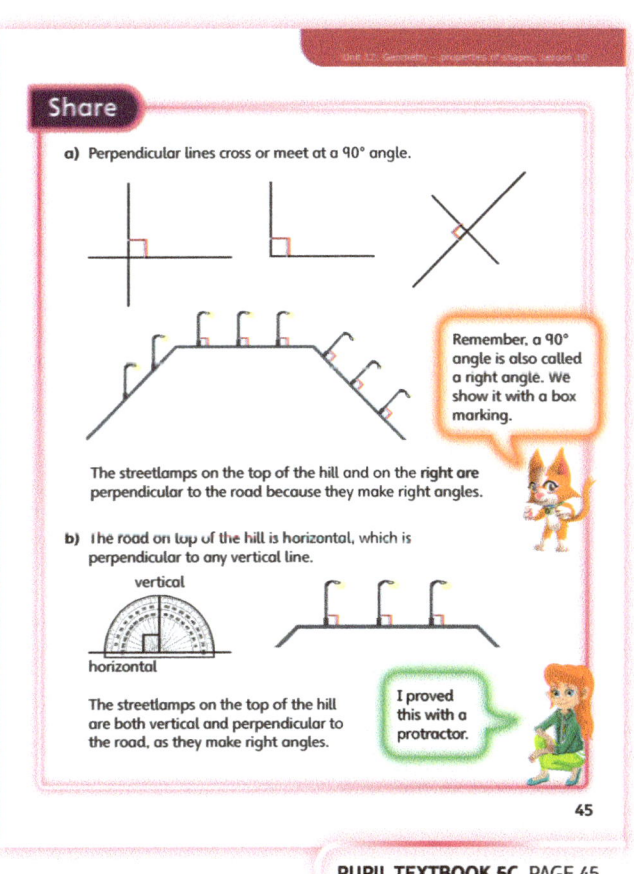

PUPIL TEXTBOOK 5C PAGE 45

Unit 12: Geometry – properties of shapes, Lesson 10

Think together

WAYS OF WORKING Whole class teacher led (I do, We do, You do)

ASK

- Question ❶: *How will you check which lines are perpendicular?*
- Question ❷: *Have you found all the perpendicular lines? Remember, they do not have to touch to be perpendicular to one another.*
- Question ❷: *Do you remember how labelling the vertices makes it easier to identify different lines?*
- Question ❸: *What do you know about the properties of rectangles that could help you?*

IN FOCUS Question ❶ prompts children to recognise perpendicular lines by using the properties of a grid (illustrated on a geoboard).

Question ❷ discusses how two perpendicular lines do not necessarily meet or cross, but would cross at a right angle if they were extended.

STRENGTHEN Work with children to explore perpendicular lines through manipulating sticks, string or geoboards. For question ❷, children may find it helpful to draw out the shapes on squared paper and extend the lines and draw right angle markings to find which are perpendicular.

DEEPEN Question ❸ challenges children to apply their understanding of perpendicular lines to the properties of rectangles, and to consider how to use the properties of a rectangle to construct perpendicular lines. Ask children to discuss different ways to alter the shapes so that they are rectangles. Can children show how to draw rectangles in different orientations?

ASSESSMENT CHECKPOINT Can children explain clearly how they worked out which lines are perpendicular in question ❷?

ANSWERS

Question ❶: C and D show perpendicular lines.

Question ❷ a): AF and CD are perpendicular to AB.

Question ❷ b): GH, KL and IJ are perpendicular to HI.

Question ❸ a): Bella is incorrect. She has not made rectangles because all the angles should be 90° and the adjacent sides should be perpendicular to each other.

Question ❸ b): Perpendicular lines to those shown are needed.

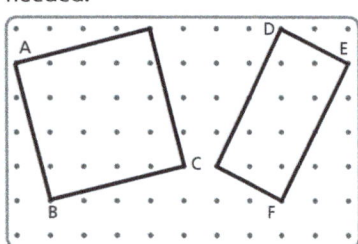

Think together

❶ Which of these diagrams show perpendicular lines?

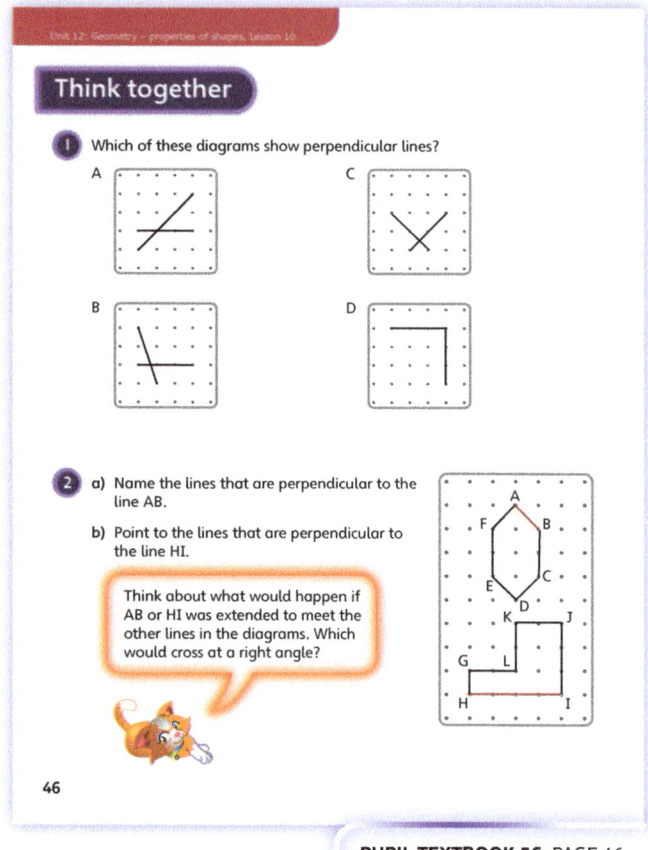

❷ a) Name the lines that are perpendicular to the line AB.

b) Point to the lines that are perpendicular to the line HI.

Think about what would happen if AB or HI was extended to meet the other lines in the diagrams. Which would cross at a right angle?

PUPIL TEXTBOOK 5C PAGE 46

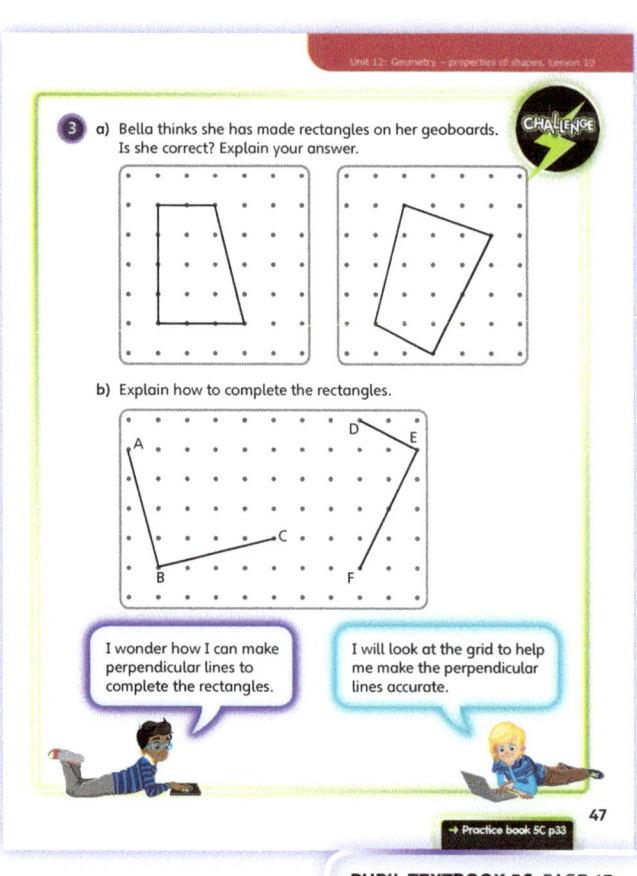

❸ a) Bella thinks she has made rectangles on her geoboards. Is she correct? Explain your answer.

b) Explain how to complete the rectangles.

I wonder how I can make perpendicular lines to complete the rectangles.

I will look at the grid to help me make the perpendicular lines accurate.

PUPIL TEXTBOOK 5C PAGE 47

Unit 12: Geometry – properties of shapes, Lesson 10

Practice

WAYS OF WORKING Independent thinking

IN FOCUS Question ❶ focuses on marking perpendicular lines that cross in different orientations and recognising examples of lines that are not perpendicular. It also helps to ensure children are confident marking right angles. Note that not all the lines meet at right angles.

Question ❷ practises the drawing of perpendicular lines, using the properties of the grid to support reasoning.

Question ❸ has been included to support learning in identifying perpendicular lines in shapes.

Question ❹ challenges children to use vertex notation to communicate about perpendicular lines.

STRENGTHEN Use construction materials to manipulate lines and shapes physically, and make versions of the shapes in the problems.

DEEPEN Question ❻ challenges children to construct rectangles in different orientations by forming perpendicular lines as adjacent sides. Ask children to draw their own part-drawn rectangles for a partner to complete.

THINK DIFFERENTLY Question ❺ challenges children to construct a shape given information about the sides that are perpendicular, and ensures they understand the vertex notation.

ASSESSMENT CHECKPOINT Questions ❸ and ❹ require a good understanding of the key concepts of the lesson. Check that children have marked all of the perpendicular lines in question ❸. Their true or false selections for question ❹ will show their understanding of the key concepts and highlight any misconceptions they may have.

ANSWERS Answers for the **Practice** part of the lesson can be found in the *Power Maths* online subscription.

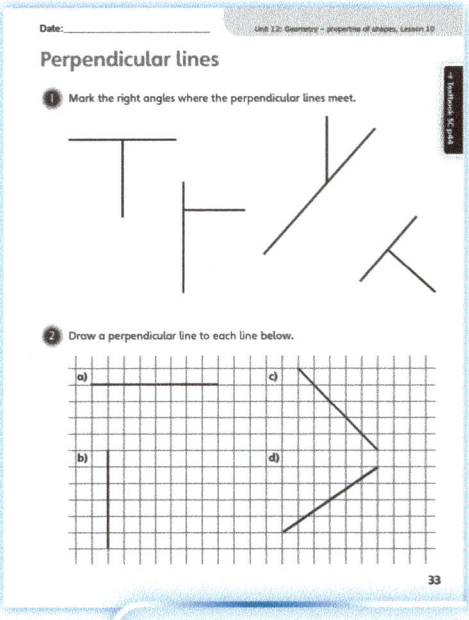

PUPIL PRACTICE BOOK 5C PAGE 33

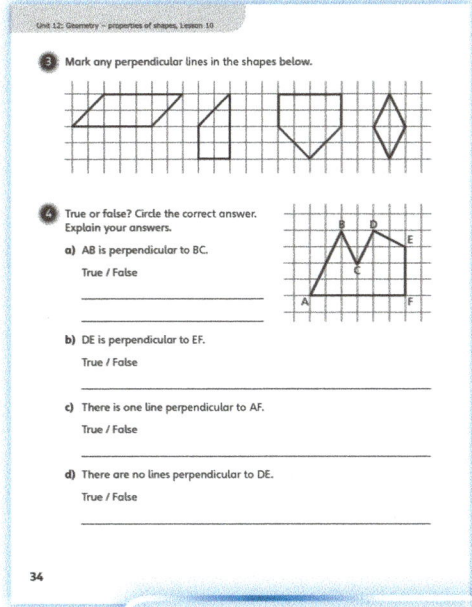

PUPIL PRACTICE BOOK 5C PAGE 34

Reflect

WAYS OF WORKING Independent thinking

IN FOCUS The **Reflect** question provides children with the opportunity to discuss the similarities and differences between the key vocabulary from Lessons 9 and 10 of this unit.

ASSESSMENT CHECKPOINT Can children confidently explain the difference in meaning between parallel and perpendicular?

ANSWERS Answers for the **Reflect** part of the lesson can be found in the *Power Maths* online subscription.

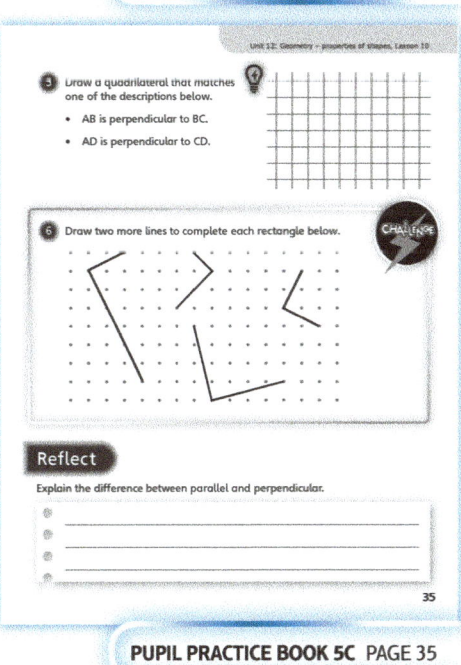

PUPIL PRACTICE BOOK 5C PAGE 35

After the lesson ⏸

- Can children recognise perpendicular lines and justify them in terms of right angles?
- Are children able to use the properties of a grid to form perpendicular lines in diagonal orientations?

Unit 12: Geometry – properties of shapes, Lesson 11

Investigate lines

Learning focus
In this lesson, children will develop their reasoning about parallel and perpendicular lines in relation to one another in shapes and patterns. Children will use angle and length properties to support their judgements.

Before you teach
- Can children identify different examples of parallel lines?
- Can children identify perpendicular lines?
- Can children describe different quadrilaterals? Do they know what is the same and what is different about them?

NATIONAL CURRICULUM LINKS
Year 3 Geometry – properties of shapes
Identify horizontal and vertical lines and pairs of perpendicular and parallel lines.

ASSESSING MASTERY
Children can discuss, investigate and evaluate mathematical statements about parallel and perpendicular lines in shapes and in patterns.

COMMON MISCONCEPTIONS
Children may use single examples to prove a general statement, rather than considering all possible examples. Ask:
- Is that always, sometimes or never true?

STRENGTHENING UNDERSTANDING
Children should manipulate lines in shapes, using geoboards or by drawing on square dotted paper, to help visualise the question they are being asked.

GOING DEEPER
Children can investigate the angles created when different quadrilaterals are overlapped and can discuss which angles can and cannot be created.

KEY LANGUAGE
In lesson: parallel, perpendicular, diagonals (of a quadrilateral), quadrilateral, protractor

Other language to be used by the teacher: angle notation, right angle, parallel line notation, arrow marking

STRUCTURES AND REPRESENTATIONS
2D shapes

RESOURCES
Mandatory: protractor, ruler

Optional: paper strips, card strips, split pins, geoboards, square dotted paper

 In the eTextbook of this lesson, you will find interactive links to a selection of teaching tools.

Quick recap
Check that children can measure and draw angles of given degrees. Write four angles on the board and ask children to use a protractor to draw one of them. Ask them to pass it on to a partner, who measures it and works out which one they have drawn.

Unit 12: Geometry – properties of shapes, Lesson 11

Discover

WAYS OF WORKING Pair work

ASK

- Question 1 a): *What do you notice about the angle these lines cross at?*
- Question 1 b): *Which lines are parallel?*
- Question 1 b): *Are there any perpendicular lines?*

IN FOCUS The main idea of questions 1 a) and b) is to explore the relationship between lines that cross a set of parallel lines. Children consider how, when a straight line crosses the parallel lines, it does so at the same angle each time, and that the parallel lines are always the same distance apart.

PRACTICAL TIPS The task can easily be modelled practically, where children use strips of coloured paper and arrange them to match the image in the **Discover** context. This will also easily allow them to demonstrate their thinking when answering the questions.

ANSWERS

Question 1 a): The plain red strip crosses each dotted strip of paper at an angle of 150°.
The angle is the same each time.

Question 1 b): The red strip is now perpendicular to the dotty strip and parallel to the striped strip.

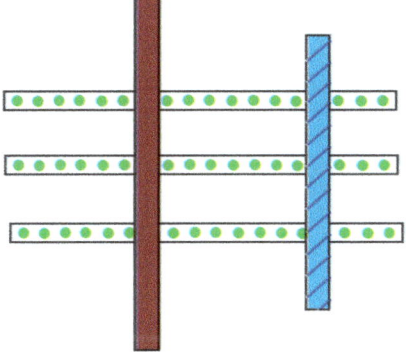

Share

WAYS OF WORKING Whole class teacher led

ASK

- Question 1 a): *Can you measure the angles using your protractor? Can you show me how you use a protractor?*
- Question 1 a): *Are all of the angles the same?*
- Question 1 b): *Can you use a ruler to measure the distance of the plain red strip from the striped blue strip as it crosses each dotted strip?*

IN FOCUS The focus in question 1 is for children to notice that a line crosses the parallel lines at the same angle.

Children may wish to explore this further, to test if it works in other cases, rather than simply accept it as true from one case.

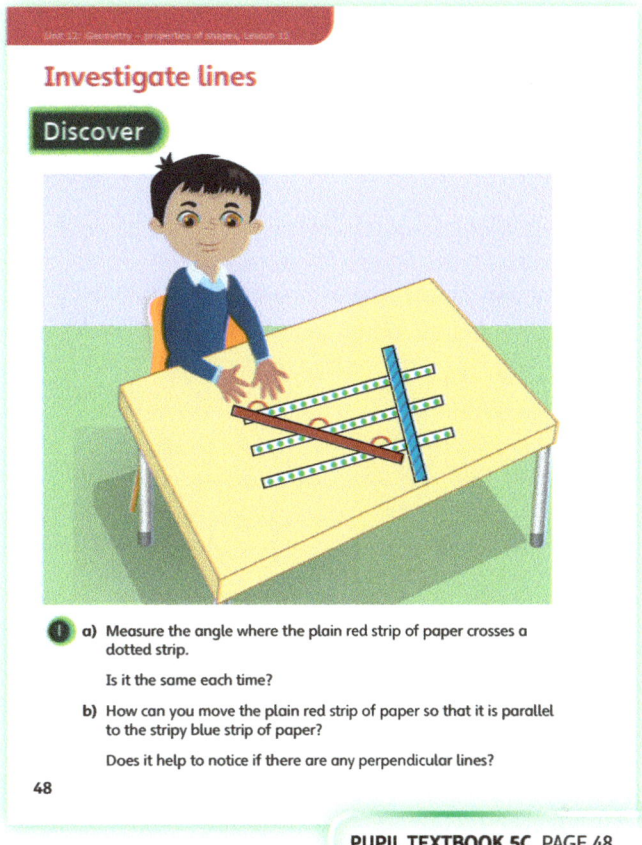

PUPIL TEXTBOOK 5C PAGE 48

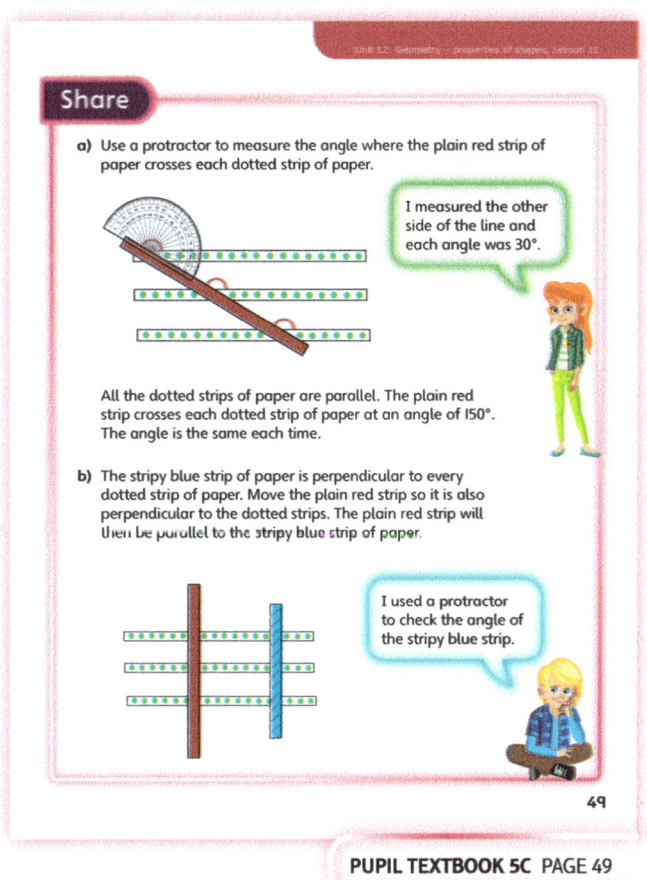

PUPIL TEXTBOOK 5C PAGE 49

85

Think together

WAYS OF WORKING Whole class teacher led (I do, We do, You do)

ASK
- Question ❶: *How will you check your lines are parallel? Can you find more than one answer?*
- Question ❷: *Can you explore the question further by folding your own rectangular piece of paper and predicting the number of parallel and perpendicular lines?*
- Question ❸: *Compare your statements with a partner's. Are they similar?*

IN FOCUS Question ❶ focuses on testing other examples of lines that cross parallel lines at a constant angle.

Question ❷ considers parallel and perpendicular lines in the context of folded paper. This will allow some children to make reasoned predictions.

STRENGTHEN Explore parallel and perpendicular lines by folding paper. Encourage children to make reasoned predictions and decisions based on the shape of the paper and the distances folded; to look for known angles and to use a ruler to measure equal lengths for the fold.

DEEPEN Question ❸ begins to explore general statements about diagonals of quadrilaterals. Children should try to investigate and formulate statements about which quadrilaterals do or do not have perpendicular diagonals. Children could use square dotted paper or geoboards to create a range of quadrilaterals in different orientations.

ASSESSMENT CHECKPOINT Can children justify their answers with geometrical reasoning, rather than relying on a case-by-case test? Children with a solid grasp of the concepts will be able to explain their answers based on geometrical reasoning: known angles, the properties of shapes and angles on lines.

ANSWERS

Question ❶: Place a strip at the same angle to the dotted strip as the red one.

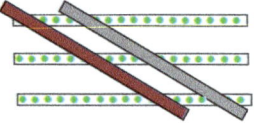

Question ❷: Children should correctly identify the parallel and perpendicular lines.

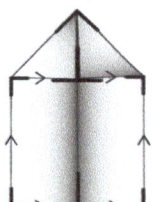

Question ❸ a): A, B and D have perpendicular diagonals.

Question ❸ b): Agree. The diagonals of all squares are perpendicular to each other. The sides of a square are all equal so the diagonals of a square make 4 identical right-angled isosceles triangles.

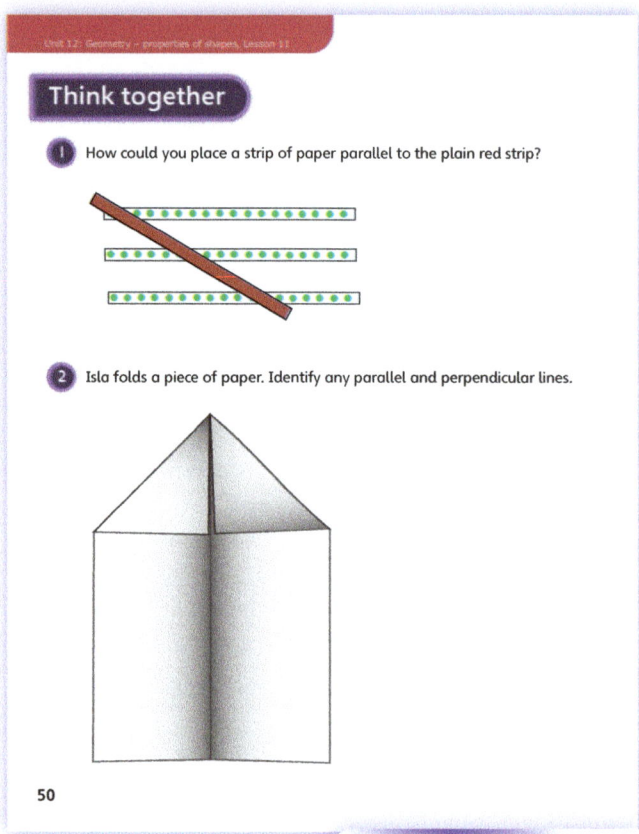

PUPIL TEXTBOOK 5C PAGE 50

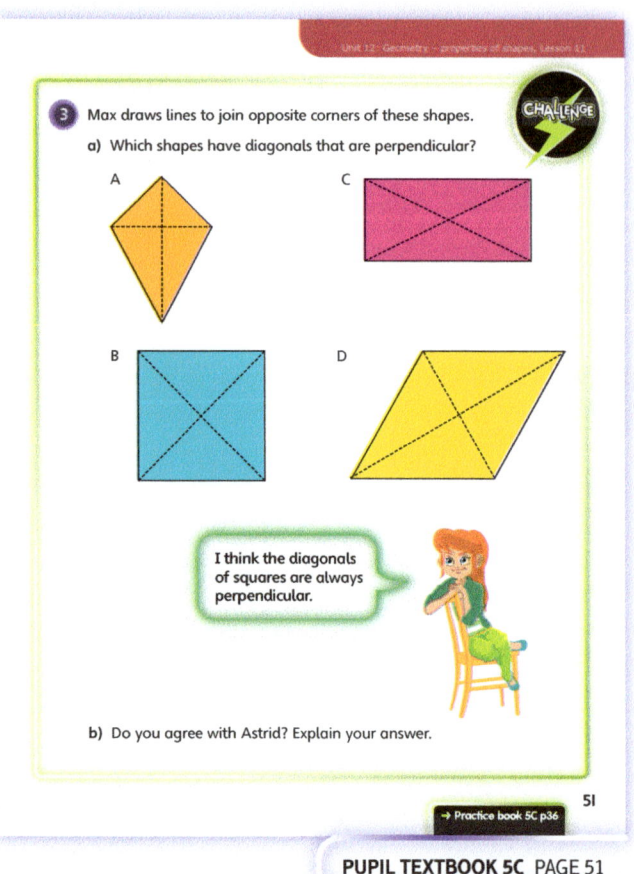

PUPIL TEXTBOOK 5C PAGE 51

Practice

WAYS OF WORKING Independent thinking

IN FOCUS Question ① continues to investigate the angles where straight lines intersect sets of parallel lines. This cements understanding that parallel lines cross at a constant angle. Question ② extends this further by asking children to investigate their own angles.

STRENGTHEN Encourage children to create and manipulate physical representations of the shapes, perhaps by using construction materials such as strips of card joined with split pins.

DEEPEN Question ⑤ challenges children to use the properties of the given dots to form parallel and perpendicular lines. To challenge them further, you could ask children to explore forming parallel and perpendicular lines from different numbers of dots equally spaced around a circle.

THINK DIFFERENTLY Question ③ gives pairs of diagonals from which quadrilaterals need to be formed. Children are therefore working in reverse, constructing a shape from its diagonals. This allows them to develop their visualisation skills whilst thinking about the problem differently.

ASSESSMENT CHECKPOINT Questions ③ and ④ require a good understanding of the distinction between parallel and perpendicular lines. Look for children who are able to articulate the distinction between the two. Can children quickly spot that the diagonal lines in question ③ d) are not perpendicular and can they therefore name which quadrilateral shape d) must be?

ANSWERS Answers for the **Practice** part of the lesson can be found in the *Power Maths* online subscription.

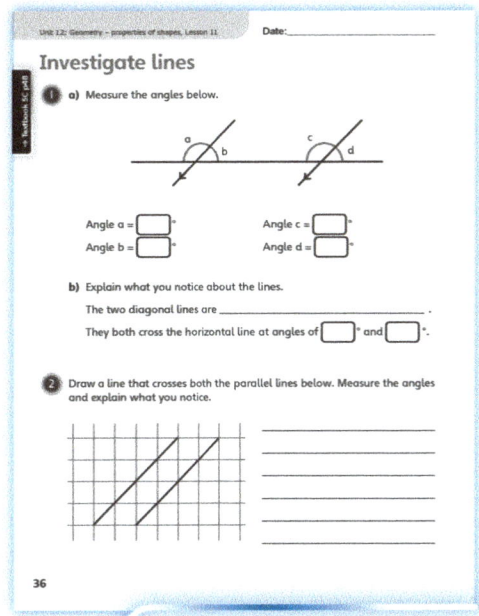

PUPIL PRACTICE BOOK 5C PAGE 36

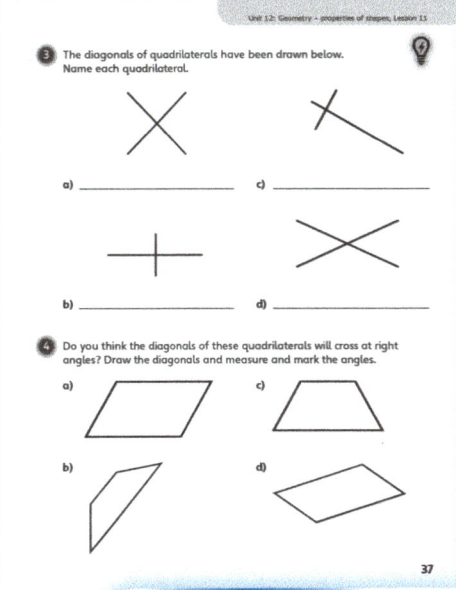

PUPIL PRACTICE BOOK 5C PAGE 37

Reflect

WAYS OF WORKING Independent thinking

IN FOCUS Children are given the opportunity to discuss their ideas, support them with reasons based on the properties of the paper, and then test them by folding. Children can then discuss the reasons behind whether their ideas produced perpendicular lines or not.

ASSESSMENT CHECKPOINT Can children explain the results based on the properties of shape and the structure of the folds they chose?

ANSWERS Answers for the **Reflect** part of the lesson can be found in the *Power Maths* online subscription.

After the lesson

- Can children justify their ideas using geometric reasoning?
- Are children confident and accurate when using the terms parallel and perpendicular in mathematical dialogue?

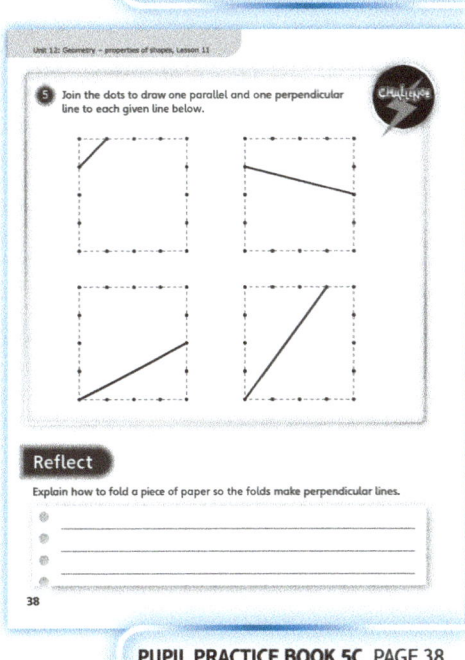

PUPIL PRACTICE BOOK 5C PAGE 38

Unit 12: Geometry – properties of shapes, Lesson 12

3D shapes

Learning focus

In this lesson, children will learn to recognise the different views of 3D shapes or collections when viewed from different positions.

Before you teach

- Which 3D shapes can children name and identify?
- Can children explain the definition of the face of a shape?

NATIONAL CURRICULUM LINKS

Year 5 Geometry – properties of shapes

Identify 3D shapes, including cubes and other cuboids, from 2D representations.

ASSESSING MASTERY

Children can recognise and identify the different viewpoints of common 3D shapes and can reproduce the different views. Children are able to justify the views based on the properties of the 3D shapes themselves, and by visualising the properties of the solid shape from a 2D representation.

COMMON MISCONCEPTIONS

Some children may need support to identify the different viewpoints and to understand that a certain elevation creates a 2D shape. Ask:

- *What would this shape look like from the side or from above?*

STRENGTHENING UNDERSTANDING

Children could explore the concept further by shining a torch onto a shape and noticing the shapes of the shadows cast and how these shadows vary as the angle of the torch or the orientation of the shape varies.

GOING DEEPER

Challenge children to explore and categorise all the different viewpoints of the common 3D shapes that they have access to in class.

KEY LANGUAGE

In lesson: position, face, cylinder, prism, sphere, cuboid, cube, surface, triangle, square, viewpoint, **top view**, **plan view**, **side view**

Other language to be used by the teacher: square-based pyramid, cone

STRUCTURES AND REPRESENTATIONS

3D shapes

RESOURCES

Optional: models of 3D shapes, a torch

 In the eTextbook of this lesson, you will find interactive links to a selection of teaching tools.

Quick recap

Show a variety of 2D and 3D shapes and ask children to name them. You may also want to ask children what shapes they can see on 3D shapes.

Unit 12: Geometry – properties of shapes, Lesson 12

Discover

WAYS OF WORKING Pair work

ASK

- Question 1 a): *What are the faces of the cylinder?*
- Question 1 a): *Would the cylinder look the same from every angle?*
- Question 1 b): *What is the same and what is different about the viewpoints marked A, B and C?*

IN FOCUS This is an introduction to the idea of viewing a shape from different positions and recording the view as a 2D shape. Children should visualise and discuss their ideas, then check these ideas for themselves.

PRACTICAL TIPS Children could explore this question using 3D shapes and shifting their own perspective to view a shape from the side or above. However, the learning will be more powerful if they spend time discussing what they expect to see before actually testing it for themselves.

ANSWERS

Question 1 a): Position A is a top view.

Question 1 b): Bella and Aki both have a side view.

Share

WAYS OF WORKING Whole class teacher led

ASK

- Question 1 b): *How can Bella and Aki see a rectangle when the cylinder has a curved surface?*
- Question 1 b): *Why do they not also see circles?*
- Question 1 b): *What would change if the cylinder was positioned differently? What would stay the same?*

IN FOCUS The focus here is for children to explore the idea that different viewpoints will show different elevations, and that these are represented as 2D shapes. Children could explore this for themselves, using models of 3D shapes and physically moving around the shapes, while they describe the different 2D elevations they see from different angles.

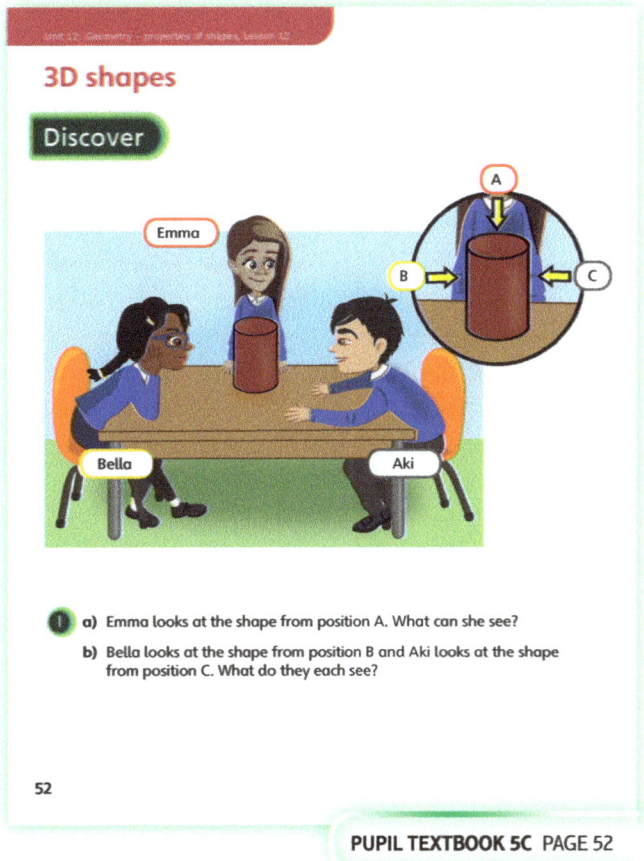

PUPIL TEXTBOOK 5C PAGE 52

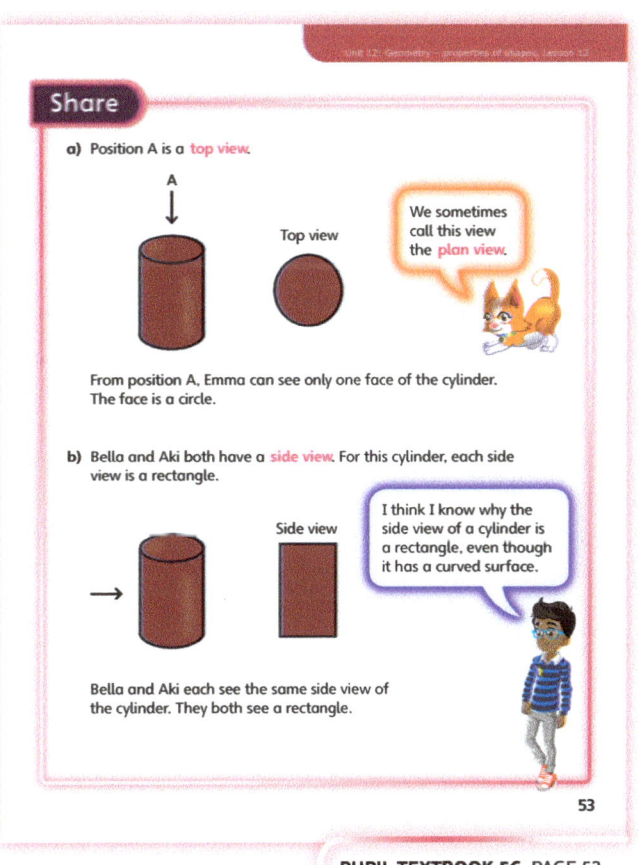

PUPIL TEXTBOOK 5C PAGE 53

Unit 12: Geometry – properties of shapes, Lesson 12

Think together

WAYS OF WORKING Whole class teacher led (I do, We do, You do)

ASK
- Question ❶: *What would this shape look like from above?*
- Question ❷: *What are the different faces of this shape?*
- Question ❷: *Which faces can you see from this angle?*
- Question ❷: *Will the side views be the same?*

IN FOCUS Question ❶ considers the views of a shape made from interlocking cubes.

Question ❷ challenges children to draw the different elevations of a prism by predicting and visualising what they expect the viewpoint to show.

Question ❸ challenges children to consider what an arrangement of different items would look like from above.

STRENGTHEN Children could explore variations on question ❶ by constructing different shapes from interlocking cubes and then viewing them from different positions.

DEEPEN Challenge children to identify a shape given only its views. Ask: *From above it looks square, from each side it looks triangular. What shape can this be?*

ASSESSMENT CHECKPOINT Correctly answering question ❷ shows a good understanding of the properties of 3D shapes. Use children's responses to gauge their ability to recognise 3D shapes from different viewpoints.

ANSWERS

Question ❶: C because all the views are 2 by 2 squares.

Question ❷: Luis and Jamie see rectangles.
Children should correctly draw a rectangle.
Andy sees a triangle.
Children should correctly draw a triangle.

Question ❸: B

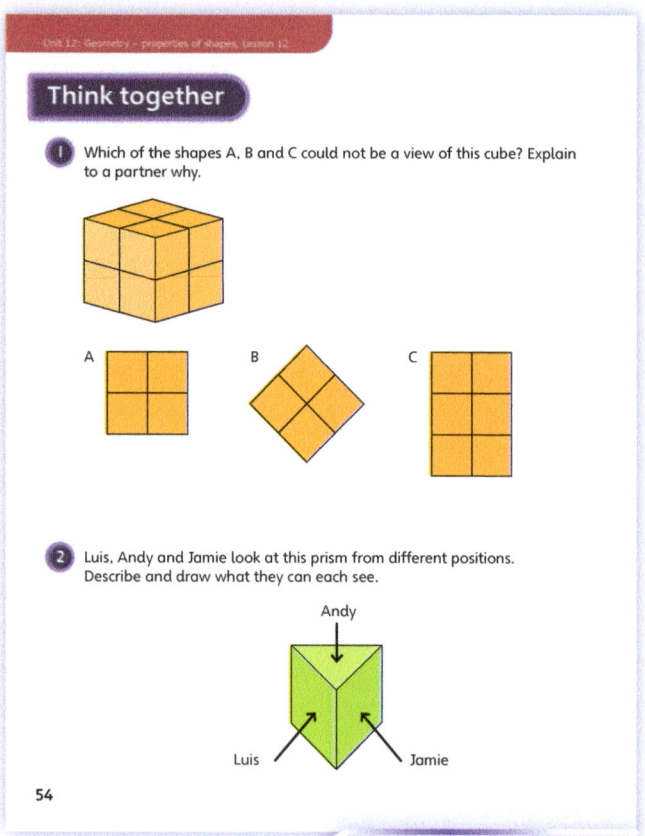

PUPIL TEXTBOOK 5C PAGE 54

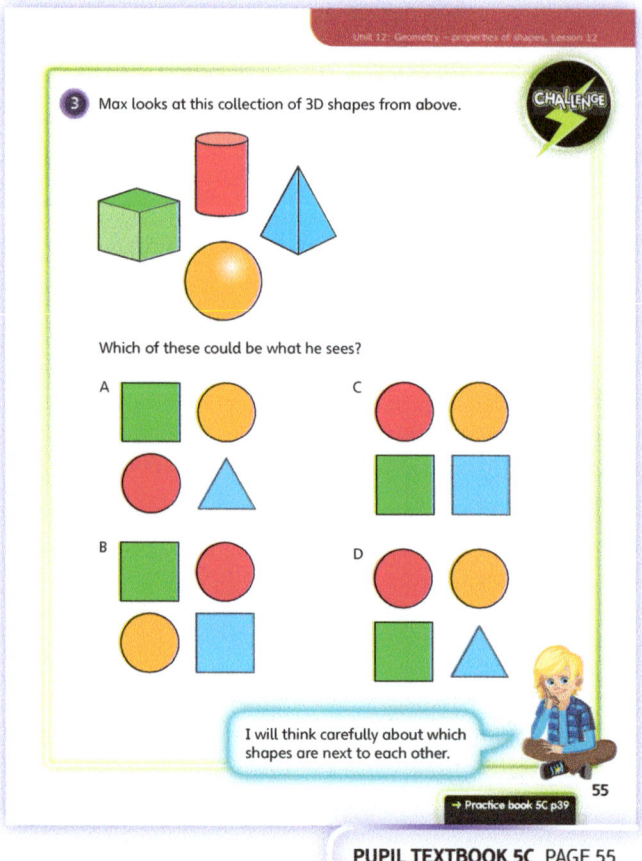

PUPIL TEXTBOOK 5C PAGE 55

90

Unit 12: Geometry – properties of shapes, Lesson 12

Practice

WAYS OF WORKING Independent thinking

IN FOCUS Question 2 looks at examples of possible views of cuboids as well as those that are not possible. This is important as children need to understand which views are not possible in order to fully understand the required properties.

Questions 3 and 4 require children to reason about the properties and faces of common 3D shapes, and require children to consider the different possible aspects from different angles.

Question 5 challenges children to reproduce a series of shapes based on what they would look like from a top view.

STRENGTHEN Children should explore different viewpoints by manipulating models of 3D shapes and shifting their perspective to notice how each view changes.

DEEPEN Question 5 can be deepened by asking children to draw the 3D shapes if viewed from the side. Alternatively, give children a group of 2D shapes and ask children to predict which 3D shapes these could represent if viewed from above.

THINK DIFFERENTLY Question 3 requires children to visualise the whole shape mentally by considering the different views from different positions.

ASSESSMENT CHECKPOINT Children can use their knowledge of the properties of 2D shapes to visualise the faces of 3D shapes. Look at children's answers to question 2 to assess their visualisation, and their answers to questions 3 and 4 to assess their knowledge of the properties of different shapes.

ANSWERS Answers for the **Practice** part of the lesson can be found in the *Power Maths* online subscription.

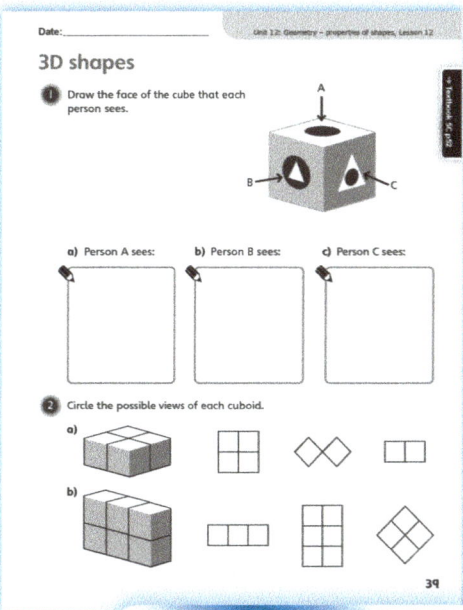

PUPIL PRACTICE BOOK 5C PAGE 39

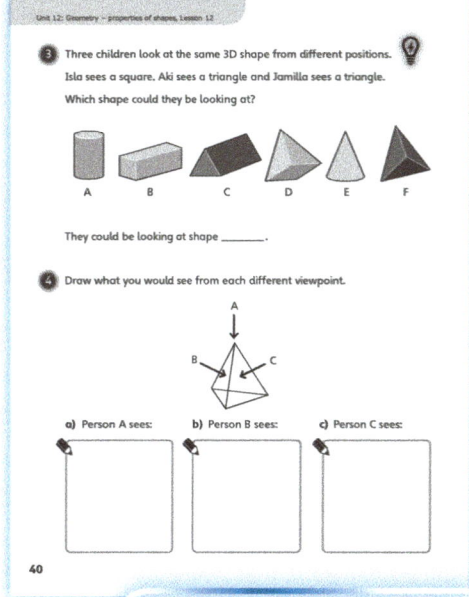

PUPIL PRACTICE BOOK 5C PAGE 40

Reflect

WAYS OF WORKING Independent thinking

IN FOCUS Children should perform this without using a model of the prism. The key skill is that they visualise the shape, rather than simply copy what they can see.

ASSESSMENT CHECKPOINT Have children drawn two rectangles and one triangle? Were they able to visualise the correct 3D shape without extra help?

ANSWERS Answers for the **Reflect** part of the lesson can be found in the *Power Maths* online subscription.

After the lesson

- Are children able to visualise what a shape will look like from different viewpoints?
- Can children justify their visualisations based on the properties of the 3D shapes?

PUPIL PRACTICE BOOK 5C PAGE 41

91

Unit 12: Geometry – properties of shapes

End of unit check

Don't forget the unit assessment grid in your *Power Maths* online subscription.

WAYS OF WORKING Group work teacher led

IN FOCUS

- Question ② requires children to realise that they need to measure accurately using a protractor.
- Question ③ identifies a common mistake of not lining the centre of the protractor up with the turn.
- Question ④ asks children to calculate angles on a straight line and around a point. To answer this question, they will need to apply what they have learnt about the sum of angles on straight lines.
- Question ⑤ assesses whether children understand 3D shapes and the different shaped faces and views they may see.
- Question ⑥ assesses children's understanding of the words parallel and perpendicular.

ANSWERS AND COMMENTARY

Children who have mastered this unit will be able to work out confidently missing angles on a line and around a point without needing to use a protractor. Children can use their knowledge of the properties of squares, triangles and circles to calculate missing angles. They can use a protractor accurately. They can also reason about 2D and 3D shapes using language such as parallel and perpendicular.

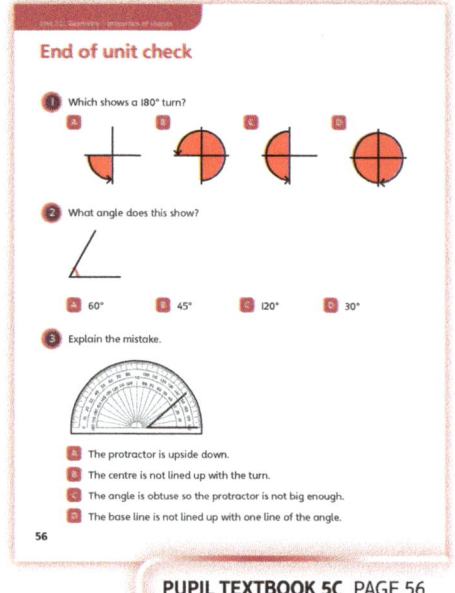

PUPIL TEXTBOOK 5C PAGE 56

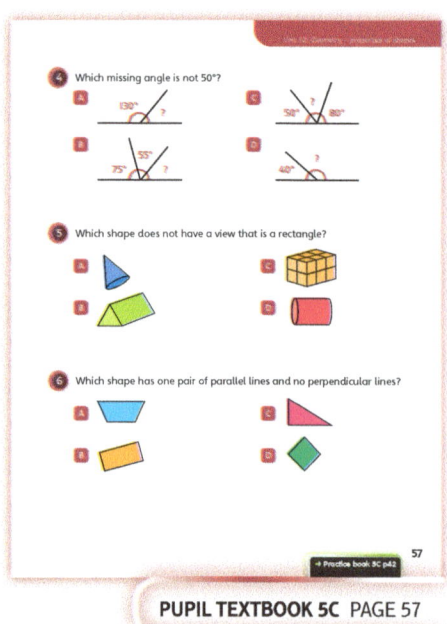

PUPIL TEXTBOOK 5C PAGE 57

Q	A	WRONG ANSWERS AND MISCONCEPTIONS	STRENGTHENING UNDERSTANDING
1	C	D shows they are unaware that a full turn is 360°.	Encourage children to sketch their ideas to support their reasoning. Model the different turns using manipulatives or by role-playing with a group of children. Revisit the three steps for measuring, using a protractor from Lesson 2. Revisit 90-degree and 180-degree angles as quarter and half turns.
2	A	C suggests that children have used the incorrect scale.	
3	B	A shows children do not understand that a protractor may be oriented differently. C shows that children think angle is a measure of the size between two lines.	
4	D	B suggests children may not have added the 1s digits given.	
5	A	C suggests children have read the question incorrectly. D suggests that children cannot see that a rectangle can be made by the curved surface.	
6	A	If children answer B or D, they may not be able to see the perpendicular sides because the shapes are in an angled orientation.	

Unit 12: Geometry – properties of shapes

My journal

WAYS OF WORKING Independent thinking

ANSWERS AND COMMENTARY In question 1 a), children must not only put their new skills of using a protractor accurately to work but also call upon their understanding of how many degrees are in a whole turn. Look to see whether children use addition or subtraction.

For question 2, children should notice that the angles formed where the internal square meets the external square are related. Encourage children to justify this reasoning based on the properties of the angles in a square, and link it with their understanding of finding angles around a point. This knowledge is also relevant for question 1 b).

Children should feel that they can measure to check, or measure first and then explain any patterns they have noticed.

Question 1 a): Groups of angles that fit together to form a whole turn are b, c, d or a, b, g.

Question 1 b): The groups of angles that fit together to form a straight line are a, b, c, f.

Question 2: a = 70° b = 20°
They add up to 90°.

Power check

WAYS OF WORKING Independent working

ASK
- What new skills have you learnt?
- Can you explain angles in a new way or using new vocabulary now?
- How confident do you feel about using a protractor?

Power puzzle

WAYS OF WORKING Pair work

IN FOCUS Use this **Power puzzle** to give children the opportunity to explore the use of a protractor in more detail. Some children may enjoy creating their own patterns for people to follow as a deepening activity.

ANSWERS AND COMMENTARY Children may find that this task requires a few drafts and redrafts. Encourage them by explaining that the protractor is one of the most challenging measuring tools that they will have to learn to use, but that once they have mastered it, they will be able to create accurate designs for their projects. They will also be able to use this skill in secondary school. Can children make a sensible decision on the length of each side so that it fits on the page?

After the unit

- Can children explain how to calculate missing angles on a line or around a point and justify their reasoning?
- Can children discuss the key challenges and common errors when measuring with a protractor?

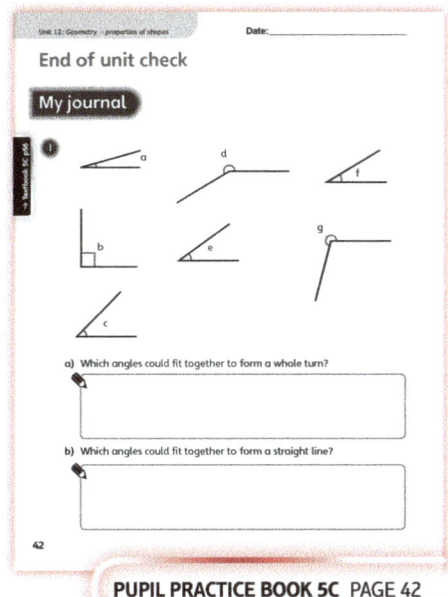

PUPIL PRACTICE BOOK 5C PAGE 42

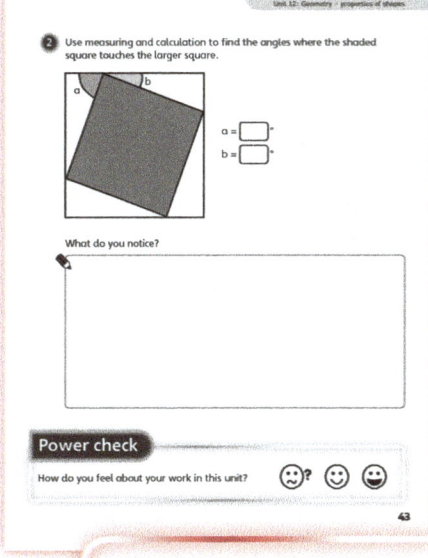

PUPIL PRACTICE BOOK 5C PAGE 43

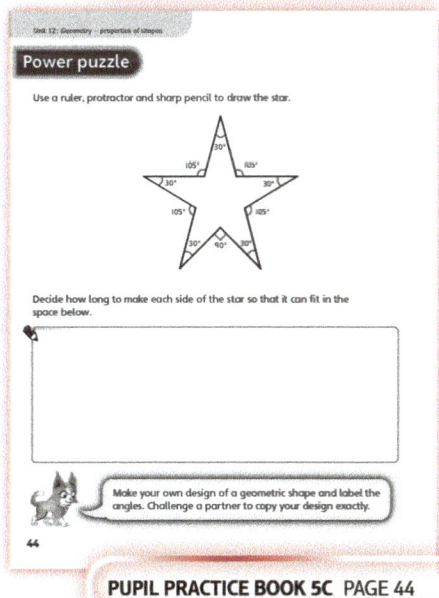

PUPIL PRACTICE BOOK 5C PAGE 44

Strengthen and **Deepen** activities for this unit can be found in the *Power Maths* online subscription.

Unit 13
Geometry – position and direction

Mastery Expert tip! 'I used the ideas of reflection and translation in art. The children created reflective patterns and made pictures using one shape translated and reflected!'

Don't forget to watch the Unit 13 video!

WHY THIS UNIT IS IMPORTANT

This unit teaches children about the position and orientation of shapes and how to reflect and translate points and shapes efficiently using coordinates. Children are reminded of their work from Year 4 on plotting coordinates in the first quadrant. They then build on these skills and apply them to different transformations of shapes. Understanding this Year 5 unit on position and direction will help children progress to problem solving involving position and direction in Year 6.

WHERE THIS UNIT FITS

→ Unit 12: Geometry – properties of shapes
→ **Unit 13: Geometry – position and direction**
→ Unit 14: Decimals

This unit builds on skills children will have gained in Year 4 – using coordinates in the first quadrant to plot points on a grid – and in Year 5 Unit 12, where children learned about the properties of shapes. It enables them to use these skills to plot reflections and translations. In Year 6, children will use coordinates in all four quadrants to complete reflections and translations, using them to solve problems.

Before they start this unit, it is expected that children:
- have experience of plotting coordinates in the first quadrant
- can describe 2D shapes using appropriate vocabulary relating to their properties (vertex, parallel, vertical, horizontal)
- are confident knowing which direction is left and which is right
- can describe the positions of 2D shapes, using language that includes left, right, up and down.

ASSESSING MASTERY

Children can plot and find coordinates of a reflected shape on a grid. They are able to use coordinates to find translations. Children can confidently identify reflections and translations and describe them.

COMMON MISCONCEPTIONS	STRENGTHENING UNDERSTANDING	GOING DEEPER
Children may translate a shape vertically or horizontally instead of reflecting it.	Provide children with mirrors to explore reflections. Ask them to predict the reflection and then check by placing the mirror on the mirror line.	Children can explore drawing and plotting reflections in vertical and horizontal mirror lines on different types of paper, such as square dotted paper, isometric paper and hexagonal grid paper.
Children may think that translation instructions (for example, left 5) refer to the distance between the original shape and the translated image, rather than the distance between each vertex on the original shape and the corresponding vertex on the translated image.	Ask children to cut out 2D shapes from squared paper so they can physically move them on a grid and place them in the new position. They should practise counting how far each vertex has moved: does this match the translation instruction?	Look at translations of more complex 2D shapes and translations involving more than one step. Use coordinates without a grid for support: a translation of 4 left, 3 up on (5,10) results in the coordinate (5 − 4, 10 + 3) = (1,13).

Unit 13: Geometry – position and direction

UNIT STARTER PAGES

Use these pages to discuss with children the vocabulary they will use, and what they can remember about reading coordinates and about reflections. Point out that Ash's diagram shows a reflection of a rectangle and that the mirror line is shown by the dotted line. Revise reading and plotting coordinates, reminding children that a coordinate tells you how far away a point is from the origin, which has the coordinate (0,0). For example, (2,3) is 2 right, 3 up from the origin. Briefly consider Dexter's diagram, which uses arrows to show the distance between two points. You may wish to use the terms right, left, up and down to compare the position of the two points in relation to each other.

STRUCTURES AND REPRESENTATIONS

Coordinate grids: Various coordinate grids are used in these lessons. They show the first quadrant and use both squared paper and blank paper.

i)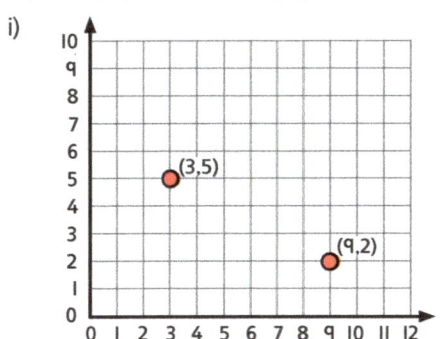

ii)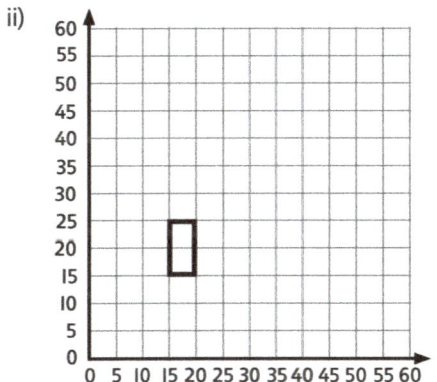

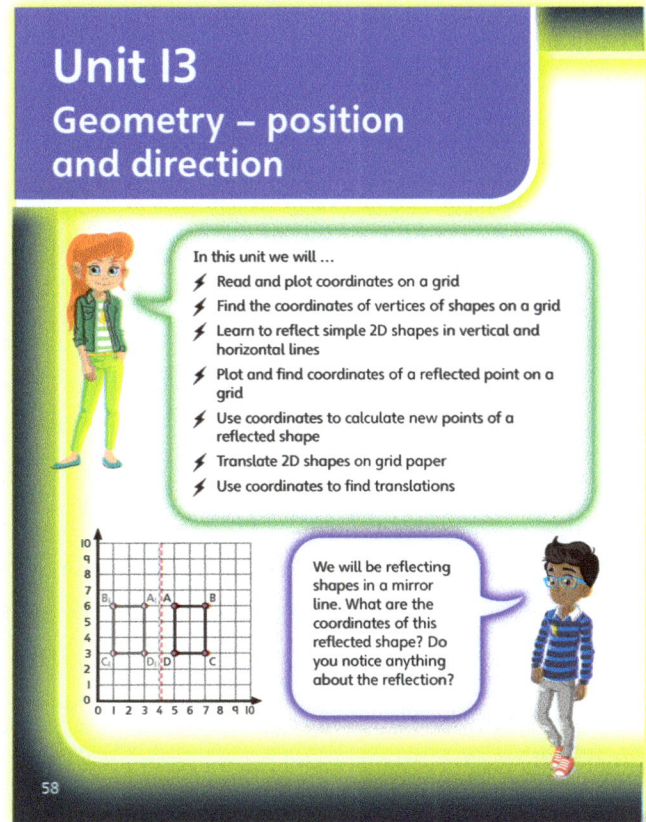

PUPIL TEXTBOOK 5C PAGE 58

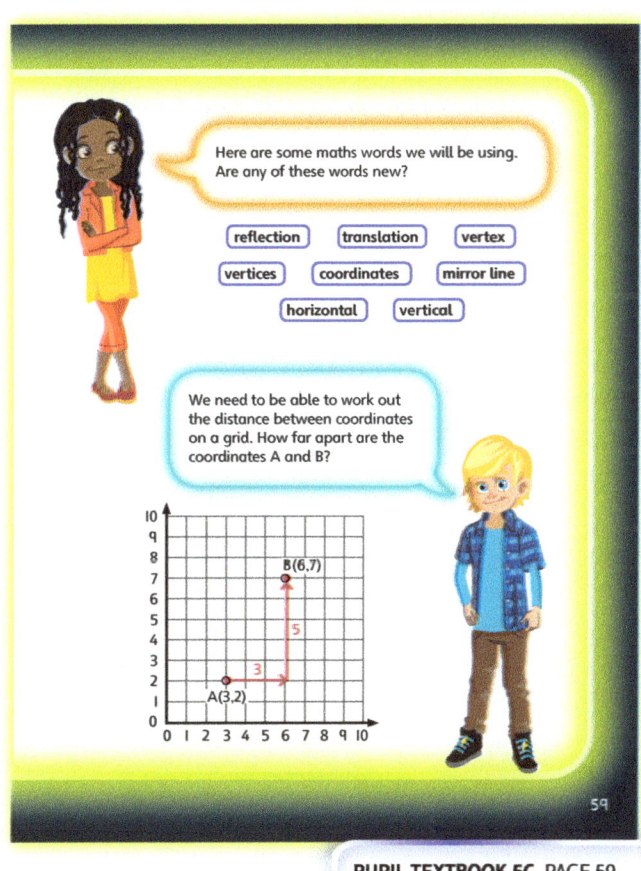

PUPIL TEXTBOOK 5C PAGE 59

KEY LANGUAGE

There is some key language that children will need to know as part of the learning in this unit.

➔ reflection, translation
➔ vertex, vertices
➔ mirror line
➔ coordinates, horizontal coordinate, *x*-coordinate, vertical coordinate, *y*-coordinate
➔ horizontal, vertical, axis, axes

Unit 13: Geometry – position and direction, Lesson 1

Read and plot coordinates

Learning focus
In this lesson, children will use coordinates to accurately read and plot points in the first quadrant.

Before you teach
- Do children already know how to plot coordinates?
- Have grids available for children to practise plotting points.

NATIONAL CURRICULUM LINKS

Year 4 Geometry – position and direction

Describe positions on a 2D grid as coordinates in the first quadrant.

Plot specified points and draw sides to complete a given polygon.

ASSESSING MASTERY

Children can confidently read and plot points in the first quadrant. They know that the order of the coordinate points matters and can plot them in the correct order. They are able to use the language of *x*- and *y*-coordinates and axes, which is new learning and builds on their understanding from Year 4.

COMMON MISCONCEPTIONS

Children may still interpret coordinates in the wrong order. Children need to know that the order matters and that it would be better that everyone agrees to use the conventional order – horizontal coordinate (*x*) first, vertical coordinate (*y*) second – otherwise children will continually encounter difficulties as they continue schooling. Children also need to know what the conventional order is. To support children indicate the point (4,2) and ask:
- *Does it matter whether you give the across coordinate first, or the up coordinate first? How will I know which you are giving first?*

Explain to children that the first number tells them how many units along the horizontal axis they need to go and that the second number tells them how many units up (or down) the vertical axes they need to go.

STRENGTHENING UNDERSTANDING

All of the plotting exercises in this lesson could usefully be carried out (or repeated) using a computer geometry package. Seeing the same coordinate system used in as many different contexts as possible should help to strengthen understanding of the system.

GOING DEEPER

Children could start to reason using coordinates. For example, ask them to plot three coordinates for the vertices of a square and then to reason where the other one could be.

KEY LANGUAGE

In lesson: read, plot, **coordinates**, horizontal, vertical, point, axis, *x*-coordinate, *y*-coordinate

Other language used by the teacher: axes, quadrant, triangle, rhombus, treasure chest

STRUCTURES AND REPRESENTATIONS

Coordinate grid

RESOURCES

Mandatory: none

Optional: Blank first quadrant grids

 In the eTextbook of this lesson, you will find interactive links to a selection of teaching tools.

Quick recap

Display a coordinate grid on the board. Ask children what they already know about coordinates. Check what children remember from plotting coordinates in Year 4.

Discover

WAYS OF WORKING Pair work

ASK

- Question 1 a): *What are the coordinates of the (centre of the) triangle? How do you know?*
- Question 1 b): *Is Reena correct? What mistake do you think she has made? What are the correct coordinates of the rhombus?*

IN FOCUS In **Discover**, children read coordinates from the first quadrant. In question 1 a) children must identify where the triangle is located. Ensure that children, with a partner, can explain why the coordinate is at (8,2). Look for children reading the coordinates the wrong way round. Question 1 b) addresses a common misconception and a common mistake that children can make, when they plot the points in the wrong order. Children should be able to explain the mistake Reena has made and state the true coordinates of the rhombus.

PRACTICAL TIPS Have copies of first quadrant grids available for children to use. You may also want to have small card cut-outs of triangles and rhombuses for the children to slide over the grids.

ANSWERS

Question 1 a): The triangle is at (8,2).

Question 1 b): Reena has written the coordinates in the wrong order.
The centre of the rhombus is at (4,5).

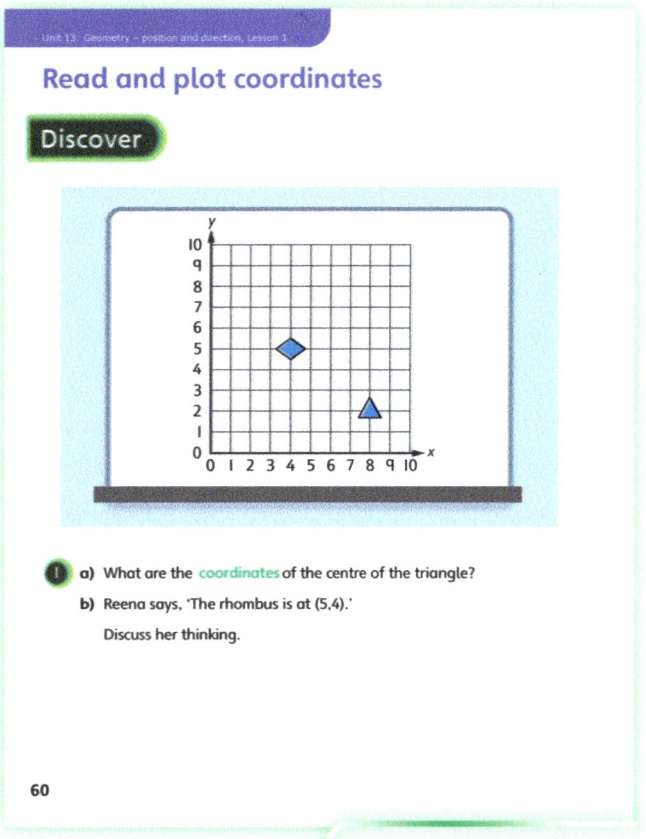

PUPIL TEXTBOOK 5C PAGE 60

Share

WAYS OF WORKING Whole class teacher led

ASK

- Question 1 a): *What do we call the horizontal axis? What about the vertical axis? What is the x-coordinate of the triangle? What about the y-coordinate?*
- Question 1 b): *What mistake has Reena made? Why is it important to write coordinates in the correct order? Which coordinate do you write first? Which coordinate do you write second?*

IN FOCUS In question 1 a), take time to discuss with children the language of x- and y-coordinates as this is new learning. It builds on the learning from Year 4, where they did not learn this formal language. Explain that the first number provides the x-coordinate (how many along from the centre of origin). Then explain that the y-coordinate tells them how many up (or down). Use question 1 b) to discuss with children the common misconception of plotting the points in the wrong order. Use the diagrams to plot both (5,4) and (4,5) and explain the difference. Ensure children understand that everyone should write the x-coordinate first and the y-coordinate second.

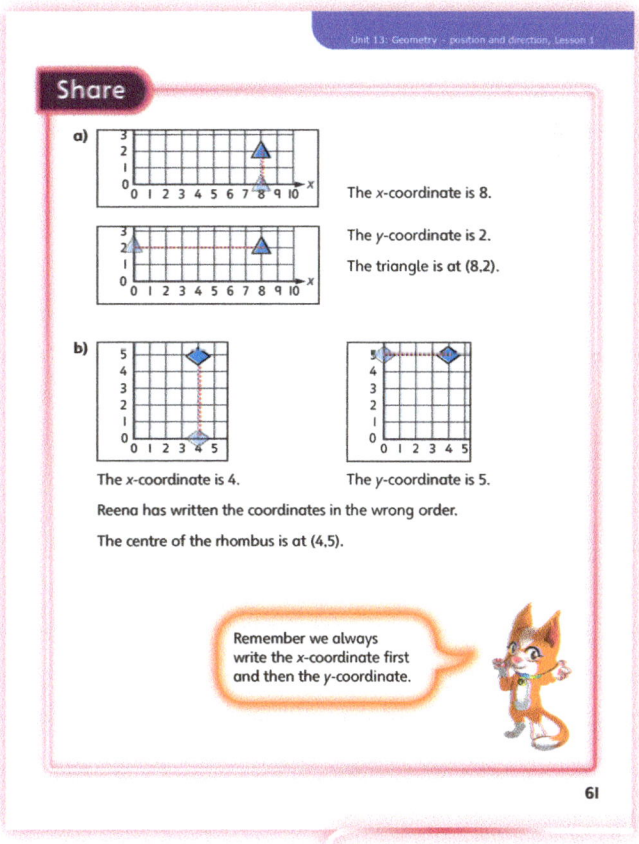

PUPIL TEXTBOOK 5C PAGE 61

Unit 13: Geometry – position and direction, Lesson 1

Think together

WAYS OF WORKING Whole class teacher led (I do, We do, You do)

ASK

- Question ①: *What does the first number in each coordinate pair tell you? Can you find this number on the x-axis? What does the second number in each coordinate pair tell you? Can you find this number on the y-axis?*
- Question ②: *What is the x-coordinate of A? Where do you write this? What is the y-coordinate of A? Where do you write this?*
- Question ③: *[for the child that drew the treasure map] What clues can you give a partner to help them guess the coordinates of your treasure? What do your partner's clues tell you about the possible coordinates of their treasure?*
- Question ③: *[for the child that is guessing] Are your guesses well-spaced apart? Are you saying the coordinates in the right order?*

IN FOCUS In question ①, children are checking that they know how to plot coordinates. Recap how to read a coordinate as well as plot one, noting that the word 'brackets', is not said. Plot the first one for children and then ask them to plot the others. Check that children are plotting in the same order. In question ②, children are reading off coordinates. Encourage them to use a ruler if they need to see the *x* or *y* values that the plotted points line up with. In question ③, children create their own treasure map on a grid and plot some treasure. In pairs, children guess where their partner's treasure is by stating coordinates. They can record their guesses on a separate map to help develop a strategy as well as make sure they don't say the same point twice. Children should be encouraged to give clues to the location (e.g. 'very close', '2 squares away', and so on).

STRENGTHEN Questions ① and ② reinforce children's understanding of the coordinate system. All the required points could fit on the grid, even if plotted in the wrong order. Remind children that the order is important and associates each coordinate with the relevant number on the axes. Children should plot other points for practice.

DEEPEN In question ③, you could ask children to use language such as 'It is 2 squares left and 1 square up'. Their partner then has one guess to work out where it is.

ASSESSMENT CHECKPOINT Use questions ① and ② respectively to assess whether children are plotting coordinates fluently and reading coordinate points confidently.

ANSWERS

Question ①: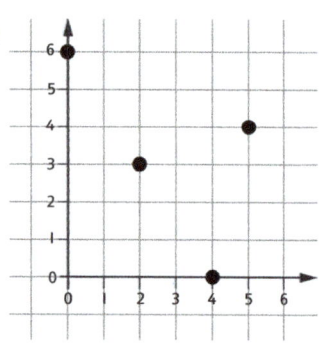

Question ②: A (1,9)   B (9,1)   C (6,2)   D (2,6)

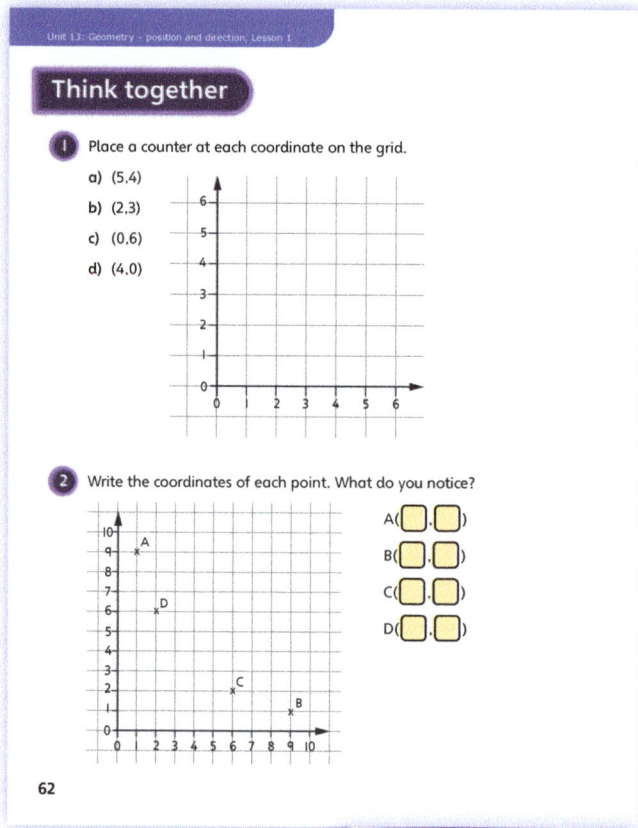

PUPIL TEXTBOOK 5C PAGE 62

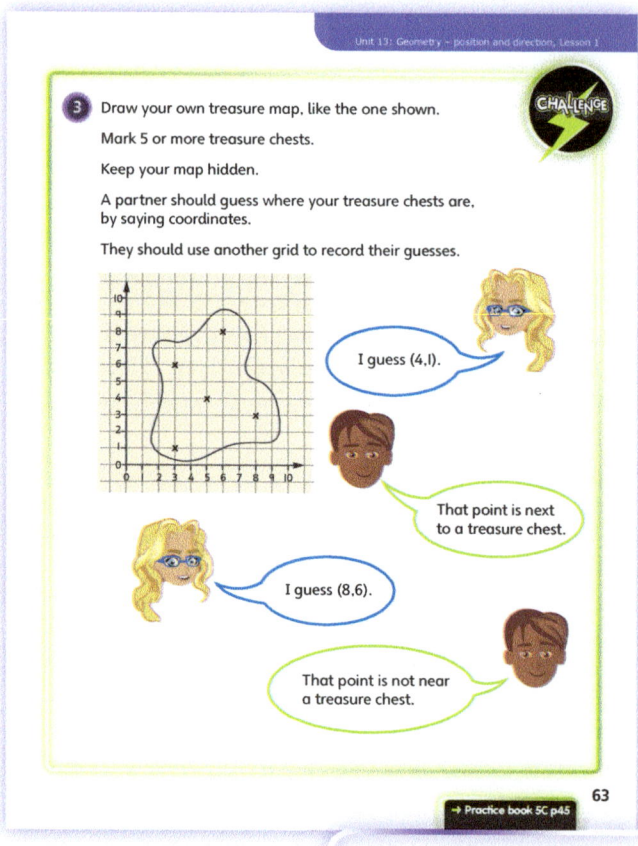

PUPIL TEXTBOOK 5C PAGE 63

98

Unit 13: Geometry – position and direction, Lesson 1

Practice

WAYS OF WORKING Independent thinking

IN FOCUS In question ❶, children write the coordinates of plotted points and, in question ❷, children plot the coordinates themselves. It is important to check that children are plotting and reading the points in the right order before they get too far into the practice. In question ❸, children can create their own mystery island. Some locations have been given but they can also make up their own. Ask children to share their islands with a partner and check each other's locations. Question ❹ b) asks children to try to find a connection between the points. Help children to elicit that, starting at (0,10), the sequence of triangles moves 2 units right and 2 units down each time.

STRENGTHEN Questions ❶ and ❷ provide an opportunity to reinforce children's understanding of the coordinate system. All of the required points could fit on the grid, even if plotted in the wrong order. Where children find this difficult, remind children that the order is important and and associates each coordinate with the relevant number on the axes. Ask children to plot other points so they become more fluent and confident.

DEEPEN To deepen the learning in question ❹, children could try to write a rule which would take them to the next plotted triangle. For example, if the coordinates of any triangle were (x,y), then the coordinates of the next triangle would be (x + 2, y – 2).

Another activity could be that children play a game with a partner where they plot five points on a coordinate grid and then try to guess where each other's coordinates are. This is a version of battleships. Children could give each other clues (like close, far away or even more accurate locations).

ASSESSMENT CHECKPOINT Questions ❶ and ❷ provide the reassurance that children can read and plot coordinates successfully and without exhibiting any common misconceptions.

ANSWERS Answers for the **Practice** part of the lesson can be found in the *Power Maths* online subscription.

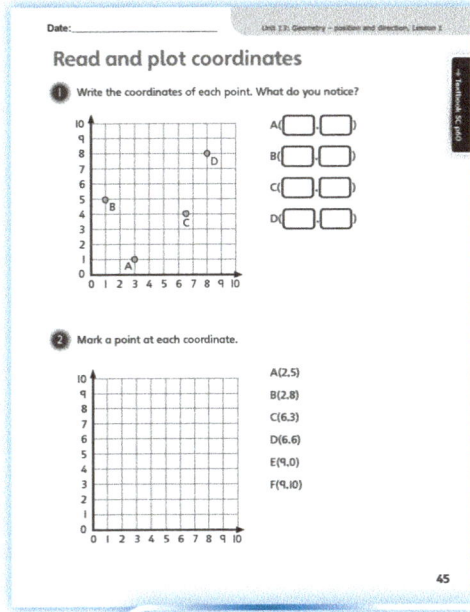

PUPIL PRACTICE BOOK 5C PAGE 45

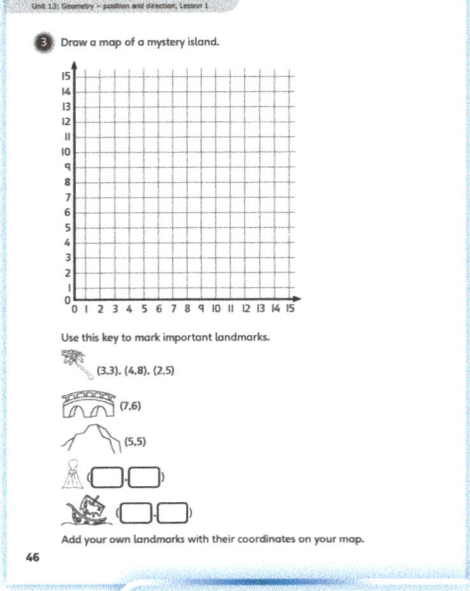

PUPIL PRACTICE BOOK 5C PAGE 46

Reflect

WAYS OF WORKING Independent thinking

IN FOCUS Ask children to reflect on the question individually and then share their thoughts with a partner. Then discuss as a class what everyone is thinking. Some children may say things like 'along first and then up'.

ASSESSMENT CHECKPOINT Check that children remember the correct way to plot coordinates.

ANSWERS Answers for the **Reflect** part of the lesson can be found in the *Power Maths* online subscription.

After the lesson ⏸

- Can children read and write coordinates in the first quadrant, getting the points the correct way around?
- Can children accurately plot points in the first quadrant?

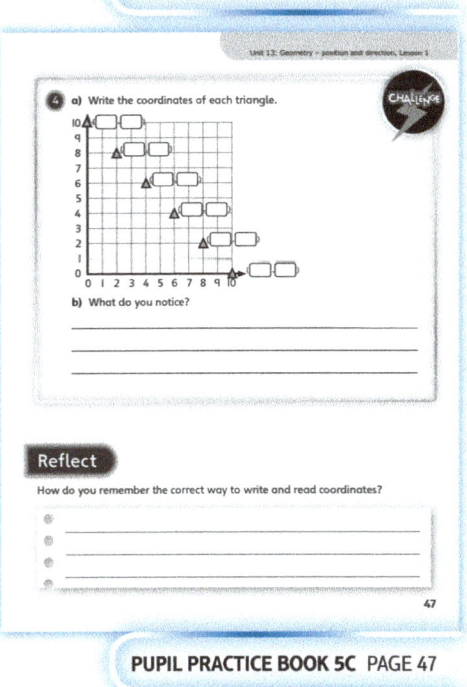

PUPIL PRACTICE BOOK 5C PAGE 47

Unit 13: Geometry – position and direction, Lesson 2

Problem solving with coordinates

Learning focus

In this lesson, children will solve problems where they find the missing vertices of 2D shapes drawn on a grid. They will give their answers using coordinates.

Before you teach

- Ensure that children can accurately read and plot coordinates in the first quadrant.
- Check that children know the names of simple 2D shapes, including different types of quadrilaterals.

NATIONAL CURRICULUM LINKS

Year 4 Geometry – position and direction

Describe positions on a 2D grid as coordinates in the first quadrant.

ASSESSING MASTERY

Children can reason about the locations of coordinates, such as finding the missing vertices of shapes on a coordinate grid. They can also solve problems involving coordinates and shapes.

COMMON MISCONCEPTIONS

Children may still interpret coordinates in the wrong order. Children need to know that the order matters and that they therefore need to use the conventional order that the *x*-coordinate is listed first and the *y*-coordinate is listed second. They also need to know what the conventional order is. To support children, indicate the point (4,2) and ask:

- *Does it matter whether you give the across coordinate first, or the up coordinate first? How will I know which you are giving first?*

Explain to children that the first number indicates how many along the horizontal axis they need to go and that the second number indicates how many up (or down) the vertical axis they need to go.

STRENGTHENING UNDERSTANDING

Use a computer geometry package to provide additional practice with the material covered in this lesson and to check solutions to the exercises. You may also want to support children by first discussing the properties of different shapes (e.g. squares and rectangles). Encourage children to draw their answers on a coordinate grid rather than trying to do it all in their heads.

GOING DEEPER

Ask children to explore more complex problems, perhaps using a computer geometry package. For example, they could try to identify the remaining two vertices of squares where (5,4) and (9,4) are opposite vertices, or where (3,5) and (9,6) are adjacent vertices.

KEY LANGUAGE

In lesson: grid, coordinates, vertices, vertex, line, square, rectangle, rhombus, horizontal, vertical, plotted, shape

Other language to be used by the teacher: symmetry

STRUCTURES AND REPRESENTATIONS

Coordinate grid

RESOURCES

Optional: Printed coordinate grids

 In the eTextbook of this lesson, you will find interactive links to a selection of teaching tools.

Quick recap

Check that children can accurately plot and read points by giving them each a coordinate grid. Ask them to mark a point on the grid (first quadrant only). They should then pass it around the class and ask each other questions to try to work out each other's points.

Unit 13: Geometry – position and direction, Lesson 2

Discover

WAYS OF WORKING Pair work

ASK

- Question 1 a): *What is a vertex? What are the coordinates of the three plotted vertices? What shape can we make with them?*
- Question 1 b): *What are the properties of a square? Point to where the missing vertex would go. How do you know where it should go?*

IN FOCUS In question 1 a), children practise reading coordinates and realise that the three points make up a triangle. Ask children to be as descriptive as possible, writing as many things about the shape as they can on a mini-whiteboard (e.g. triangle, three vertices, at (2,2), (2,8) and (8,8), right-angled and so on). Some children may want to draw their own version and this should be encouraged. In question 1 b), children reason where the missing vertex would be if the three points are part of a square. Encourage children to draw their own version so that they can see the square.

PRACTICAL TIPS If devices are available, encourage the use of geometrical software to plot the points.

ANSWERS

Question 1 a): This is a right-angled triangle.
It is isosceles because two sides are the same length.

Question 1 b): A (2,8)
B (2,2)
C (8,8)
D (8,2)

Share

WAYS OF WORKING Whole class teacher led

ASK

- Question 1 a): *How can you show the triangle? What are the coordinates of the vertices? What type of triangle have you made?*
- Question 1 b): *What do you know about a square? How can you make sure the point you plotted is the same distance from both B and C?*

IN FOCUS Work through the explanation of the triangle in question 1 a), and then start again with a fresh grid for question 1 b) – it could become confusing if you add additional working to the grid in question 1 a) to try to make a square.

In question 1 a), discuss the different things children may have seen. They should talk about the coordinates of the vertices. Discuss how the coordinates make a right-angled triangle. Children may also realise that the triangle has two sides the same length and some children may know that this is called an isosceles triangle. For question 1 b), encourage children to plot the three points again and then place the fourth vertex, checking that they know the final coordinate needs to be the same distance from the two adjacent points (B and C) to make a square.

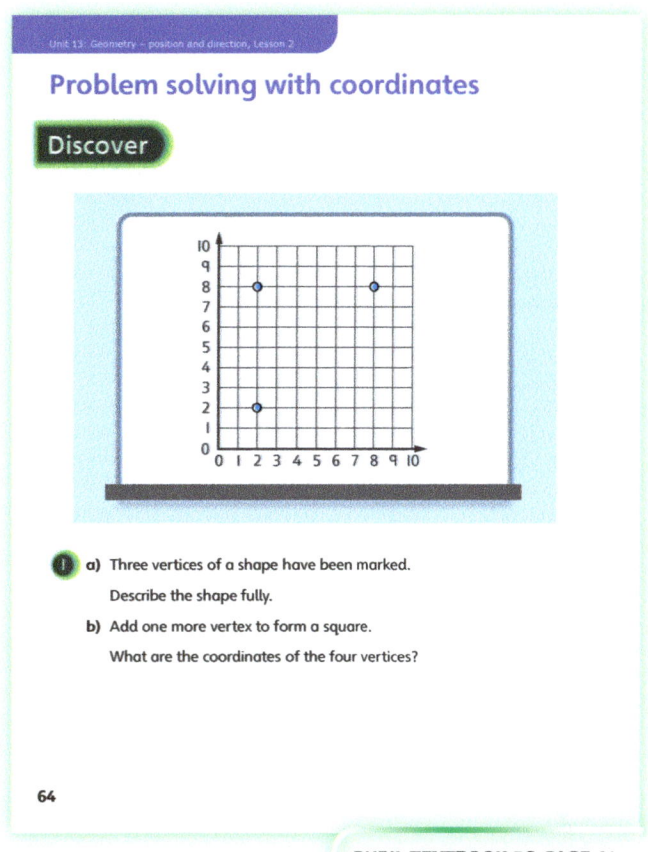

Problem solving with coordinates

Discover

1. a) Three vertices of a shape have been marked. Describe the shape fully.
   b) Add one more vertex to form a square. What are the coordinates of the four vertices?

PUPIL TEXTBOOK 5C PAGE 64

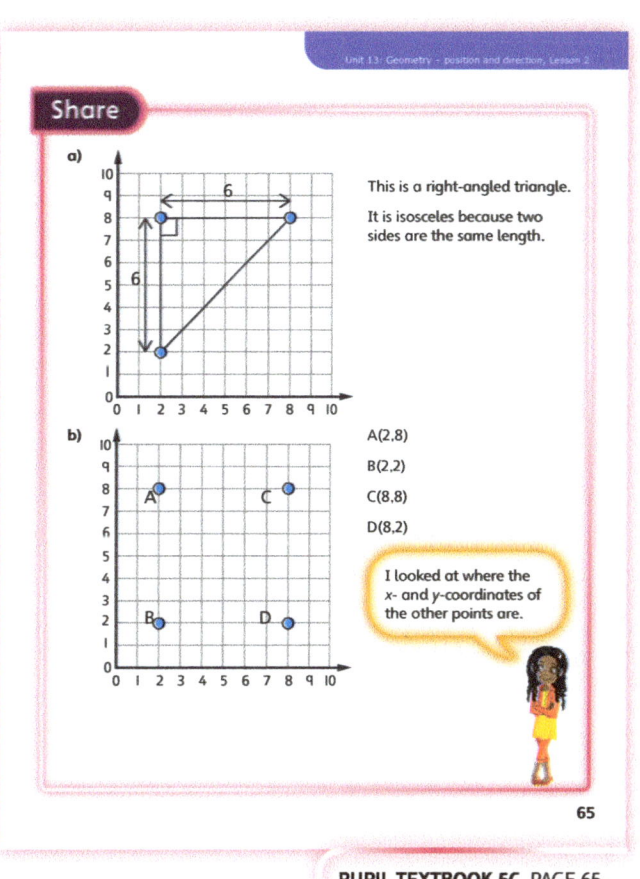

Share

a) This is a right-angled triangle. It is isosceles because two sides are the same length.

b) A(2,8)
B(2,2)
C(8,8)
D(8,2)

I looked at where the x- and y-coordinates of the other points are.

PUPIL TEXTBOOK 5C PAGE 65

101

Unit 13: Geometry – position and direction, Lesson 2

Think together

WAYS OF WORKING Whole class teacher led (I do, We do, You do)

ASK

- Question 1: *What shape are these two points the vertices of? Where could the other two go? How can you prove that the shape you have plotted is a rectangle?*
- Question 2: *What is a rhombus? Can you draw one? What are the properties of the shape? Where do you think the missing point goes?*
- Question 3: *Can you label any points on the x- or y-axis? How do you know?*

IN FOCUS In **Think together** children gradually build their reasoning and problem-solving skills with coordinates. In question 1, children are given two points of a rectangle and they work out where the other two points are. Question 2 gives children three points of a rhombus for them to plot the final point. Discuss with children the properties of a rhombus and show them other examples if needed for support. Question 3 may be the first time children have seen coordinates that are not on a grid. You may consider asking them copy the diagram onto a grid, but the grid will be too large unless children use a different scale. Instead ask them to think carefully about what a coordinate tells them about points on the *x*- and *y*-axis.

STRENGTHEN To support children throughout, use geometry software or provide copies of grids so that they can experiment with plotting points. It's vital that children don't have to do all the visualisation in their heads.

DEEPEN Provide children with more challenging examples, if needed, that extend each of the questions. For example, extend question 1 by asking children to find possible coordinates of other vertices if the two points given are actually adjacent coordinates of a rectangle. In question 3, ask children to reason whether the points (12,40), (40,12) and (10,30) lie on, inside or outside the square.

ASSESSMENT CHECKPOINT Use the questions to determine children's ability to reason with coordinates and to find missing coordinates and information.

ANSWERS

Question 1: (1,5) and (5,2)

Question 2: (5,9)

Question 3: (10,30) and (25,45)

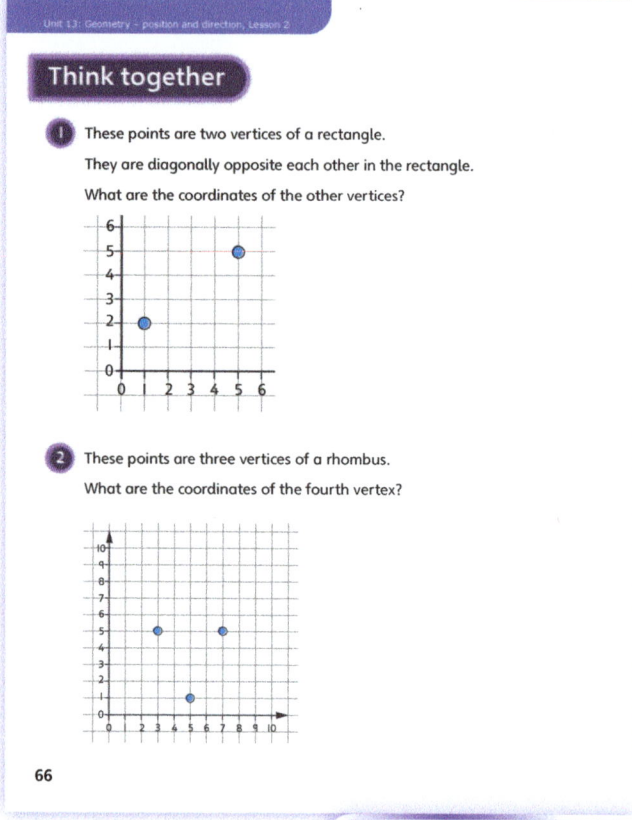

PUPIL TEXTBOOK 5C PAGE 66

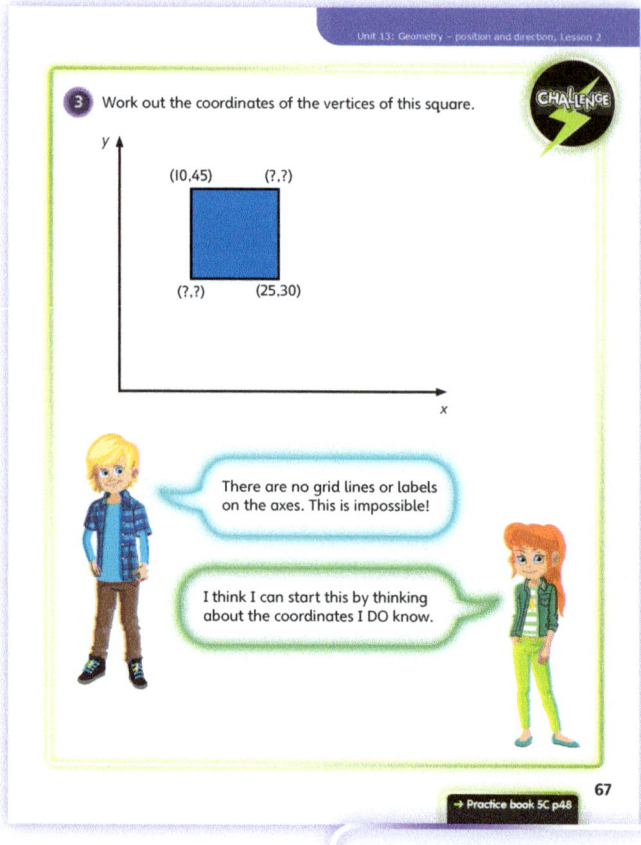

PUPIL TEXTBOOK 5C PAGE 67

102

Unit 13: Geometry – position and direction, Lesson 2

Practice

WAYS OF WORKING Independent thinking

IN FOCUS In questions 1 and 2, children use their knowledge about rectangles and squares to find the two missing vertices. Grids are provided for the first three questions so that children can draw the shapes. In question 3, children are given information about three of the vertices of a parallelogram and then they need to work out the final vertex. Questions 4 and 5 provide a set of coordinate problems where the grids are not given. You may consider asking children to draw a grid, but the grid will be too big to draw unless children use a different scale. Instead get them to think carefully about what a coordinate tells them about points on the x- and y-axis.

STRENGTHEN To support children throughout, use geometry software or provide copies of the grids so that they can experiment with plotting points. It's vital that children don't have to do all the visualisation in their heads. Additionally you may want to make sure children know the different types of shapes. Have relevant shapes accessible in case children are unsure what they look like. This could be a reason that children are struggling – not because they cannot reason with coordinates.

DEEPEN For question 5, ask children to reason if another rotated triangle could fit on the grid. Where could the triangle go and what could the coordinates of the vertices be?

ASSESSMENT CHECKPOINT Use the questions to determine children's ability to reason with coordinates and find missing coordinates and information.

ANSWERS Answers for the **Practice** part of the lesson can be found in the *Power Maths* online subscription.

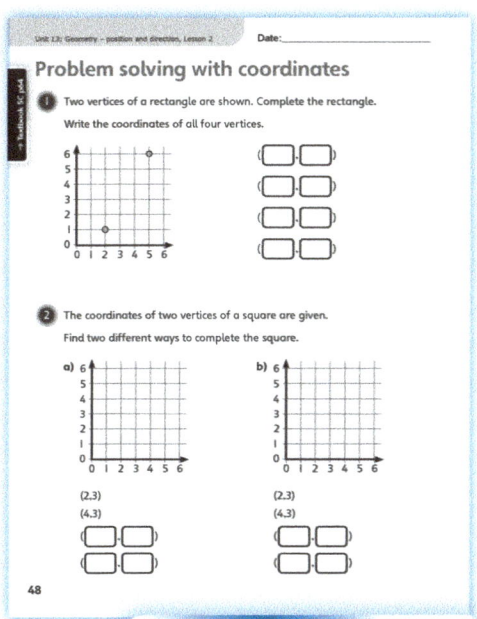

PUPIL PRACTICE BOOK 5C PAGE 48

PUPIL PRACTICE BOOK 5C PAGE 49

Reflect

WAYS OF WORKING Independent thinking

IN FOCUS Children are provided with axes without a grid. They may want to make some grid lines if they need further support. Children should plot all the points to make a square. They should reason that the lengths of the sides are all the same and that the distance between adjacent coordinates must therefore be equal.

ASSESSMENT CHECKPOINT Can children explain how to calculate a reflected point using coordinates? Can they give an example when explaining?

ANSWERS Answers for the **Reflect** part of the lesson can be found in the *Power Maths* online subscription.

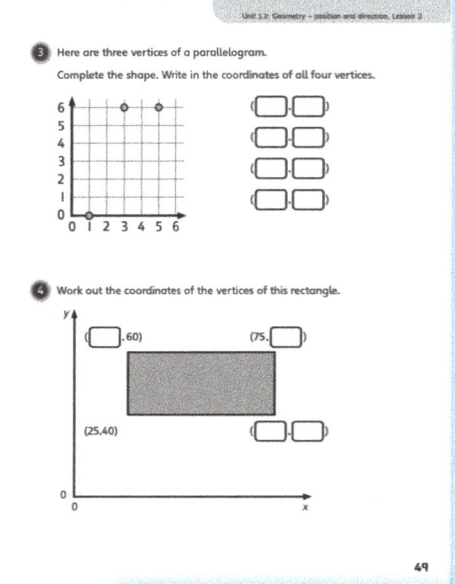

PUPIL PRACTICE BOOK 5C PAGE 50

After the lesson
- Can children find the missing vertices of simple shapes drawn on a coordinate grid?
- Where grid lines are not given, can children start to work out the missing information?

Unit 13: Geometry – position and direction, Lesson 3

Translate shapes

Learning focus

In this lesson, children will explore translations. They will learn how to translate simple 2D shapes on grid paper by moving one vertex at a time. Children will understand that the shape has not changed.

Before you teach

- Can children use the correct language to describe and identify 2D shapes: vertex, sides and parallel?
- Can children describe the change in position of a point on grid paper using left, right, up and down, *without* using coordinates?

NATIONAL CURRICULUM LINKS

Year 5 Geometry – position and direction

Identify, describe and represent the position of a shape following a reflection or translation, using the appropriate language, and know the shape has not changed.

ASSESSING MASTERY

Children can use appropriate language to describe translations: translation, direction, position, vertex, left, right, up and down. They can confidently complete translations of 2D shapes on a grid by moving one vertex at a time. Children understand and can explain that the shape has not changed in any way.

COMMON MISCONCEPTIONS

Children may slide the shape so that the translation instruction is applied to the distance between the shape and its translation rather than the distance between corresponding vertices of the two shapes. Ask:
- *How many squares has this vertex moved? Does this match the instruction?*

STRENGTHENING UNDERSTANDING

Provide 2D shapes, or cut out 2D shapes from squared paper, so that children can physically move them on a grid and place them in a new position. Marking one of the vertices may also help. This will help children to understand that the shape does not change. They will be able to see clearly how far and in which direction the shape has moved.

If you have patio slabs in the school grounds, you could model translations on a large scale outdoors, using the patio squares as a grid and moving shapes cut out from cardboard. Alternatively, draw a large grid in chalk.

GOING DEEPER

Look at moving more complex 2D shapes. Complete translations involving more than one step, first by plotting each vertex in turn, then by plotting one vertex and using the dimensions to complete the translation.

KEY LANGUAGE

In lesson: position, **translation**, triangle, vertex, coordinates, left, right, up, down, reflection, grid

Other language to be used by the teacher: vertices, 2D shape, distance, plot, translate, rectangle

STRUCTURES AND REPRESENTATIONS

2D shapes

RESOURCES

Optional: squared paper, square dotted paper, 2D shapes cut out of squared paper, chalk

 In the eTextbook of this lesson, you will find interactive links to a selection of teaching tools.

Quick recap

Ensure that children can accurately plot shapes on coordinate grids. Ask children to plot the points (5,1), (1,4). Ask: *If these are two corners of a rectangle, can you tell me the other two coordinates?*

Unit 13: Geometry – position and direction, Lesson 3

Discover

WAYS OF WORKING Pair work

ASK

• Question 1 a): *What do the shapes represent?*
• Question 1 a): *In which directions could the bed move? Can you point to the direction that Bella's bed will move? Will it be touching the table after she moves it?*
• Question 1 b): *Can you work out where the table will move to? Could Bella move her desk by the same amount?*

IN FOCUS Ensure children understand that this is a plan of a bedroom. Provide children with squared paper to draw the new position of the bed. Discuss moving other items, thinking about the maximum distance each object can be moved – left, right, up and down.

PRACTICAL TIPS Children could create a plan of their own bedroom on squared paper and cut out each item of furniture from another sheet of paper so they can physically move the items. Encourage them to explain the movements using the language of translation.

ANSWERS

Question 1 a): The new position of Bella's bed will be above the table.

Question 1 b): The table will now be in the bottom right corner of her room, near the desk.

PUPIL TEXTBOOK 5C PAGE 68

Share

WAYS OF WORKING Whole class teacher led

ASK

• Question 1 a): *What is a translation? Which movement comes first?*
• Question 1 a): *Where will each point or vertex of the shape be once you have translated the shape?*
• Question 1 b): *Has the size or orientation of the shape changed? What changes when you translate a shape?*
• Question 1: *How is a translation different to a reflection?*

IN FOCUS Use question 1 a) to discuss what a translation is. Ensure children understand that a translation is a slide left or right, up or down, on a grid and that the shape does not change in shape, size or orientation. Identify and discuss any misconceptions about what the translation instruction refers to: it is the distance each vertex moves, not the number of squares between the shape and its translation. Encourage children to use the vocabulary 'translate' and 'translation'. The convention for translations is to give the horizontal direction first, then the vertical (as with coordinates). It is important to establish this now.

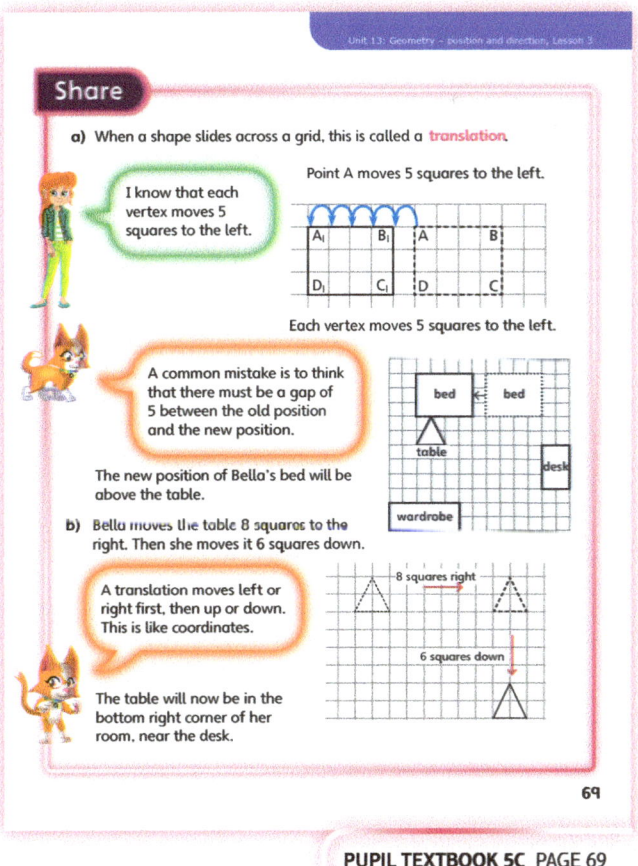

PUPIL TEXTBOOK 5C PAGE 69

105

Unit 13: Geometry – position and direction, Lesson 3

Think together

WAYS OF WORKING Whole class teacher led (I do, We do, You do)

ASK

- Question ①: *Where will each translated vertex be?*
- Question ②: *What pattern do you see when you reverse a translation?*
- Question ③: *What is the difference between a translation and a reflection?*

IN FOCUS In this section, children identify the translation between two shapes. In question ②, the translations involve movement in two directions, so establish that the horizontal movement is stated first, followed by the vertical movement. The easiest way to identify a translation is to focus on a single vertex and work out how it has moved. In question ②, discuss how to reverse a translation. Ensure children recognise they need to do the opposite movements (for example, left 1, down 2 becomes right 1, up 2). Discuss the differences between a translation and a reflection using question ③. It is important that children understand that translation does not change anything about the shape itself but just slides it.

STRENGTHEN Give each child a sheet of squared paper with a shaded rectangle, labelled A, drawn in the middle. Provide instructions for children to practise translating this rectangle, labelling the new positions B, C and so on. Begin with one direction only, then move on to translations in two directions.

DEEPEN Use the **Strengthen** activity, but give children a rectilinear shape to translate. Ask them to write the translations from one image to another and back again. Can they generalise about how the instructions change?

ASSESSMENT CHECKPOINT Can children translate a shape by plotting one vertex at a time? Can they identify the translation between two shapes using the correct vocabulary? Do children understand that a shape does not change its size or orientation when translated?

ANSWERS

Question ①: The desk has moved 3 squares up.

Question ② a): A to B: 1 left, 2 down. B to A: 1 right, 2 up.

Question ② b): C to D: 3 left, 5 down. D to C: 3 right, 5 up.

Question ② c): The number of squares is the same, but left becomes right and down becomes up.

Question ③ a): B, D: translations. C, E: reflections.

Question ③ b): A to B: 2 left, 2 up. A to D: 3 right, 4 up.

Questions ③ c) and d):

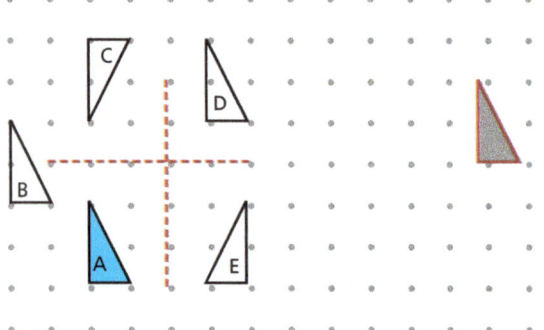

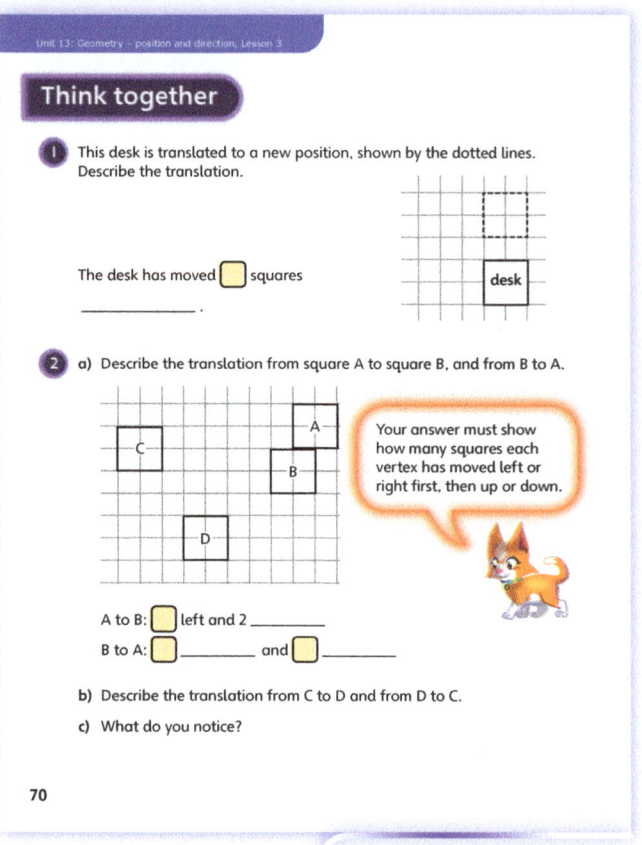

PUPIL TEXTBOOK 5C PAGE 70

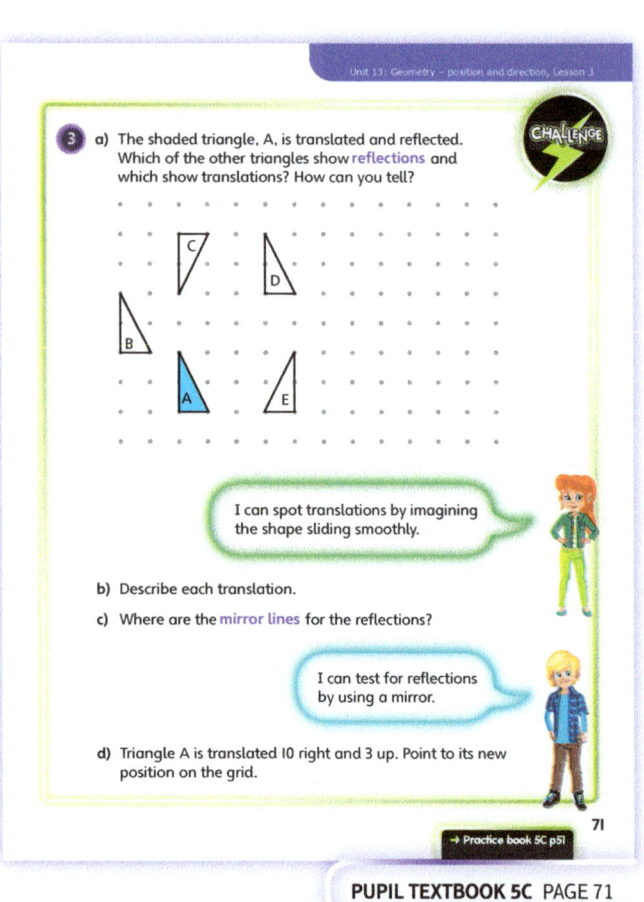

PUPIL TEXTBOOK 5C PAGE 71

Unit 13: Geometry – position and direction, Lesson 3

Practice

WAYS OF WORKING Independent thinking

IN FOCUS In questions 1 and 2, children practise translating shapes and describe a translation. In question 3, children are required to reverse a given translation to identify the starting position of the original shape. They may need to follow a step-by-step approach to interpret what they are being asked to do. Establish that, when going back to the original shape, the instructions are reversed – so right becomes left and up becomes down – but the numbers stay the same.

STRENGTHEN With question 4, provide a mirror so children can check if the shape has been reflected. Discuss whether the shape looks exactly the same (translation) or whether it has changed in orientation (reflection), and why you can or cannot tell that with this shape.

DEEPEN To extend questions 5 and 6, provide children with squared or dotted grid paper and challenge them to create their own problems where one shape is split and parts are translated to form a different shape or a new picture.

THINK DIFFERENTLY In question 4, children need to identify whether the movement is a translation or a reflection and explain how they know. The rectangle has not changed in shape, size or orientation but it has moved up and across, so it is a translation. Some children may need to check with a mirror to see where the reflection would be.

ASSESSMENT CHECKPOINT Can children describe a translation using the correct terminology and giving the directions in the correct order? Can they translate a simple shape accurately? Do children understand that a translated shape does not change in size or shape but moves left or right, up or down? Can they describe the difference between a reflection and a translation?

ANSWERS Answers for the **Practice** part of the lesson can be found in the *Power Maths* online subscription.

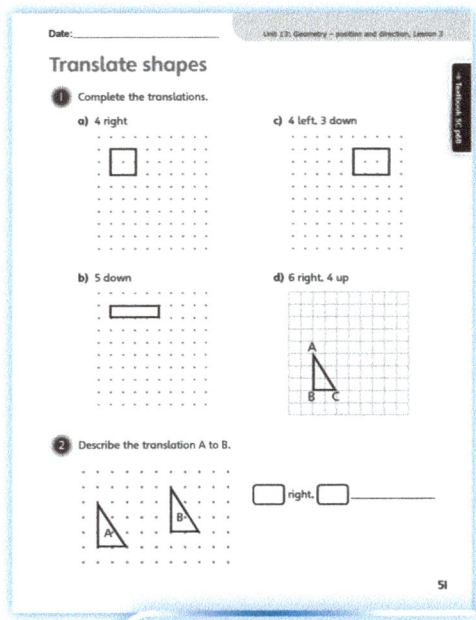

PUPIL PRACTICE BOOK 5C PAGE 51

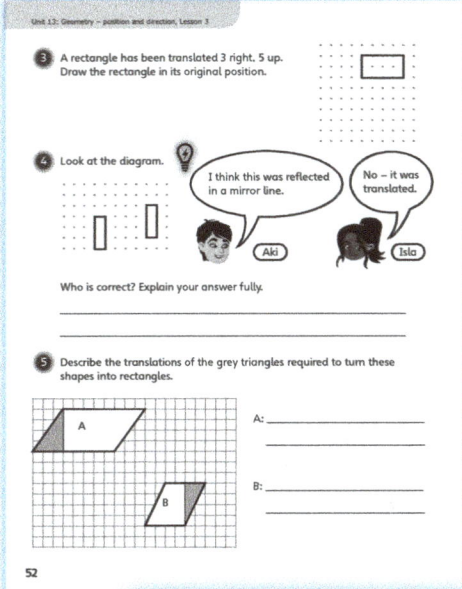

PUPIL PRACTICE BOOK 5C PAGE 52

Reflect

WAYS OF WORKING Independent thinking

IN FOCUS The challenge here is for children to think generally and only use the instructions to answer the question – demonstrating their understanding that, to return a shape to its original position, the translation is reversed. For example, left becomes right and up becomes down. If necessary, children can draw a shape onto grid paper and work out the translation visually.

ASSESSMENT CHECKPOINT Can children explain how the translation instructions need to be reversed to return the shape to its original position?

ANSWERS Answers for the **Reflect** part of the lesson can be found in the *Power Maths* online subscription.

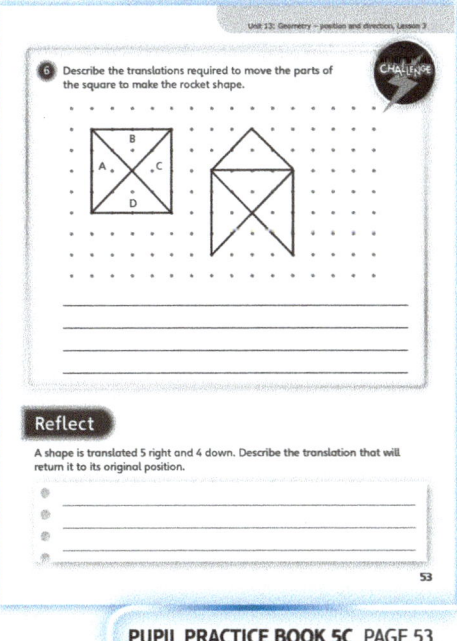

PUPIL PRACTICE BOOK 5C PAGE 53

After the lesson

- Can children explain what is meant by the term translation?
- Can children draw translations accurately, using a ruler and keeping to the grid lines?
- Do children understand the difference between reflection and translation?

Unit 13: Geometry – position and direction, Lesson 4

Translate points

Learning focus

In this lesson, children will build on their knowledge of translation from the previous lesson. They will use coordinates to find translations.

Before you teach

- Can children describe the translation of a shape (left, right, up, down)? Can they identify left from right?
- Are children confident using and plotting coordinates?

NATIONAL CURRICULUM LINKS

Year 5 Geometry – position and direction

Identify, describe and represent the position of a shape following a reflection or translation, using the appropriate language, and know the shape has not changed.

ASSESSING MASTERY

Children can use coordinates to calculate the new position after a translation. They understand that the shape has not changed in any way. Children can confidently plot the new coordinates of a point and/or shape on a coordinate grid.

COMMON MISCONCEPTIONS

Children may default to counting methods because they do not know which part of the translation to add to or subtract from each number in the coordinate pair. Ask:
- *Which number in the coordinate pair shows how far right the point lies from 0? Which part of the translation will you add this to?*
- *Which number in the coordinate pair shows how far up the point lies from 0? Which part of the translation will you add this to?*

STRENGTHENING UNDERSTANDING

Provide children with a coordinate grid and counters. (Alternatively, you could mark out a large grid on the playground and children could move themselves as 'counters'.) Ask children to find the point (2,3) to begin. Then give a list of translations (for example: right 3, up 4; left 2, down 5; right 4, down 1) and ask children to practise using coordinates to find the new position of the counter. Always begin at the original point.

GOING DEEPER

Children can practise calculating a new position using coordinates by playing the game described in the **Discover** section of the Textbook, using a larger grid and 0–9 dice. Encourage them to add to (right/up) or subtract from (left/down) the starting coordinate using the translation numbers. Can they give a combination of two or more double translations (right/left **and** up/down) to get from one point to another?

KEY LANGUAGE

In lesson: translation, translate, coordinate, position, calculate, efficient, decrease, increase, horizontal coordinate, vertical coordinate, vertices, left, right, up, down

Other language to be used by the teacher: vertex, axis, horizontal axis, vertical axis

STRUCTURES AND REPRESENTATIONS

Coordinate grids, 2D shapes

RESOURCES

Optional: squared paper, two different coloured dice

 In the eTextbook of this lesson, you will find interactive links to a selection of teaching tools.

Quick recap

On the board display a triangle or rectangle on a squared grid. Ask children to translate the shape 3 squares right and 2 up. Where does it end up? Now translate the shape 3 squares left and 2 down. What do they notice? Repeat for other translations. Can they predict the translation that will move the shape back to its original position.

Unit 13: Geometry – position and direction, Lesson 4

Discover

WAYS OF WORKING Pair work

ASK

- Question ① a): *What are the coordinates of the counters marked A and B?*
- Question ① a): *Look at the two dice. Could Andy have chosen a different translation? How many different possibilities are there with each pair of numbers?*
- Question ①: *What translation would get from the yellow counter to the red one? What about red to yellow?*
- Question ①: *Can you calculate the new coordinates by adding or subtracting instead of moving the counters?*

IN FOCUS Encourage children to explore the game practically using dice, counters and a grid. Discuss the current coordinates of the counters. Can children identify the new coordinates? What methods do they use? Do they move the counters physically and then determine the coordinates or can they use the coordinates to find the new positions?

PRACTICAL TIPS Provide children with two different coloured dice, red and yellow counters and a coordinate grid so they can play the **Discover** game. Ask them to explore and apply the eight possible translations for each roll of the two dice. They can record their moves on the coordinate grid, but encourage them to use addition or subtraction to find the new coordinates.

ANSWERS

Question ① a): The coordinates of Andy's new position are (10,8).

Question ① b): A translation of Alex's counter 3 left, 4 down would win.

PUPIL TEXTBOOK 5C PAGE 72

Share

WAYS OF WORKING Whole class teacher led

ASK

- Question ① a): *What is a translation?*
- Question ① a): *Who used Dexter's method? Who used Flo's method?*
- Question ① a): *Can you describe Flo's method to a partner?*
- Question ① b): *Is Alex's method the same as Dexter's or Flo's method?*
- Question ① b): *Use Alex's method to work out where the red counter could be with dice scores of 2 and 3.*

IN FOCUS Focus on Alex's/Flo's method by discussing how to use the coordinates to calculate the new position after a translation.

- to move right, add to the horizontal coordinate
- to move left, subtract from the horizontal coordinate
- to move up, add to the vertical coordinate
- to move down, subtract from the vertical coordinate.

Use the scores shown (6,3) to write and apply different translations to either the red or yellow counter using this method. There are eight possible translations.

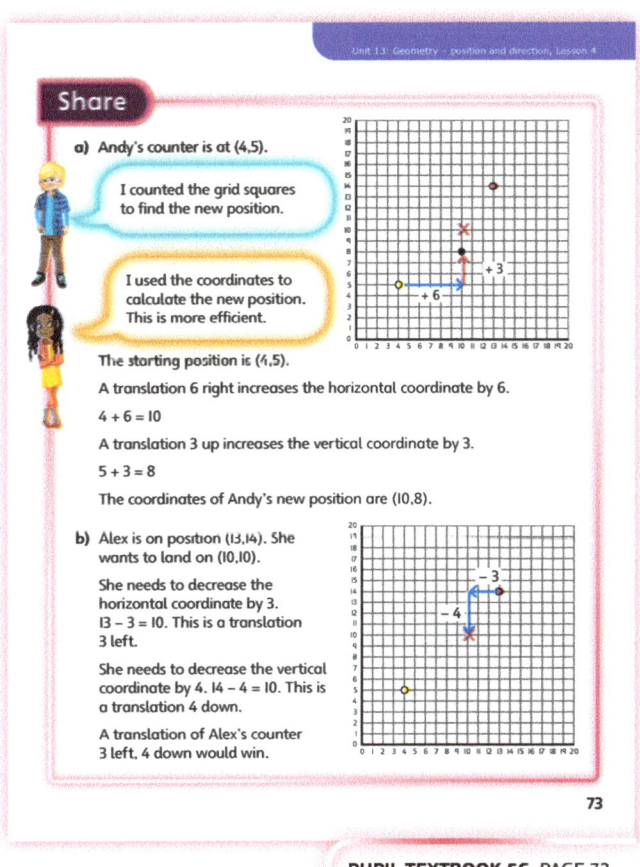

PUPIL TEXTBOOK 5C PAGE 73

109

Think together

WAYS OF WORKING Whole class teacher led (I do, We do, You do)

ASK

- Questions 1 and 2: *Do the translations affect the horizontal coordinates or the vertical coordinates?*
- Questions 1 and 2: *Do the translations increase or decrease the coordinates?*
- Question 3: *What needs to be done to the translations to reverse them?*

IN FOCUS The focus for all these questions is on children using the calculation method to find new points, rather than moving the shape on the grid. Talk through the fact that a translation right or up increases the relevant part of the coordinate, while a translation left or down decreases it. Make the link with coordinates: the horizontal coordinate is how far along (left or right) and the vertical coordinate is how far up or down.

STRENGTHEN Give children a coordinate grid marked with one point, A. Ask them to calculate the new coordinates after given translations, labelling each new point alphabetically; ensure children use the coordinate method rather than counting. Start with right/left **or** up/down before moving on to a combination. Then give examples that will allow children to practise reversing the translation and identifying translations between marked points by considering the changes in the coordinates.

DEEPEN Give children a list of six coordinates, A to F, and ask them to identify the two translations between each pair of coordinates without the support of a coordinate grid. For example, for A (6,11) and B (4,5), moving from A to B: 2 left, 6 down, and moving from B to A: 2 right, 6 up.

ASSESSMENT CHECKPOINT Can children calculate a new point using the coordinates? Do they know whether to increase or decrease the coordinate depending on the direction of the translation?

ANSWERS

Question 1: $A_1(12,8)$
$B_1(13,5)$
$A_2(12,5)$
$B_2(13,2)$

Question 2: $A_1(14,10)$
$B_1(14,12)$
$C_1(16,12)$
$D_1(16,10)$

Question 3 a): P(0,1)
Q(2,3)
R(3,0)

Question 3 b): A(13,13)
B(12,10)
C(15,11)

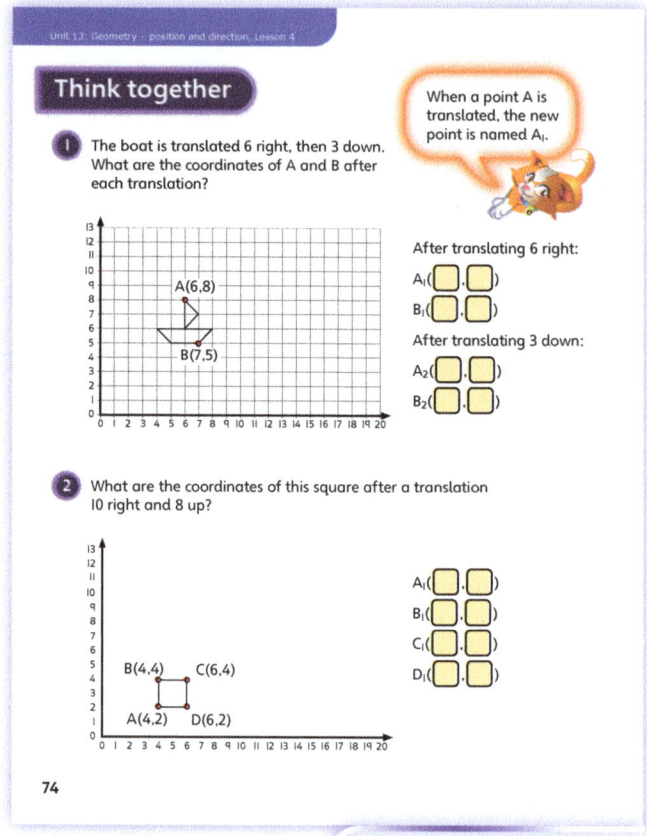

PUPIL TEXTBOOK 5C PAGE 74

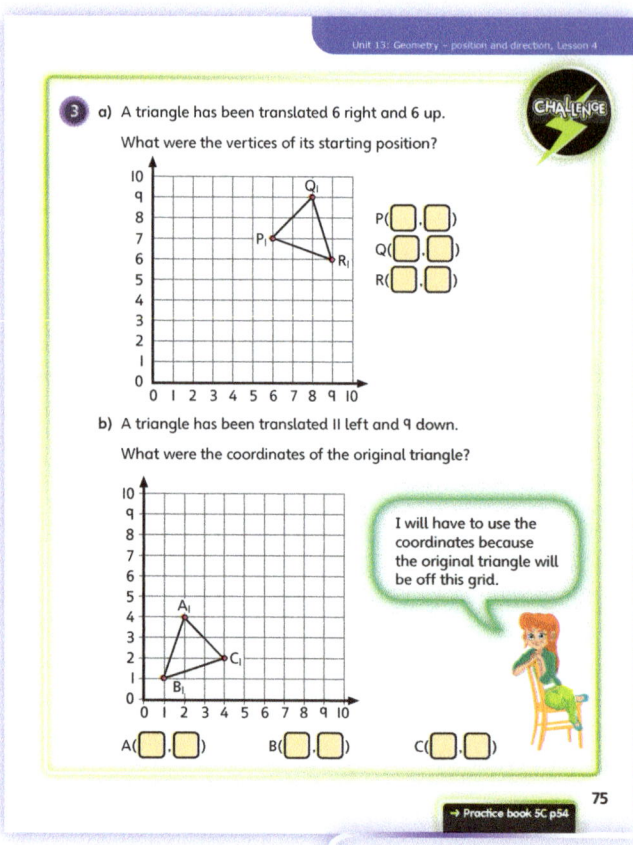

PUPIL TEXTBOOK 5C PAGE 75

Unit 13: Geometry – position and direction, Lesson 4

Practice

WAYS OF WORKING Independent thinking

IN FOCUS In question ①, ensure children realise that each translation is applied to the coordinate immediately above it, not to the starting coordinate. Children will need to work out the final translation based on the ending position of point A. When the table is completed, the ending position in all columns should show a translation of 1 left, 7 up from the previous coordinate, or 3 right, 9 up from the starting coordinate – this is a useful check on accuracy. Encourage children to add or subtract rather than using the grid. The arrow diagram will help them to see which operation they need to use.

STRENGTHEN Give children a starting coordinate of (10,10) and a series of translations to perform on it, involving all four combinations: right and up (+,+); right and down (+,–); left and up (–,+); and left and down (–,–). You may wish to provide copies of the arrow diagram from question ① for reference.

DEEPEN Ask children to work in pairs. Each child draws a shape twice on a coordinate grid then asks their partner to work out the translation from one shape to the other. What translation would move the shape back to its original position?

ASSESSMENT CHECKPOINT Can children calculate a new point after a translation by adding to or subtracting from the coordinates, rather than by counting squares? Can they explain when to increase or decrease the coordinate, depending on the direction of the translation? Can children explain which coordinate to increase or decrease for each part of the translation?

ANSWERS Answers for the **Practice** part of the lesson can be found in the *Power Maths* online subscription.

Reflect

WAYS OF WORKING Pair work

IN FOCUS Children reflect on their work in this lesson by discussing how they will translate the triangle. Listen carefully to the language they use. Encourage them to use correct language, e.g. right, left, up and down, increase and decrease.

ASSESSMENT CHECKPOINT Can children explain that 'translate' in mathematics means movement of a shape? Do they understand that they can move shapes so many squares left or right and up or down. Ask them to demonstrate their understanding by showing you an example and explaining the translation.

ANSWERS Answers for the **Reflect** part of the lesson can be found in the *Power Maths* online subscription.

After the lesson

- Do children still rely on counting squares to work out a translation? How can you further help them to add to or subtract from the coordinates, rather than count squares?
- Do children understand the terms increase, decrease, horizontal coordinate and vertical coordinate in relation to translations?

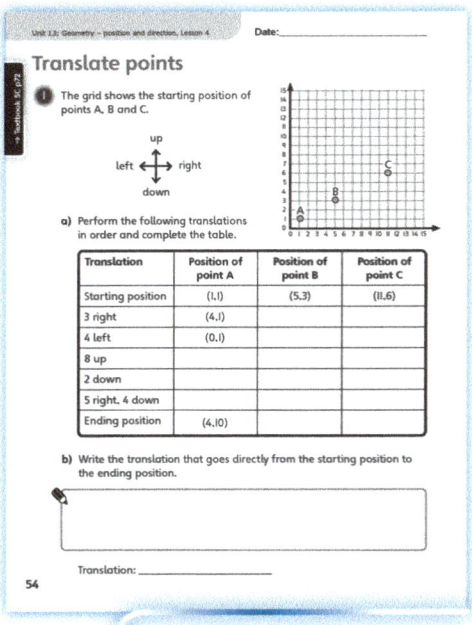

PUPIL PRACTICE BOOK 5C PAGE 54

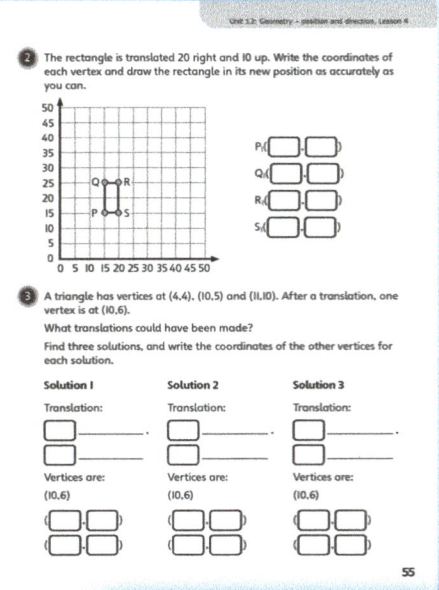

PUPIL PRACTICE BOOK 5C PAGE 55

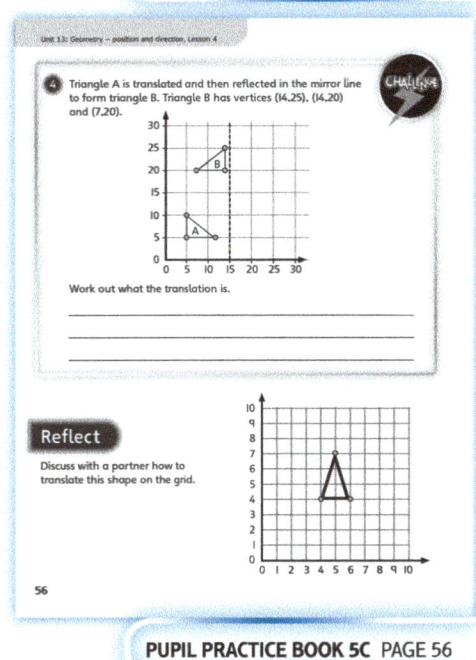

PUPIL PRACTICE BOOK 5C PAGE 56

Unit 13: Geometry – position and direction, Lesson 5

Reflection

Learning focus

In this lesson, children will explore reflection. They will learn how to reflect simple 2D shapes in vertical and horizontal lines. Children will explore drawing reflections on different types of paper (squared, square dotted and isometric).

Before you teach

- Are children able to use the correct language to describe 2D shapes: vertex, side, parallel, symmetric and right angle?

NATIONAL CURRICULUM LINKS

Year 5 Geometry – position and direction

Identify, describe and represent the position of a shape following a reflection or translation, using the appropriate language, and know that the shape has not changed.

ASSESSING MASTERY

Children can reflect simple 2D shapes in horizontal and vertical mirror lines. They can confidently use the terms 'reflect', 'reflection' and 'mirror line' and understand that a reflected shape does not change in size or shape but will face in a different direction (orientation). Children can accurately draw reflections in mirror lines on different types of paper and can identify whether a given reflection is correct. They are able to explain errors.

COMMON MISCONCEPTIONS

Children may simply transfer the shape to the other side of the mirror line, rather than reflecting it. Ask:
- *How many squares away from the mirror line is point A? Where should the reflected point be on the other side of the mirror line? Is each point the same distance away from the mirror line?*

STRENGTHENING UNDERSTANDING

Provide children with mirrors to explore reflections. Ask children to predict what the reflection will look like and then check by placing the mirror on the mirror line. Discuss, in detail, pictures and photographs of reflections. Cut 2D shapes out of thin card and model how the shape 'flips over' but does not change size or shape in a reflection. Mark a point with a dot to show where that point is when the shape has been reflected. Make sure reflected images are drawn accurately using grid lines.

GOING DEEPER

Practise reflecting more complex shapes with several diagonal sides or by using different types of paper: squared, isometric or dotted. Ask children to draw and reflect the capital letters of the alphabet and place them in two groups: 1) the reflection is different to the original; 2) the reflection is exactly the same (answers will vary depending on whether the mirror line is horizontal or vertical).

KEY LANGUAGE

In lesson: reflection, reflect, mirror line, vertex, symmetric, predict, isometric grid

Other language used by the teacher: mirror image, translation, orientation, flip over, position, direction

STRUCTURES AND REPRESENTATIONS

2D shapes

RESOURCES

Mandatory: mirrors, squared paper, square dotted paper, isometric paper, rulers

Optional: images of reflections

 In the eTextbook of this lesson, you will find interactive links to a selection of teaching tools.

Quick recap

On the board display some symmetrical shapes and ask children to identify any horizontal or vertical lines of symmetry. First use simple shapes, such as rectangles and isosceles triangles, before moving on to shapes such as circles and regular hexagons.

Unit 13: Geometry – position and direction, Lesson 5

Discover

WAYS OF WORKING Pair work

ASK

• Question ① a): *What is a reflection?*
• Question ① a): *How many mirror lines are shown in this image?*
• Question ① a): *What happens to the shape when it is reflected? Does the shape change size?*
• Question ① b): *Does it matter if the reflection is not exactly on the grid lines?*

IN FOCUS The image shows a triangle with two mirror lines. Encourage children to place a mirror on each of the mirror lines, discussing what they can see. Explore what happens to the shape when it is reflected – ask: *Does the shape change?* It is important to spend a significant amount of time using the mirrors to develop children's understanding of what reflections look like. Children should also realise the importance of accurate drawing, so ensure that they use a ruler and that the shape is drawn on the grid lines.

PRACTICAL TIPS Provide children with mirrors, irregular 2D shapes and squared paper to explore both reflections and mirror lines.

Demonstrate how to place the mirror on the mirror lines to see the reflections.

ANSWERS

Question ① a): Question ① b):

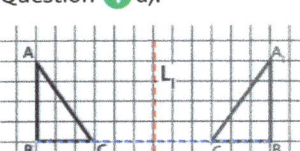 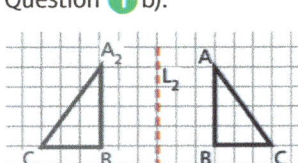

Share

WAYS OF WORKING Whole class teacher led

ASK

• Question ① a): *In which line are you reflecting the shape?*
• Question ① a): *What has happened to each vertex – where has it moved to?*
• Question ① b): *Where will each point/vertex of the shape be when you have reflected the shape?*
• Question ① b): *Can you check your answer using a mirror?*

IN FOCUS This part of the lesson shows how to reflect a shape one vertex at a time. Show children how to count squares from the vertex to the mirror line. Ask children to draw the problem onto squared paper. Can they use a mirror to check their answers? Discuss if and how the shape has changed – ensure children understand that the shape has not changed in size or shape, only in position or orientation. Use a triangle cut out of thin card to model this, marking a dot at one of the vertices: show how the shape is 'flipped over' and where the vertex with the dot has moved to.

PUPIL TEXTBOOK 5C PAGE 76

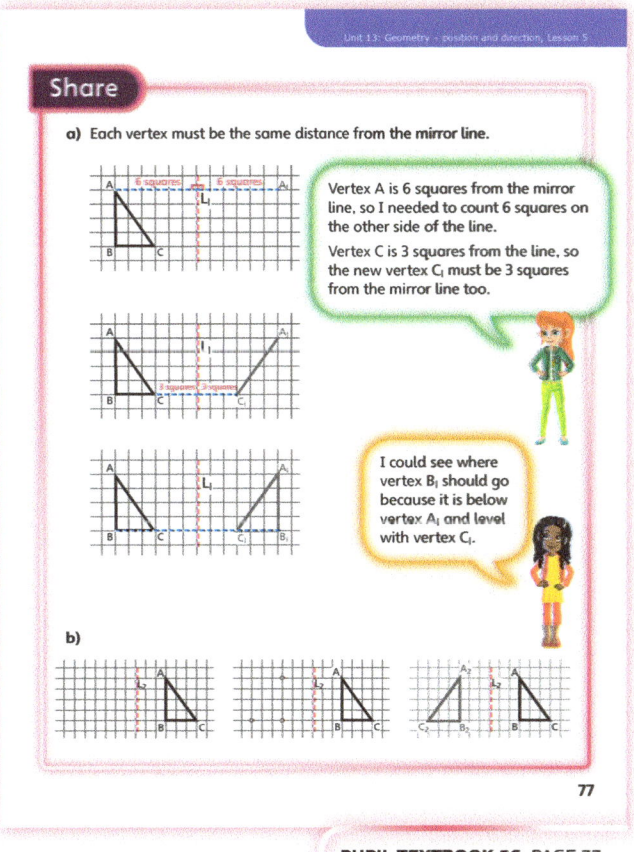

PUPIL TEXTBOOK 5C PAGE 77

Unit 13: Geometry – position and direction, Lesson 5

Think together

WAYS OF WORKING Whole class teacher led (I do, We do, You do)

ASK
- Question ❶: *How far is each vertex from the mirror line?*
- Questions ❶ and ❷: *Where will each reflected vertex be?*
- Question ❸: *Can you predict where to hold a mirror?*

IN FOCUS In questions ❶ and ❷, children draw reflections of simple 2D shapes one vertex at a time and check their drawings using a mirror. Give children square dotted and isometric paper.

STRENGTHEN Ensure children have plenty of opportunities to check reflections with a mirror and to discuss what has happened to the shape and to each vertex of the shape.

DEEPEN Encourage children to practise drawing similar 2D shapes and reflecting them.

ASSESSMENT CHECKPOINT Can children draw the reflection in a mirror line one vertex at a time? Do they know to count squares to find the position of reflected vertices?

ANSWERS

Question ❶: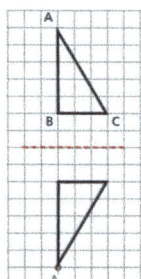

Question ❷: Bella has translated the triangle 9 squares up.
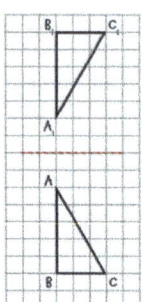

Question ❸ a): Reflect it vertically then horizontally or horizontally then vertically.
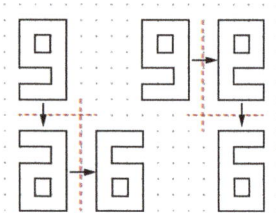

Question ❸ b):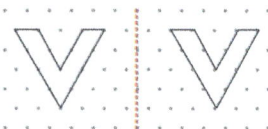

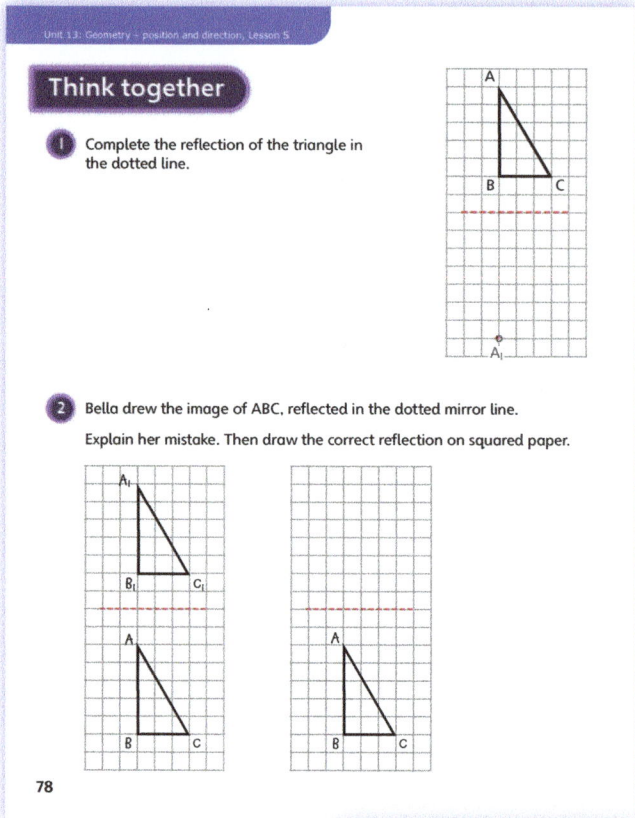

PUPIL TEXTBOOK 5C PAGE 78

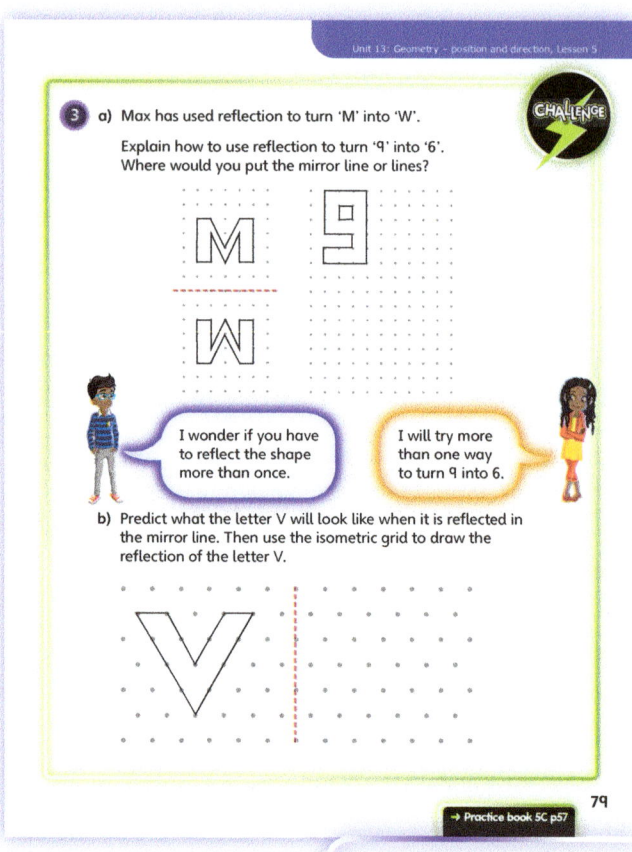

PUPIL TEXTBOOK 5C PAGE 79

Unit 13: Geometry – position and direction, Lesson 5

Practice

WAYS OF WORKING Independent thinking

IN FOCUS Question ❷ involves reflecting the numbers 5 and 3, drawn on squared or dotted paper so that they look as they would on a digital clock face. Children may see the shapes as an S and a back-to-front E, so help them to recognise the numbers if necessary. Encourage children to check each drawing by placing a mirror on the mirror line.

STRENGTHEN Spend time exploring reflections by placing a mirror on the mirror line and discussing what happens to each shape. Encourage children to write capital letters or other numbers to see what they look like when reflected. Ask them to explain what has changed (orientation) and what has not changed (size and shape).

DEEPEN Ask children to draw and reflect the numbers 1 to 9 on squared, square dotted or isometric paper. Which paper is easier to draw numbers on? Which paper do children prefer and why?

THINK DIFFERENTLY In question ❸, children are given the original shape and the reflection and asked to draw the mirror line. Provide mirrors for children to check their answers.

ASSESSMENT CHECKPOINT Do children understand that reflected shapes do not change in size or shape but are positioned differently? Can children reflect shapes accurately by measuring the distance of each vertex from the mirror line?

ANSWERS Answers to the **Practice** part of the lesson can be found in the *Power Maths* online subscription.

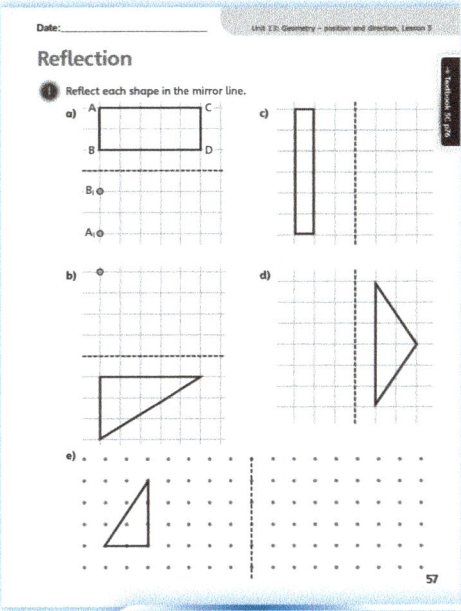

PUPIL PRACTICE BOOK 5C PAGE 57

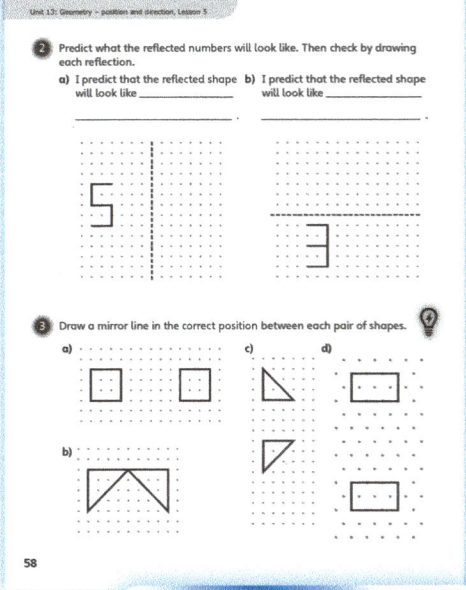

PUPIL PRACTICE BOOK 5C PAGE 58

Reflect

WAYS OF WORKING Independent thinking

IN FOCUS Prompt children by asking: *Which way is the arrow facing? Which way will the arrow be facing once reflected?* Children should identify that the arrow will point to the right (in the opposite direction) after it is reflected. Encourage children to draw the reflection on squared paper and to check with a mirror.

ASSESSMENT CHECKPOINT Can children draw and describe reflections in a vertical mirror line accurately? Can they check their answers using a mirror?

ANSWERS Answers to the **Reflect** part of the lesson can be found in the *Power Maths* online subscription.

After the lesson

- Can children draw reflections accurately?
- What misconceptions did children have relating to reflection?

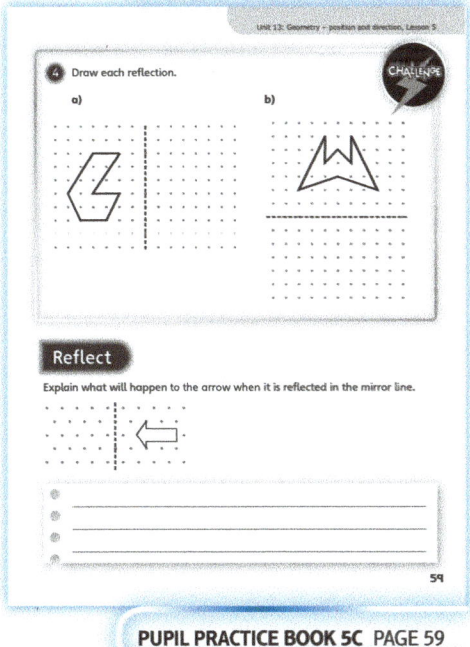

PUPIL PRACTICE BOOK 5C PAGE 59

Unit 13: Geometry – position and direction, Lesson 6

Reflection in horizontal and vertical lines

Learning focus

In this lesson, children will find the coordinates of a reflected point on a grid. They will use coordinates to calculate new points rather than counting squares.

Before you teach

- Are children secure in reflecting simple shapes in a mirror line?
- Can children plot and find coordinates in the first quadrant?

NATIONAL CURRICULUM LINKS

Year 5 Geometry – position and direction

Identify, describe and represent the position of a shape following a reflection or translation, using the appropriate language, and know that the shape has not changed.

ASSESSING MASTERY

Children can use coordinates to find a reflected point on a grid. They are able to use coordinates to calculate the new point/vertex of a shape by adding or subtracting rather than counting squares. Children can find the coordinates of a reflected 2D shape by plotting each vertex.

COMMON MISCONCEPTIONS

Children may rely on counting squares rather than using coordinates to calculate the position of points. Ask:
- *How could you work out the new coordinate without counting squares?*

When the mirror line is parallel to the axes, children may not see that that part of the coordinate stays the same. Ask:
- *How can you tell that one number in this coordinate will stay the same?*

STRENGTHENING UNDERSTANDING

Provide children with coordinate grids to practise reflecting individual points before moving on to shapes. Ask children to work with a partner: partner A should place red counters on a 10×10 coordinate grid and partner B should place yellow counters to indicate the reflection of each point in a vertical mirror line (*x* = 5); then the children should swap roles. You may wish to encourage children to draw lines to the axes and use number line jumps to show an equal distance either side of the mirror line.

GOING DEEPER

Ask children to explore the coordinates of squares and rectangles set parallel to the axes on a 10×10 coordinate grid and investigate their reflections in the axes. Ask: *Which part of the coordinate stays the same when the mirror line is parallel to the horizontal or vertical axis?* Encourage children to use addition or subtraction of the relevant coordinates to work out the reflected vertices, and to predict what will happen to these reflected vertices if the mirror line is moved.

KEY LANGUAGE

In lesson: reflection, reflect, mirror line, grid, coordinates, position

Other language to be used by the teacher: vertex, mirror image, translation, direction, orientation, horizontal axis, vertical axis

STRUCTURES AND REPRESENTATIONS

Coordinate grids, 2D shapes

RESOURCES

Optional: mirrors, squared paper

 In the eTextbook of this lesson, you will find interactive links to a selection of teaching tools.

Quick recap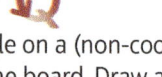

Draw a 4×2 rectangle on a (non-coordinate) square or dotted grid on the board. Draw a vertical line two squares away from the rectangle. Ask children to draw the reflection. Repeat if time allows with a horizontal line below the rectangle.

Unit 13: Geometry – position and direction, Lesson 6

Discover

WAYS OF WORKING Pair work

ASK

- Question 1 a): *How do you read coordinates? What are the coordinates of T?*
- Question 1 a): *Where is the mirror line?*
- Question 1 b): *How far away from the mirror line is coordinate (6,8)?*

IN FOCUS Encourage children to discuss the map. Can they identify the coordinates of point T and find the point (6,8) on the map? Ensure that children are able to identify coordinates confidently.

What happens when children put a mirror on the mirror line? Note that it will show an image of where point T and the secret cave will be, but will not show the new coordinates correctly.

PRACTICAL TIPS Provide children with mirrors to explore the reflection. Provide a copy of the coordinate grid with the mirror line in place so children can practise plotting and reflecting other points.

ANSWERS

Question 1 a): The true coordinates of the treasure, T_1, are (8,3).

Question 1 b): The coordinates of the secret cave are (4,8).

PUPIL TEXTBOOK 5C PAGE 80

Share

WAYS OF WORKING Whole class teacher led

ASK

- Question 1 a): *What do you need to do to answer the question?*
- Question 1 a): *What has happened to the vertical coordinate of the reflected point?*
- Question 1 b): *How have the coordinates of the reflected point been worked out? Which part of the coordinate has not changed?*

IN FOCUS Model how to use addition or subtraction to find new coordinates. For example, 5 + 3 = 8, so the horizontal coordinate is 8. Encourage children to use this method rather than counting squares. You could also discuss how adding (or subtracting) double the distance to the original horizontal coordinate will give the new horizontal coordinate. For example, in question 1 a), 2 + 6 = 8; in question 1 b), 6 − 2 = 4. Highlight that, when the mirror line is vertical, the vertical coordinate does not change. If the mirror line is horizontal, the horizontal coordinate will not change. Explore examples with the class.

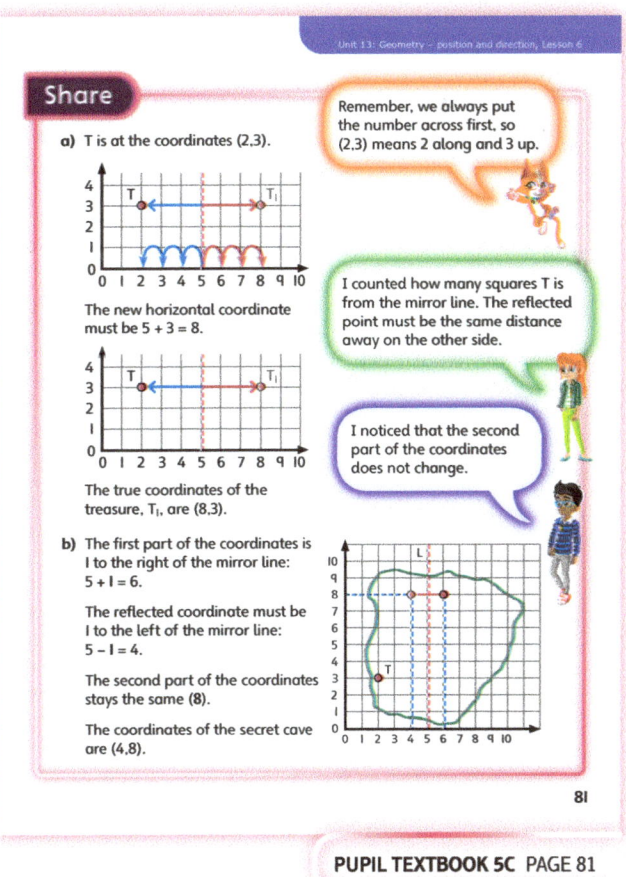

PUPIL TEXTBOOK 5C PAGE 81

117

Unit 13: Geometry – position and direction, Lesson 6

Think together

WAYS OF WORKING Whole class teacher led (I do, We do, You do)

ASK

- Question ①: *Where is the mirror line?*
- Question ②: *How far are the coordinates from the mirror line? How can you use this to work out the coordinates of the reflection?*
- Question ③: *What can you say about the coordinates of the rectangle? Which parts of the coordinates are the same/different? How will this help you to find the coordinates of the reflected rectangle?*

IN FOCUS For questions ① and ②, concentrate on the method of adding (or subtracting) the distance from the mirror line to (or from) the original coordinate to work out the reflected coordinate. Explain to children that they can also add (or subtract) double this distance to (or from) the original coordinate. Ensure children understand which coordinate does not change when the mirror line is vertical (the vertical coordinate) or horizontal (the horizontal coordinate). In question ③ a), there are no grid lines; look out for children trying to draw them in, so they can count squares rather than adding or subtracting. Discuss how to work out how long and high the rectangle is using the coordinates (height = 14 − 11 = 3; length = 11 − 6 = 5).

STRENGTHEN For questions ② and ③, provide coordinate grid paper so children can explore drawing rectangles parallel to the axes. Ask children to write the coordinates of each vertex and identify which parts of the coordinates 'match' and which parts change. Encourage children to practise reflecting one rectangle in different mirror lines (vertical and horizontal) to consolidate which parts of the coordinates change and which do not. Relate the difference between original and reflected coordinates to the distance from the mirror line to the original rectangle.

DEEPEN Enable children to explore reflecting parallelograms in horizontal and vertical mirror lines.

ASSESSMENT CHECKPOINT Can children identify the new coordinates when a point is reflected? Can they identify the new coordinates when a rectangle is reflected? Can children calculate reflected coordinates without counting the squares?

ANSWERS

Question ①: $A_1(2,5)$
$B_1(3,6)$
$C_1(7,1)$

Question ②: $A_1(8,11)$
$B_1(8,14)$
$C_1(3,14)$
$D_1(3,11)$

Question ③ a): $A_1(6,5)$
$B_1(11,5)$
$C_1(6,2)$
$D_1(11,2)$

Question ③ b): Children should estimate the coordinates.

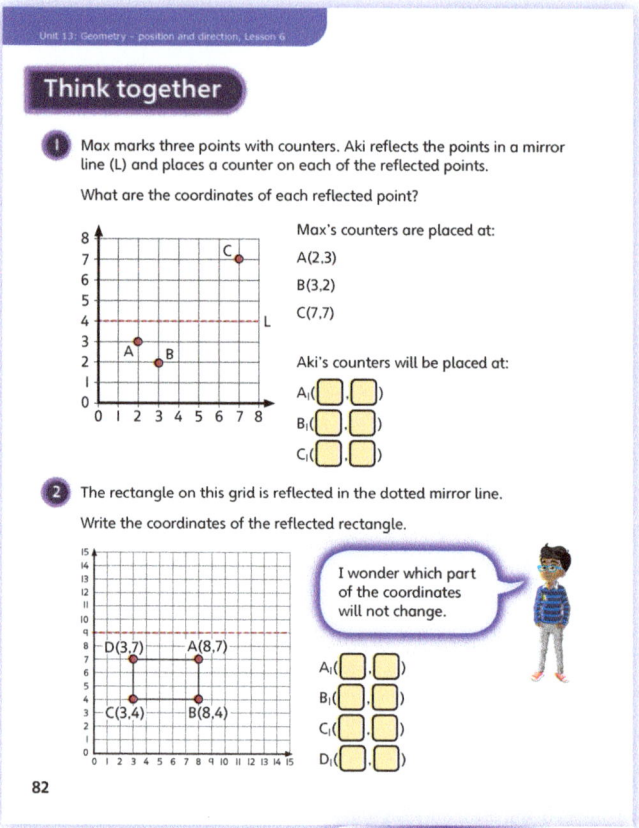

PUPIL TEXTBOOK 5C PAGE 82

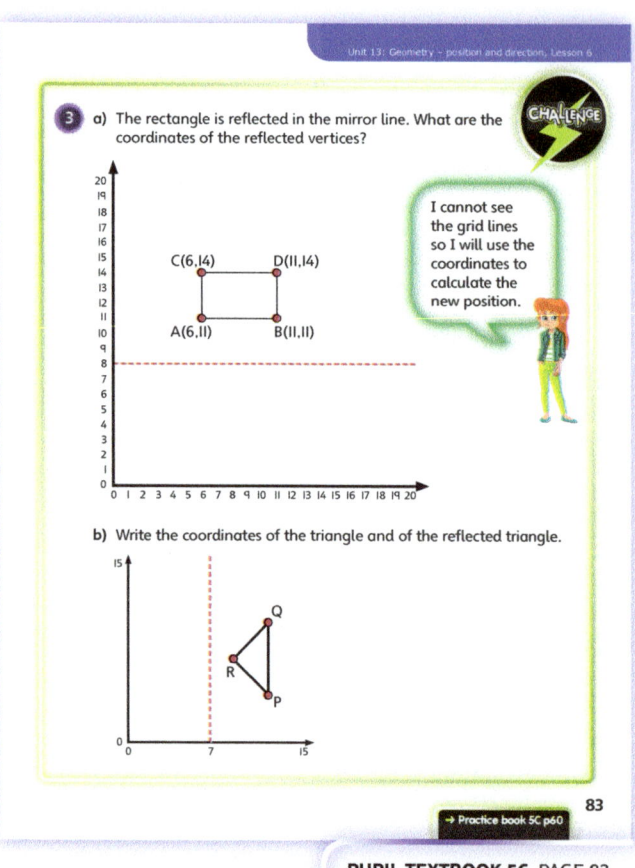

PUPIL TEXTBOOK 5C PAGE 83

Unit 13: Geometry – position and direction, Lesson 6

Practice

WAYS OF WORKING Independent thinking

IN FOCUS In question ❻, children will need to identify the coordinates of all the vertices. Two can be deduced from the markings on the axes: (22,14) and (22,20). Discuss how to find the length of the sides using the vertical coordinates given (20 – 14 = 6); this will allow children to identify the other two coordinates: (28,14) and (28,20). Some children may need to draw the reflected square and plot each point to see where it lies. Encourage more able children to simply mark on the axes where the reflection will be.

STRENGTHEN In question ❷, discuss which parts of the coordinates stay the same and which are changed and by how much, depending on the distance from the mirror line. Can children see that they should add or subtract double the distance from the mirror line to the horizontal or vertical coordinate? Do they understand that one part of the coordinate will not change when the mirror line is horizontal or vertical?

DEEPEN Give each child a 12×12 coordinate grid and ask them to mark the four vertices of a rectangle or square, plus a mirror line. Children then swap with a partner to work out the vertices of the reflected shape. Can they work out the new vertices even if the image does not lie within the grid they have?

THINK DIFFERENTLY In question ❺, fewer numbers are provided on the axes of the grid. Children should realise that they still have enough information to complete the task: they can use their knowledge of scales and number lines to find the correct coordinates for the reflected shape.

ASSESSMENT CHECKPOINT Can children calculate the reflected coordinates using addition or subtraction rather than by counting squares? Can children explain which part of the coordinate will/will not change with a given mirror line?

ANSWERS Answers for the **Practice** part of the lesson can be found in the *Power Maths* online subscription.

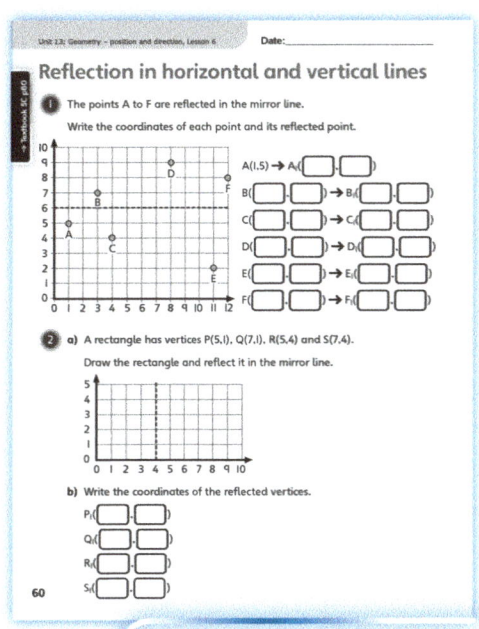

PUPIL PRACTICE BOOK 5C PAGE 60

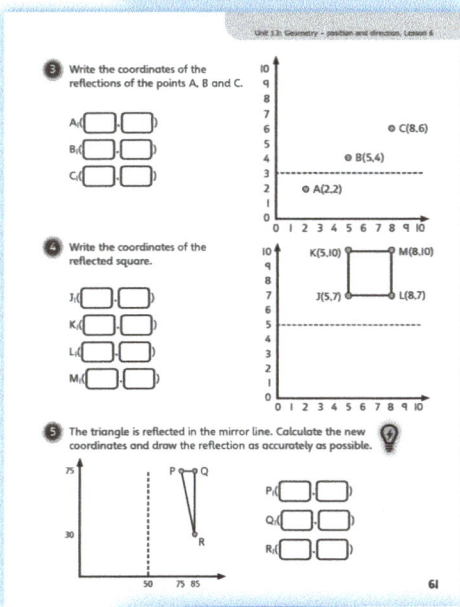

PUPIL PRACTICE BOOK 5C PAGE 61

Reflect

WAYS OF WORKING Pair work

IN FOCUS Look out for children explaining the following.
When you reflect across a vertical mirror line:
- the x-coordinate will change and the y-coordinate will stay the same
- work out the distance from the point to the mirror line
- add/subtract double that distance to/from the x-coordinate.

When you reflect across a horizontal mirror line:
- the y-coordinate will change and the x-coordinate will stay the same
- work out the distance from the point to the mirror line
- add/subtract double that distance to/from the y-coordinate.

ASSESSMENT CHECKPOINT Can children explain how to calculate a reflected point using coordinates? Can they give an example when explaining?

ANSWERS Answers for the **Reflect** part of the lesson can be found in the *Power Maths* online subscription.

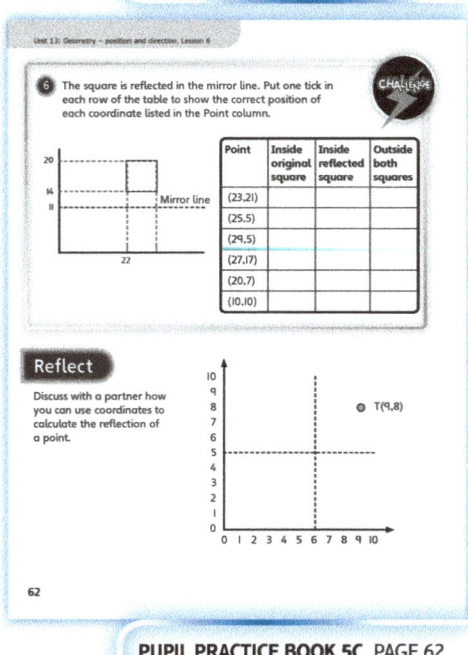

PUPIL PRACTICE BOOK 5C PAGE 62

After the lesson
- Are all children confident in reading coordinates?
- Do children understand which coordinate does not change when reflecting in a line parallel to one of the axes (vertical or horizontal)?

119

Unit 13: Geometry – position and direction

End of unit check

Don't forget the unit assessment grid in your *Power Maths* online subscription.

WAYS OF WORKING Group work adult led

IN FOCUS
- Question ❶ checks that children can accurately read a coordinate in the first quadrant.
- Question ❷ checks that children can reason about the position of missing vertices of shapes using coordinates to help them be precise in their reasoning.
- Questions ❸ and ❹ test children's ability to identify common errors in reflections and translations.
- Question ❹ ensures children are confident with left and right and know which instruction to follow first.
- Question ❺ is a SATS-style question which will assess whether children remember that the x-coordinate is unchanged when a mirror line is horizontal. They should also know that the vertical distance from the point to the mirror line is the same as the distance from the mirror line to the reflected image.

ANSWERS AND COMMENTARY Children who have mastered the concepts in this unit will be able to plot and find coordinates of a reflected shape on a grid. They will be able to use coordinates to find translations. Children will also be able to confidently identify and describe reflections and translations.

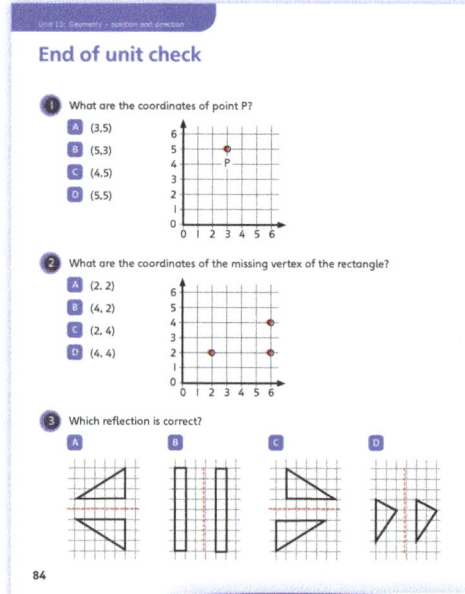

PUPIL TEXTBOOK 5C PAGE 84

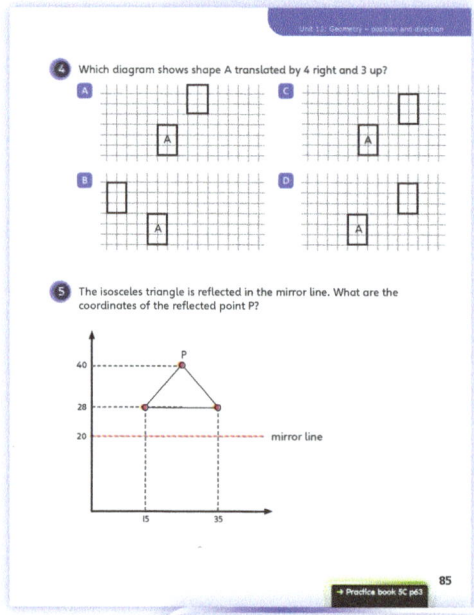

PUPIL TEXTBOOK 5C PAGE 85

Q	A	WRONG ANSWERS AND MISCONCEPTIONS	STRENGTHENING UNDERSTANDING
1	A	B suggests children are still making the mistake of reading a coordinate and getting the x and y values the wrong way round.	Provide children with mirrors to explore reflections; ask them to predict the reflection and then check by placing the mirror on the mirror line. Ask children to cut out 2D shapes from squared paper so they can physically move them on a grid and place them in the new position. This will help children to understand that the shape does not change. They will be able to see clearly how far and in which direction the shape has moved. Practise using coordinates to find translations involving more than one move.
2	C	B again suggests children are still reading a coordinate the wrong way round.	
3	A	D suggests children do not understand the difference between reflection and translation.	
4	C	A shows 3 right, 4 up; B shows movement left instead of right; D shows 3 squares between the rectangles, not between the corresponding vertices.	
5	(25,0)	The mirror line is horizontal, so only the vertical coordinate is affected, but children also need to work out the horizontal coordinate (half-way between 35 and 15).	

Unit 13: Geometry – position and direction

My journal

WAYS OF WORKING Independent thinking

ANSWERS AND COMMENTARY

Question ❶: A (53,25), B (67,25), C (77,25), D (82,13), E (72,13)

As the grid is blank, children will need to calculate the new coordinates using the distance from the mirror line. The width of the parallelogram is 10. The mirror line is vertical, so the vertical coordinates of the reflection will be the same as those of the original shape. The mirror line is on the right-hand side of the shape, so the horizontal coordinates of the reflection will be greater than those of the original shape.

Question ❷: Children may need to think a bit more carefully about how to draw the vertical mirror line.

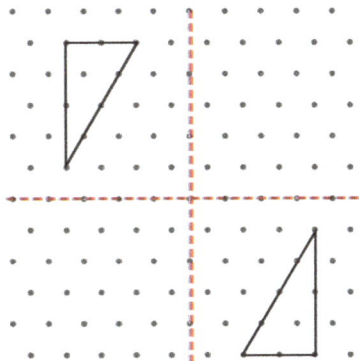

Power check

WAYS OF WORKING Independent thinking

ASK
- What have you learnt about reflections?
- What have you learnt about translations?
- Are you confident with using coordinates?

Power puzzle

WAYS OF WORKING Independent thinking

IN FOCUS Use this **Power puzzle** to see if children can independently create and reflect a design in four quadrants, using coordinates to plot accurately. Encourage children to sketch their designs first. Begin with designs based on rectilinear shapes, then suggest designs involving diagonal lines and more complex rectilinear shapes.

ANSWERS AND COMMENTARY If children complete the task successfully, this shows that they can plot the coordinates for a reflection in both horizontal and vertical mirror lines.

After the unit ⏸

- Are children confident with plotting coordinates?
- Do children use coordinates to plot a translation or reflection, or do they rely on counting squares?
- What misconceptions did children have and how can you address these before the next unit?

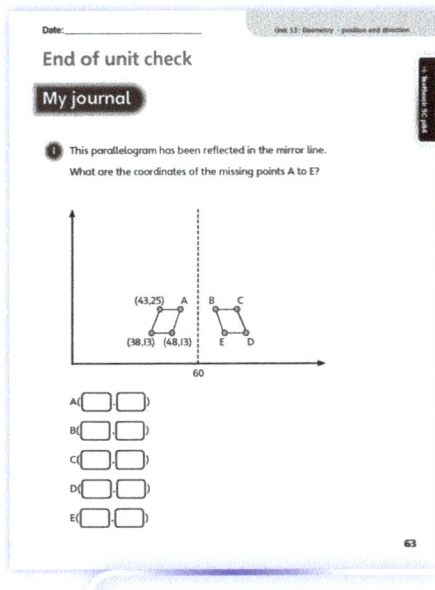

PUPIL PRACTICE BOOK 5C PAGE 63

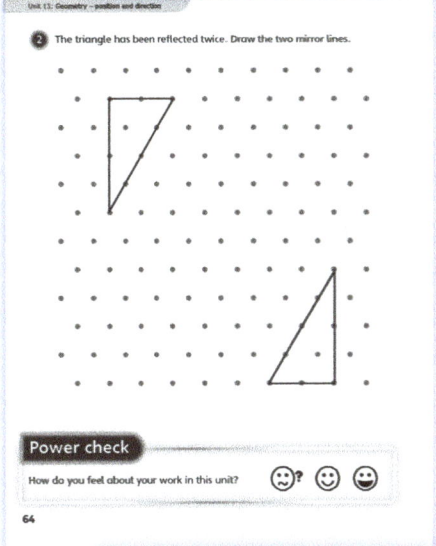

PUPIL PRACTICE BOOK 5C PAGE 64

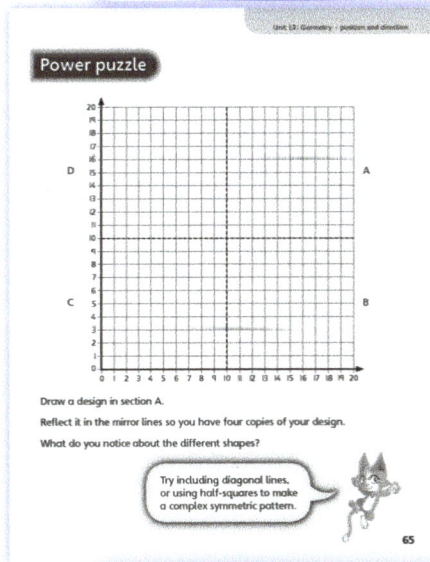

PUPIL PRACTICE BOOK 5C PAGE 65

Strengthen and **Deepen** activities for this unit can be found in the *Power Maths* online subscription.

Unit 14
Decimals

Mastery Expert tip! 'To build confidence with decimals, we invented a board game called "93". This was because there were 93 children in Year 5. They each created one question where the answer was a decimal number. Children applied their learning from this unit to the game. We challenged children to add, subtract, multiply, divide and solve problems with decimals to win points. They tried to make as many points as possible before reaching the "Finish". The game really engaged children and was a great way to build confidence and fluency in working with decimals.'

Don't forget to watch the Unit 14 video!

WHY THIS UNIT IS IMPORTANT

This unit is important because it applies the formal methods of addition and subtraction to numbers with up to three decimal places. It also teaches children to multiply and divide decimal numbers by 10, 100 and 1,000. The range of problem-solving questions in this unit will develop confidence and flexibility when exploring the most efficient ways to work with decimal numbers.

WHERE THIS UNIT FITS

→ Unit 13: Geometry – position and direction
→ **Unit 14: Decimals**
→ Unit 15: Negative numbers

This unit builds on children's work in Years 4 and 5 of adding and subtracting whole numbers, and multiplying and dividing whole numbers by 10, 100 and 1,000. It also extends children's work with number patterns. By considering the place value of each digit, children will broaden their understanding of adding and subtracting using formal written methods, and of multiplying and dividing decimal numbers.

Before they start this unit, it is expected that children:
- can add and subtract numbers with up to 4 digits
- are able to solve addition and subtraction word problems
- can multiply and divide whole numbers by 10, 100 and 1,000
- understand what place value means and can use a place value grid to partition a decimal number.

ASSESSING MASTERY

Children who have mastered this unit can add and subtract numbers with up to three decimal places using a variety of methods, including formal written methods. They can multiply and divide decimal numbers by 10, 100 and 1,000. They can confidently apply their knowledge of decimal numbers to solve word problems.

COMMON MISCONCEPTIONS	STRENGTHENING UNDERSTANDING	GOING DEEPER
When using column addition or subtraction, children may line numbers up from right to left rather than according to their place value.	Use a place value grid alongside column addition or subtraction so that children can link the concrete and abstract representations.	Compare calculations laid out correctly and incorrectly. What is the difference between the correct and incorrect answers? Why?
Children may add the digits when adding decimals that make one. For example, with 0·3 + 0·7, children may give the answer 0·10.	Use a bar model or number line to help children understand whether the answer is 1 or not. Discuss that 0·3 means 3 tenths, and that 0·7 means 7 tenths.	Ask increasingly difficult missing-number questions. Give children a total and ask them to find the different ways they can make the total.

Unit 14: Decimals

UNIT STARTER PAGES

As a whole class, discuss the unit starter pages. Talk through the key learning points and vocabulary mentioned by the characters. Use Ash's comment to remind children of the column method, noting the importance of setting out column addition neatly.

STRUCTURES AND REPRESENTATIONS

Place value grid: This model uses counters to show the value of each column. It supports the column method layout.

O	•	Tth	Hth	Thth
①		⓪⋅¹ ⓪⋅¹ ⓪⋅¹ ⓪⋅¹ ⓪⋅¹ ⓪⋅¹	⓪⋅⁰¹ ⓪⋅⁰¹	⓪⋅⁰⁰¹ ⓪⋅⁰⁰¹ ⓪⋅⁰⁰¹ ⓪⋅⁰⁰¹

Column addition and subtraction: This model demonstrates the place value of each digit in addition and subtraction calculations and shows exchanges between columns.

	O	•	Tth	Hth	Thth
	2	•	4	3	1
+	0	•	0	8	0
	2	•	5	1	1
			1		

	O	•	Tth	Hth	Thth
	⁷8̸	•	¹2	⁶7̸	¹0
−	6	•	6	5	3
	1	•	6	1	7

Bar model: This model can be used to compare numbers and identify missing information. It can be used to represent the information in some addition and subtraction word problems.

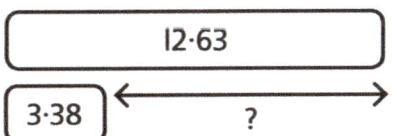

KEY LANGUAGE

There is some key language that children will need to know as part of the learning in this unit:

➡ add, subtract, multiply, divide
➡ ones, tenths, hundredths, thousandths
➡ difference, group, share, compare, represent
➡ decimal, decimal point, decimal place, digit, whole
➡ column, place value, exchange
➡ mass, weight, length, width, cost, height
➡ complement

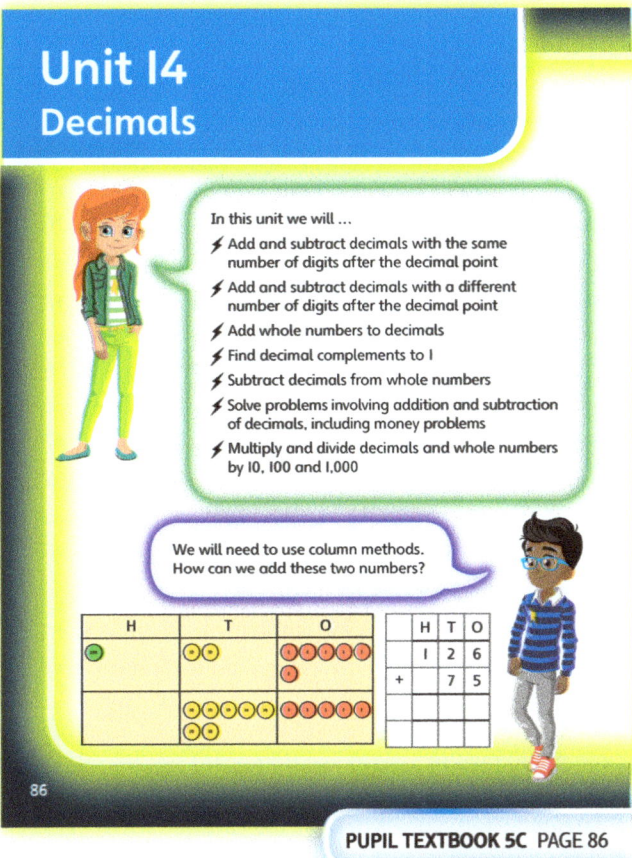

PUPIL TEXTBOOK 5C PAGE 86

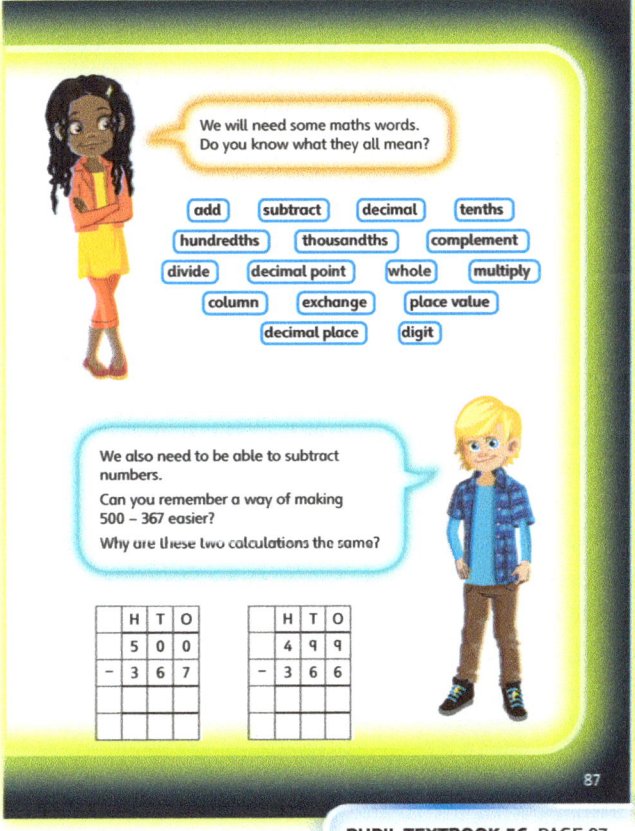

PUPIL TEXTBOOK 5C PAGE 87

Unit 14: Decimals, Lesson 1

Add and subtract decimals within 1

Learning focus
In this lesson, children will learn to add decimals, recognising the importance of place value.

Before you teach
- Do children know how to add and subtract 1- and 2-digit numbers?
- Do children understand what a decimal is?

NATIONAL CURRICULUM LINKS

Year 5 Number – fractions (including decimals and percentages)

Solve problems involving number up to three decimal places.

ASSESSING MASTERY

Children can add and subtract decimals where the answer is less than one. Children can answer abstract questions or those presented in a context and draw diagrams to express their thinking.

COMMON MISCONCEPTIONS

Children may not understand the importance of place value or what each digit represents. When adding decimals that total 1, such as 0·3 + 0·7, children often add the digits 3 and 7 and give the answer as 0·10. Ask:
- *What does the 3 represent? Can you represent each number with place value counters?*
- *Can you simplify 0·10? Can you show 0·10 on a number line? Can you use a number line to add 0·3 and 0·7?*

STRENGTHENING UNDERSTANDING

Children should practise adding 2 or 3 whole numbers before moving on to decimals. They can use a number line to better visualise the calculations they do. Discuss that 0·3 means 3 tenths, and 0·7 means 7 tenths. Ask: *What is the total of 3 tenths and 7 tenths?* If children say 10 tenths, revisit the concept of fractions. Encourage them to clearly describe the place value of each column, and ensure they understand the importance of the decimal point, particularly in an exchange.

GOING DEEPER

Give children a total and ask how many different ways they can make the total, for example: ☐ + ☐ = 0·9. This could also be represented on a part-whole model to help children to see the link between adding and subtracting. Change each decimal to a fraction and link both concepts together. Children know that $\frac{3}{10} + \frac{7}{10} = 1$. What other decimals and fractions can they find with a total of 1?

KEY LANGUAGE

In lesson: add, subtract, difference, total, tenths, part-whole, metres (m)

Other language to be used by the teacher: ones, hundredths, digit, column, place value, decimal point

STRUCTURES AND REPRESENTATIONS

Bar model, number line, part-whole model

RESOURCES

Mandatory: place value counters

Optional: toy train tracks, paper, glue

 In the eTextbook of this lesson, you will find interactive links to a selection of teaching tools.

Quick recap

Support children to recap their bonds to 10 and 100. Focus on bonds such as 20 + 80 first, then move on to examples such as 53 + ? = 100.

Unit 14: Decimals, Lesson 1

Discover

WAYS OF WORKING Pair work

ASK

- Question 1 a): *What lengths are the different track pieces? What does 0·8 m show?*
- Question 1 b): *How many track pieces can be used to make a track of 0·8 m length? What does this depend on?*

IN FOCUS Question 1 a) is important as it requires children to consider and identify three decimal numbers that have a total of 0·8. Question 1 b) requires children to explore alternative ways that 0·8 can be made.

PRACTICAL TIPS Consider allowing children to assemble their own train tracks so that they feel fully involved in the context of the lesson. Bring in different length train tracks or make paper tracks. Children could link or glue the tracks together. This is a good opportunity to discuss the vocabulary needed for this lesson, including terms such as total, longer, shorter, difference and sum.

ANSWERS

Question 1 a): 0·4 m, 0·3 m and 0·1 m sections

Question 1 b): There are several possible answers:
$$0·4 + 0·2 + 0·2 = 0·8$$
$$0·3 + 0·3 + 0·2 = 0·8$$

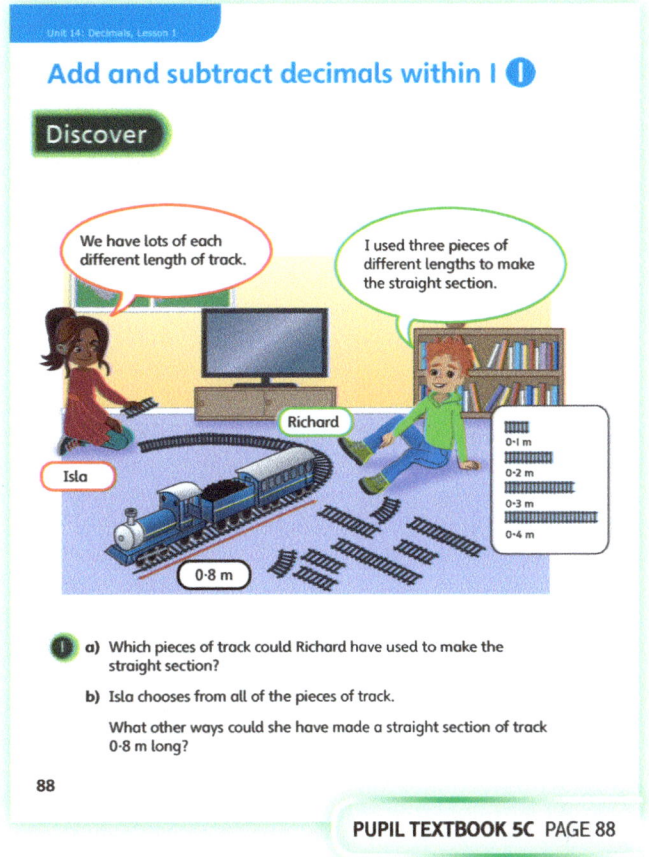

PUPIL TEXTBOOK 5C PAGE 88

Share

WAYS OF WORKING Whole class teacher led

ASK

- Question 1 a): *How many decimal numbers do you need? What should they add up to (sum to)? What method could you use to add the three numbers together?*
- Question 1 a): *Does it matter which number you start with?*
- Question 1 b): *How many decimal numbers are you looking for? What method could you use to work out which numbers make the total? Is there a quicker way?*

IN FOCUS For question 1 a), take the opportunity to discuss the use of the bar model above the number line. Demonstrate how a bar model helped Dexter work out the possible answers. Place value grids can also be used to reinforce the place value of each digit. For example, ask: *What is 3 tenths add 4 tenths add 1 tenth?*

For question 1 b), discuss the use of adding whole numbers that make 8, emphasising the need to not miss any calculations. Discuss Flo's comment. To extend, children can investigate what their strategy would be if the total was 0·9 m or 1·0 m.

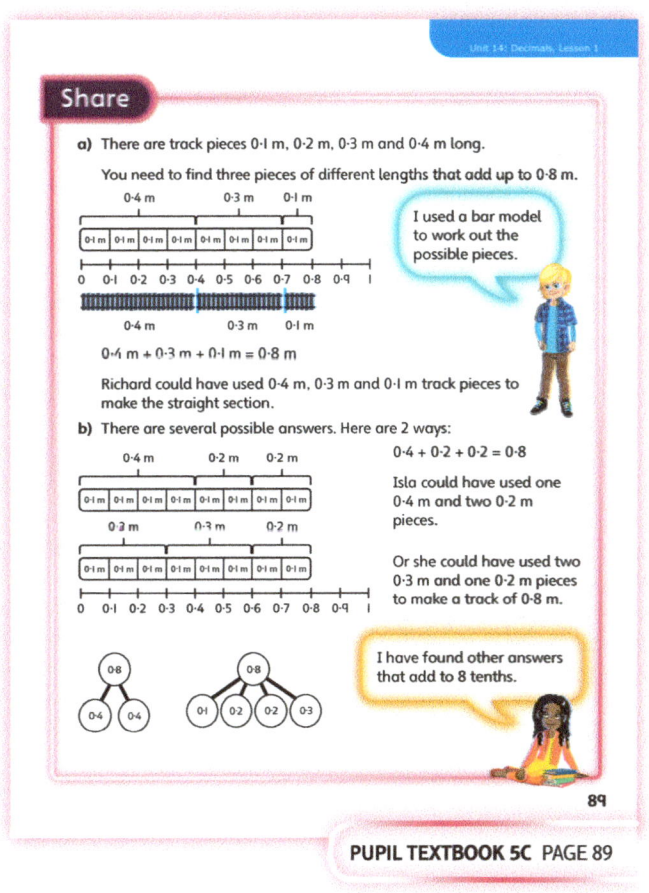

PUPIL TEXTBOOK 5C PAGE 89

125

Unit 14: Decimals, Lesson 1

Think together

WAYS OF WORKING Whole class teacher led (I do, We do, You do)

ASK

- Questions ① a) and b): *What is the same and what is different about this question and the question in Discover?*
- Question ② b): *Does the question only have one answer or is there more than one possible answer?*
- Question ③ b): *What strategy will you use? Can you use a number line to show the calculation?*

IN FOCUS Question ③ highlights the most common misconception in this lesson. Give children the opportunity to discuss what mistake Isla has made. Encourage them to use a bar model with a number line to support their answer. Discuss the methods Astrid and Flo used to find the difference between the decimal numbers in question ③ b). Decimals are numbers, so children should realise that all the strategies used to add and subtract numbers can be applied to decimals.

STRENGTHEN Support understanding by representing calculations using counters on a place value grid. Revisit the vocabulary that children have already encountered when adding and subtracting numbers. Ask children to explain what each word means – for example, difference, total, more, less and sum.

DEEPEN Explore all the possible ways of making each of the numbers given in question ②. Do children work systematically to find all the possible combinations? Children could explore more open-ended questions. For example, ask: *Can you find five decimals with a sum of 1?*

ASSESSMENT CHECKPOINT Can children explain how to add two or more decimals with a total less than or equal to 1? Can children find the difference between two decimal numbers? Can they draw a diagram to support their calculation?

ANSWERS

Question ① a): 0·6 m + 0·2 m = 0·8 m
The track is 0·8 m in total.

Question ① b): 0·7 m – 0·1 m = 0·6 m
Track piece C is 0·6 m longer than track piece D.

Question ② a): 0·4 + 0·5 = 0·9

Question ② b): 0·5 – 0·4 = 0·1
0·4 – 0·3 = 0·1
0·3 – 0·2 = 0·1
0·2 – 0·1 = 0·1

Question ② c): 0·2 and 0·4

Question ③ a): Isla has put the decimal point in the wrong position.
0·1 + 0·2 + 0·7 = 1·0

Question ③ b): 1 m – 0·7 m = 0·3 m

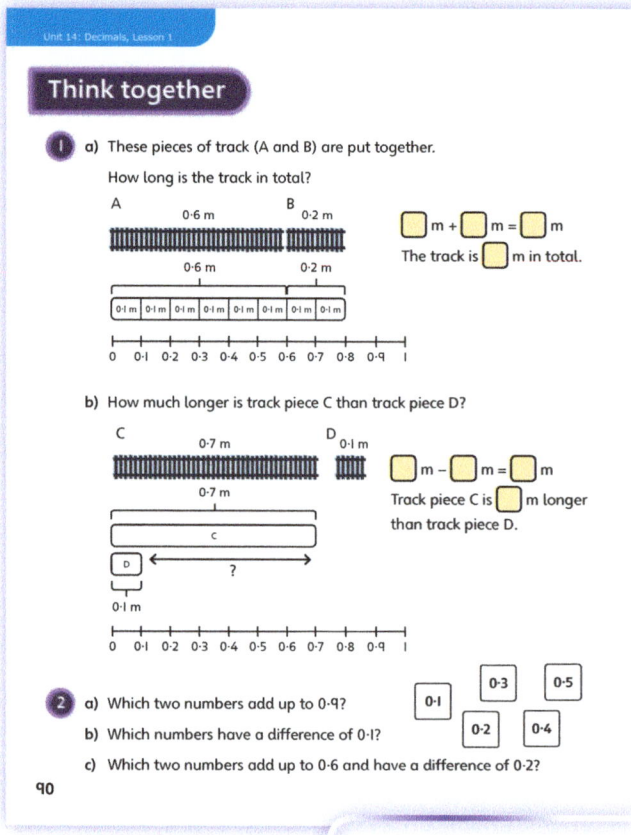

PUPIL TEXTBOOK 5C PAGE 90

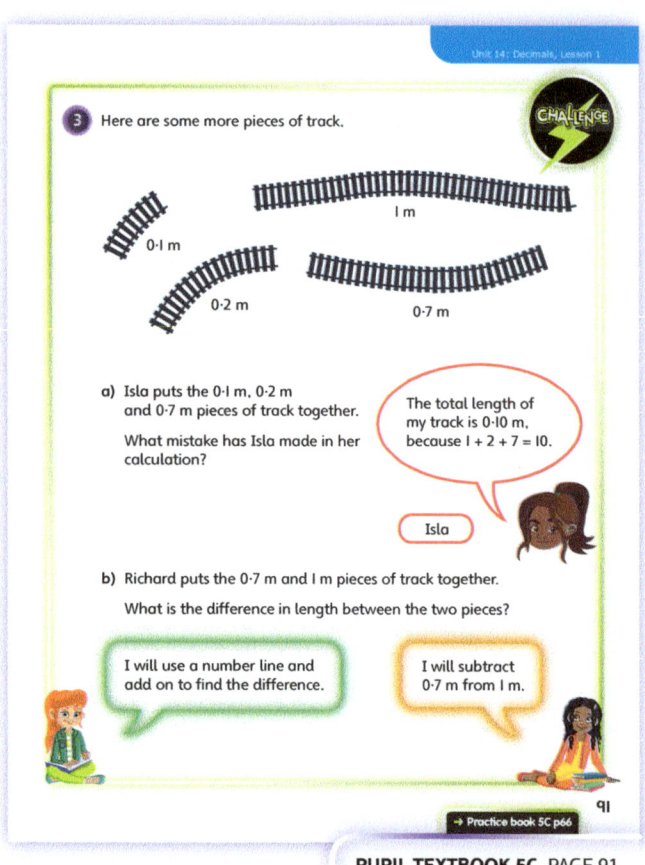

PUPIL TEXTBOOK 5C PAGE 91

Unit 14: Decimals, Lesson 1

Practice

WAYS OF WORKING Independent thinking

IN FOCUS Question ❶ consolidates children's understanding of adding two or more decimal numbers.

Question ❷ consolidates children's understanding of subtracting decimals and revisits the vocabulary that they have used previously to compare numbers.

For question ❻, children need to identify whether to add or subtract to find the missing numbers. To solve these questions, they need to understand that addition and subtraction are inverse operations. Ask children to write the number sentence they used to find each missing number.

STRENGTHEN Encourage children to use counters on a place value grid to support understanding, and to use the bar model to check their answers are correct. Ask: *Are number bonds important when adding decimals? Why?*

DEEPEN If children are comfortable solving question ❽, then ask them to solve the question in a different way. Ask: *What is your strategy? Do you start by finding the numbers in each vertex or do you complete one side at a time? What addition and subtraction statements can you write?*

THINK DIFFERENTLY Question ❼ will require children to think differently and work backwards from the answer to determine what the possible question might be. Discuss whether there might be more than one answer and then challenge children to find all the possible answers.

ASSESSMENT CHECKPOINT Children are confident adding and subtracting decimal numbers less than 1.

ANSWERS Answers for the **Practice** part of the lesson can be found in the *Power Maths* online subscription.

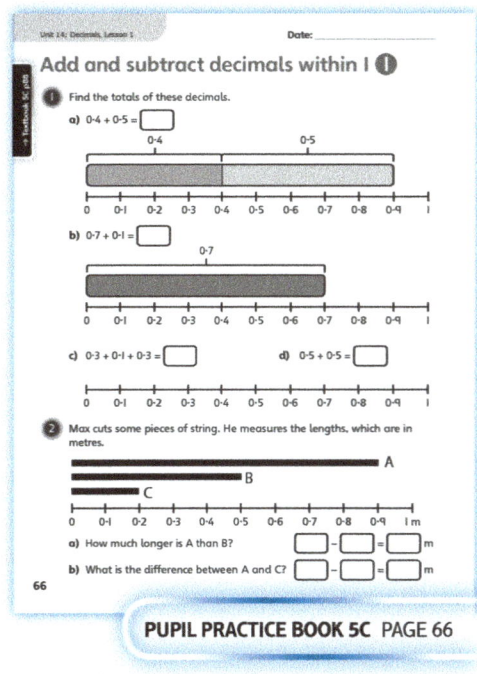

PUPIL PRACTICE BOOK 5C PAGE 66

PUPIL PRACTICE BOOK 5C PAGE 67

Reflect

WAYS OF WORKING Independent thinking

IN FOCUS This **Reflect** activity checks that children are recognising and adding decimals correctly. Encourage them to explain (without doing any calculations) why Emma is wrong. Children should recognise that Emma has added 0·1 and 0·4 rather than 1 and 0·4. Look for children who are able to spot the mistake without any prompting.

ASSESSMENT CHECKPOINT Assess if children can correctly explain how to find two decimal numbers that have a total of less than 1, emphasising the importance of place value and what each digit represents.

ANSWERS Answers for the **Reflect** part of the lesson can be found in the *Power Maths* online subscription.

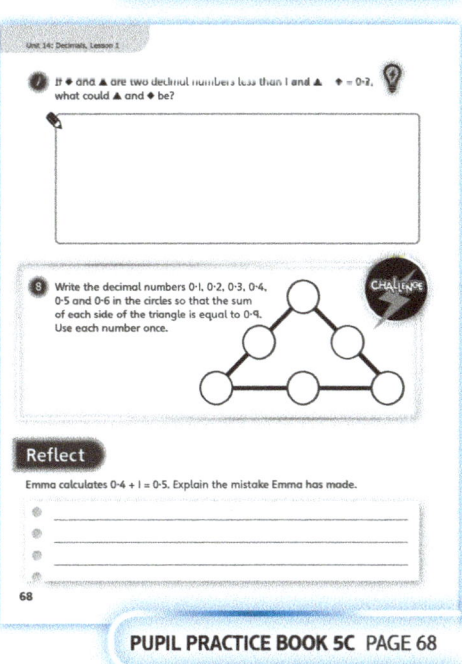

PUPIL PRACTICE BOOK 5C PAGE 68

After the lesson

- Can children add and subtract decimals less than 1?
- Can children find two or more decimals that equal a given total?
- Which children needed to use counters on a place value grid for support? How can you continue to support these children with their calculations?
- Can children accurately draw or use a bar model to support their answers?

Unit 14: Decimals, Lesson 2

Add and subtract decimals within 1 ②

Learning focus
In this lesson, children will add and subtract decimals less than one. They will use the written column method to add or subtract decimals.

Before you teach
- Are children confident making an exchange when using the column addition method?
- Can children set out a column addition when one of the numbers is zero?

NATIONAL CURRICULUM LINKS

Year 5 Number – fractions (including decimals and percentages)
Solve problems involving number up to three decimal places.

ASSESSING MASTERY
Children can add or subtract decimals less than one using the written column method.

COMMON MISCONCEPTIONS
Children may not set the column method out correctly when adding decimals with a zero as one of the digits. They may ignore the zero and add the wrong number. For example, instead of adding 0·03 to a number, they add 0·3. Ask:
- What does each digit represent? What is the value of each digit? Do digits in the same column have the same value?

STRENGTHENING UNDERSTANDING
Encourage children to use counters on a place value grid. This can be done alongside the column addition or subtraction so that children can see how the concrete and abstract methods link together. You could also ask children to check their answers by looking at the approximate size of the numbers in each calculation and the relative size of the answer. For example, if children add 0·35 + 0·21, they should know that their answer must be less than 2 because both numbers are less than 1. If they get an answer of 5·6, then they should recognise that they must have made a mistake.

GOING DEEPER
Ask children to create their own questions that involve adding or subtracting decimals within 1 that have both tenths and hundredths. They could find two numbers where the sum or difference is given, for instance: *Find two numbers that total 0·87* or *Find two numbers where one is 0·12 more than the other*. Encourage children to make the link between addition and subtraction when solving these kinds of calculations.

KEY LANGUAGE
In lesson: add, subtract, total, altogether, difference, sum, exchange, decimal, column, how much more?, less than, tenths, hundredths, litres (l)

Other language to be used by the teacher: place value, ones

STRUCTURES AND REPRESENTATIONS
Column addition, column subtraction, place value grid, bar model, number line, part-whole model

RESOURCES
Mandatory: place value counters

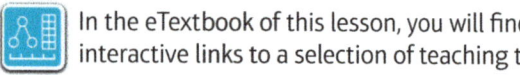 In the eTextbook of this lesson, you will find interactive links to a selection of teaching tools.

Quick recap
Check that children can complete simple column addition with 3- and 4-digit numbers. Ask questions that will involve children completing an exchange.

Discover

WAYS OF WORKING Pair work

ASK

- Question ① a): *Do you think this question is about addition or subtraction? Can you answer this question without doing a calculation? How do you know you are correct?*
- Question ① b): *What strategy could you use to find the answer?*

IN FOCUS Discuss with children the language used in the **Discover** section. Usually, children link the words 'how much' and 'more' with addition; however, in question ① b), they need to subtract. If children are unsure how to proceed, give real-life examples and use smaller numbers.

PRACTICAL TIPS Consider linking this **Discover** to an art class. Discuss with children why it is important to know the measurements of red and yellow paint when mixing them. Children need to think about the container that they will use to store the paint and how much paint they might use. Use words like 'total', 'more' and 'less' when discussing the amount of paint that they will mix.

ANSWERS

Question ① a): Olivia and Luis can make 0·68 l of orange paint.

Question ① b): 0·89 l − 0·68 l = 0·21 l
Olivia and Luis need to make 0·21 l more orange paint.

PUPIL TEXTBOOK 5C PAGE 92

Share

WAYS OF WORKING Whole class teacher led

ASK

- Question ① a): *Look at the columns in the place value grid. What does each column represent? How can you use column addition of whole numbers to add decimal numbers?*
- Question ① a): *Why do you need to be careful with how you write the numbers so that they are correctly lined up?*
- Question ① b): *To answer this question, should you add or subtract? How do you know? What part of the question tells you this? What calculation could you do to prove your answer is correct?*

IN FOCUS In question ① b), children should be able to articulate why they have performed a subtraction and not an addition. Children can use bar models and number lines to prove their answers are correct. Show children the column method for the addition and subtraction of decimal numbers. Use this question to reinforce the place value of each digit when carrying out the calculation.

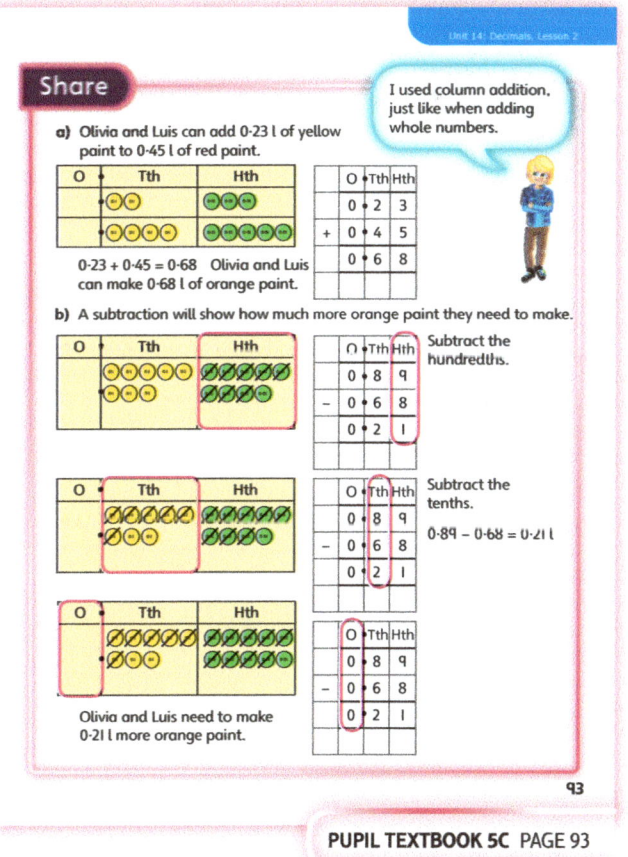

PUPIL TEXTBOOK 5C PAGE 93

129

Think together

WAYS OF WORKING Whole class teacher led (I do, We do, You do)

ASK

- Question ① a): *What is the same and different about this question and the question in Discover?*
- Question ① b): *What is the question asking? What is the unit of measurement?*
- Question ③ a): *What do you need to know to find the amount of slime altogether? Do you know how much slime each child has? Can you work it out?*

IN FOCUS Questions ① a) and b) apply learning from **Discover**. Use question ① b) to reinforce the place value of each digit when adding the numbers together. Emphasise the use of the correct units of measurement. Question ② asks children to find the missing numbers in column additions. Question ③ introduces a two-step question. Children need to find how much slime Kate has before adding the amounts together to find the total. Some children may just add the two numbers given without fully reading the question. Encourage them to describe the question before finding a solution.

STRENGTHEN To support understanding, use counters to represent each calculation on a place value grid and then display these alongside the abstract calculations. For question ③, ask children to use a bar model to find the answer.

DEEPEN Challenge children to create their own addition and subtraction problems using the words 'more than', 'less than' and 'how much more?'. What language will they use to indicate a subtraction? Can they use a variety of different words? If children need support, provide them with an addition and subtraction example.

ASSESSMENT CHECKPOINT Can children use a bar model and a number line to explain their answers? Can children explore the best method to use to add or subtract two decimal numbers? If so, can they describe their strategy?

ANSWERS

Question ① a): 0·83 l

Question ① b): 0·47 l

Question ① c): 0·23 l more

Question ② a): 0·63 + 0·05 = 0·68

Question ② b): 0·316 + 0·263 = 0·579

Question ② c): 0·53 + 0·46 = 0·99

Question ③ a): They have 0·526 l of slime altogether.

Question ③ b): 0·322 l

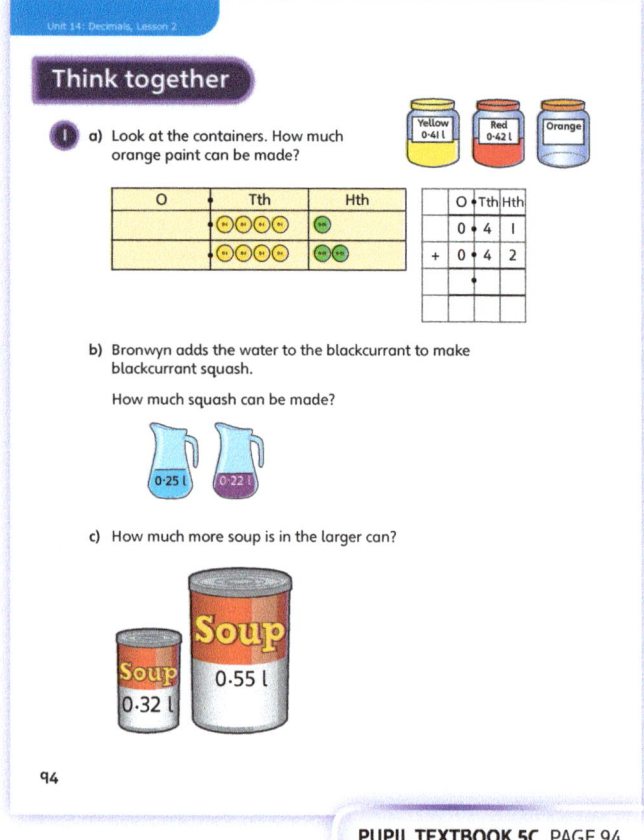

PUPIL TEXTBOOK 5C PAGE 94

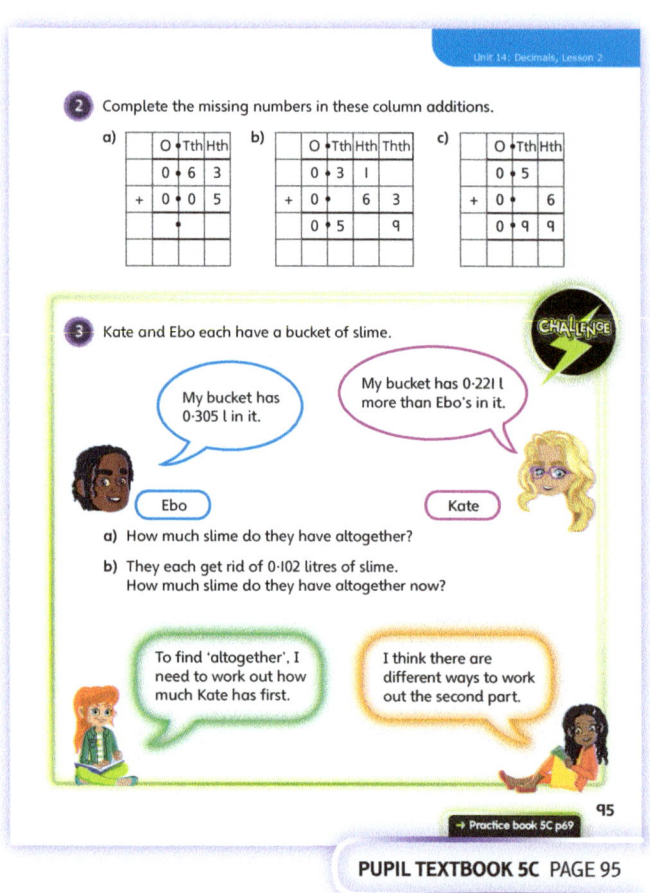

PUPIL TEXTBOOK 5C PAGE 95

Unit 14: Decimals, Lesson 2

Practice

WAYS OF WORKING Independent thinking

IN FOCUS Questions 1 and 3 aim to consolidate children's understanding of adding or subtracting two decimal numbers using the column method. The numbers for the additions and subtractions have already been set out in column format.

In some of the other questions, children need to set up the calculations. Questions 2 and 5 provide a part-whole model where children have to work out whether they are finding the whole or the part. Question 6 provides a challenge where children have to work out which decimals add up to a given number in a scenario where multiple combinations are possible.

STRENGTHEN Ensure children have access to concrete materials and number lines throughout the exercise. Encourage all children to continuously explain their thought processes to a teacher, a parent or a partner. Urge children to use counters on a place value grid to support their understanding.

DEEPEN Ask children to discuss how they tackled the **Challenge** question. Is there only one way to approach it? Did they start with the hundredths column or the tenths column? Why?

ASSESSMENT CHECKPOINT Are children confident in using the addition and subtraction column methods to add or subtract decimals? Do they understand how they can use the link between addition and subtraction to find missing numbers?

ANSWERS Answers for the **Practice** part of the lesson can be found in the *Power Maths* online subscription.

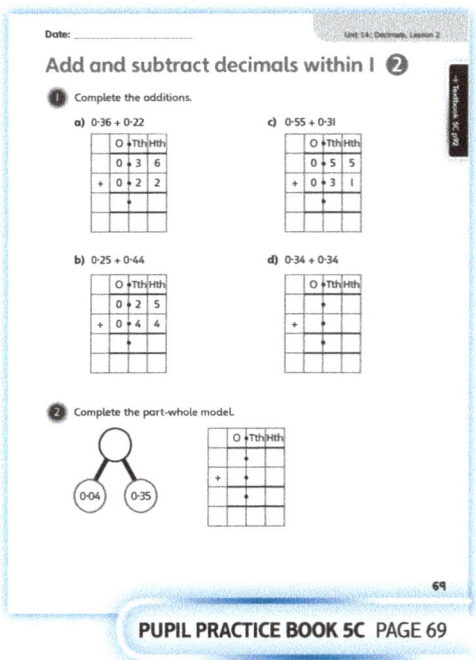

PUPIL PRACTICE BOOK 5C PAGE 69

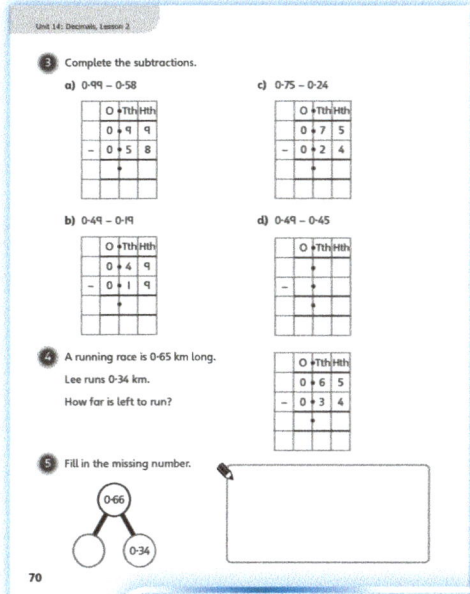

PUPIL PRACTICE BOOK 5C PAGE 70

Reflect

WAYS OF WORKING Independent thinking

IN FOCUS This **Reflect** activity checks that children can use the addition of two whole numbers to calculate the addition of two decimal numbers. Identify children who can use the link between the additions and are able to add decimal numbers mentally. Identify those who need further support in understanding the link between the additions, then explain the links by using concrete resources.

ASSESSMENT CHECKPOINT Check that children are able to explain the similarities between adding decimal numbers and whole numbers. Do children recognise that the methods for adding and subtracting whole numbers can be adapted and used when working with decimal numbers?

ANSWERS Answers for the **Reflect** part of the lesson can be found in the *Power Maths* online subscription.

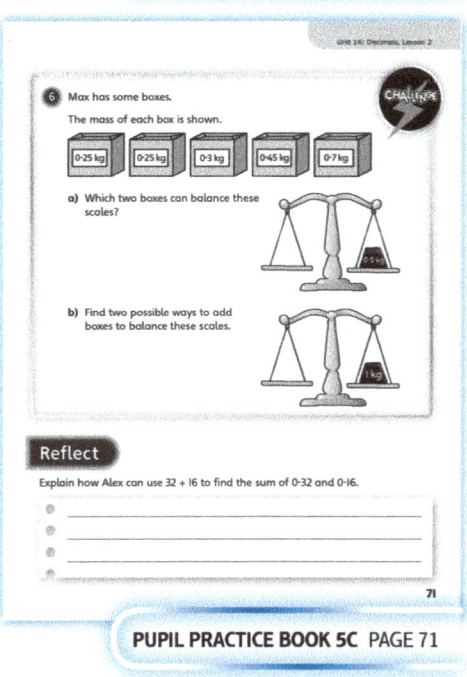

PUPIL PRACTICE BOOK 5C PAGE 71

After the lesson

- Can children use column addition or subtraction to add two decimal numbers with confidence?
- Do children understand the importance of place value when adding decimal numbers?

131

Unit 14: Decimals, Lesson 3

Complements to 1

Learning focus
In this lesson, children will work out how much needs to be added to another decimal to make the whole: to find the complement to 1.

Before you teach
- Do children know how to make an exchange when subtracting?
- Can children confidently add two or more whole numbers with two or more decimal places?

NATIONAL CURRICULUM LINKS

Year 5 Number – fractions (including decimals and percentages)

Solve problems involving number up to three decimal places.

ASSESSING MASTERY

Children can use number bonds to 10 and to 100 and column addition and subtraction to find decimal numbers that add to make 1. Children can answer abstract questions or those presented with a context and draw diagrams to represent their thinking.

COMMON MISCONCEPTIONS

Children may not set the column method out correctly when subtracting decimals from 1. For example, they may think of 1 as 0·1 and subtract the smaller digit from the bigger digit, so when calculating 1 – 0·23 they may actually calculate 0·23 – 0·1. Ask:
- What digit do you need to subtract? How many tenths are there in 1? How many hundredths are there in one tenth?

STRENGTHENING UNDERSTANDING

Children should first practise subtracting whole numbers from 100 or 1,000 before progressing to subtracting decimals with two and three decimal places from 1. Children should be given the opportunity to use a hundredths grid (or a similar resource) to link visually what happens when different parts of the whole (that complement to make 1) are added together. To embed this process, move parts away from the whole to see where the two complementary parts come from.

GOING DEEPER

Challenge children to find three or four decimal numbers that add to make 1. They should write these as addition equations and derive the subtraction equations from them.

KEY LANGUAGE

In lesson: complement, whole, place value, ones (1s), tenths, hundredths, exchange, number bonds, column, add, subtract

Other language to be used by the teacher: parts, split, partitioned, thousandths, decimal point, two decimal places, three decimal places

STRUCTURES AND REPRESENTATIONS

Bar model, part-whole model, hundredths grid, number line, column addition, column subtraction, place value grid

RESOURCES

Mandatory: place value counters

 In the eTextbook of this lesson, you will find interactive links to a selection of teaching tools.

Quick recap

With children, practise the number bonds to 100. Provide children with 10 numbers and ask them to work out the number bonds to 100.

Unit 14: Decimals, Lesson 3

Discover

WAYS OF WORKING Pair work

ASK

- Question 1 a): *How long are the two lengths of mirror Aki needs to decorate? What lengths of paper does he have?*
- Question 1 b): *Will you add or subtract to find the answer?*

IN FOCUS Question 1 a) asks children to find the pairs of numbers that have a total of 1. Ensure that children understand what the question requires. Question 1 b) requires children to calculate how much ribbon is left from 1 m, so they will need to subtract a number with three decimal places from 1.

PRACTICAL TIPS Consider providing children with three different lengths of paper or ribbon, where two of the pieces add up to 1 m, and also provide a 1 m stick. Ask children to measure each piece, write the measurements and put together the pieces that make 1 m. Provide children with a hundredths grid and discuss how it can be used.

ANSWERS

Question 1 a): Aki can use the 0·7 m and 0·3 m pieces to decorate one border of the mirror, and 0·57 m and 0·43 m pieces to decorate the other.

Question 1 b): Aki has 0·765 m of ribbon left.

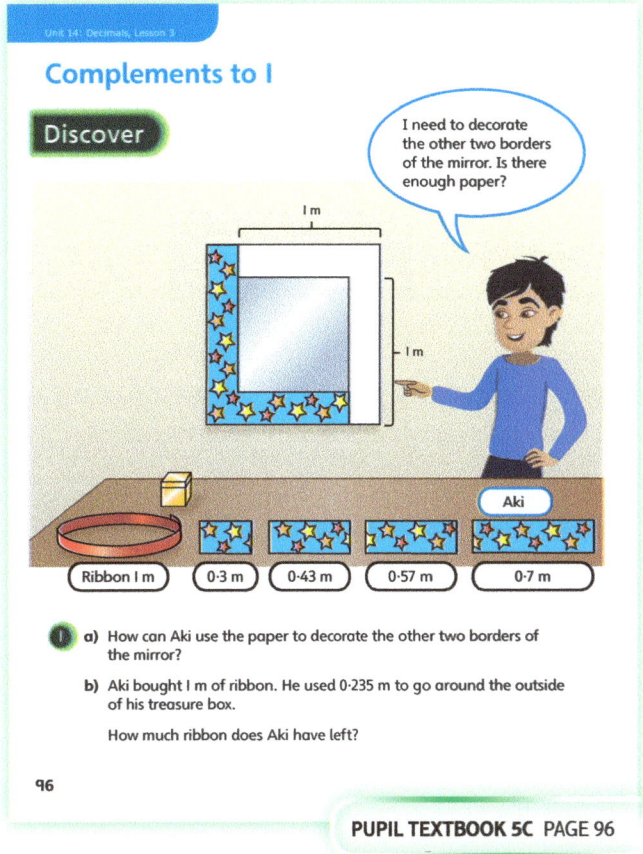

PUPIL TEXTBOOK 5C PAGE 96

Share

WAYS OF WORKING Whole class teacher led

ASK

- Question 1 a): *How can you use number bonds to help? How did Astrid use the hundredths grid to help her?*
- Question 1 b): *Why is it helpful to write 1 as 1·000 here? What exchanges do you need to make to find the difference?*

IN FOCUS In this section, children are shown how to find decimal numbers that have a total of 1. They are also introduced to subtracting a decimal number from 1. Show children the column method of subtraction. They should be able to explain which place value column they need to start with and how to exchange when subtracting a 3-digit number from 100. Draw similarities between subtracting 2- and 3-digit numbers from 100 and 1,000 and subtracting decimal numbers from 1.

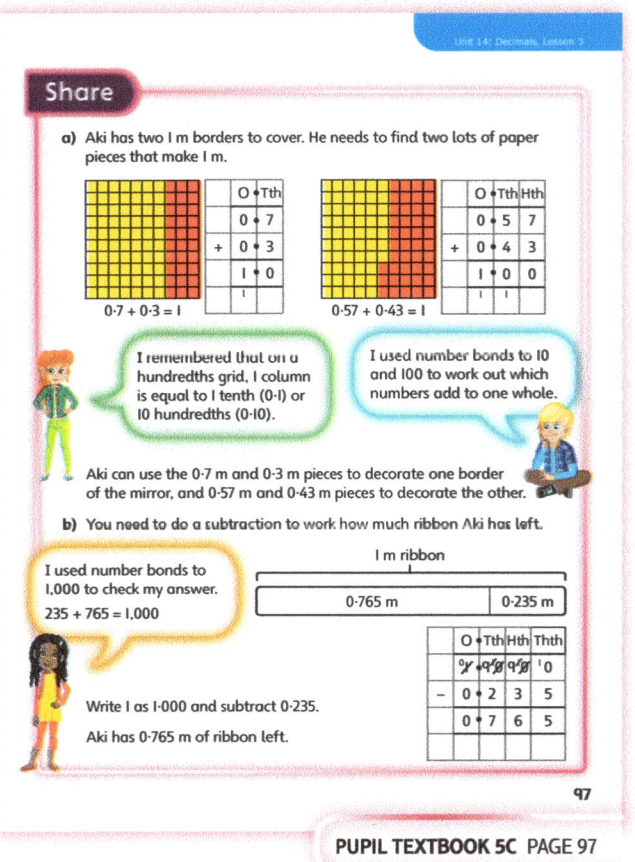

PUPIL TEXTBOOK 5C PAGE 97

133

Think together

WAYS OF WORKING Whole class teacher led (I do, We do, You do)

ASK
- Question ❶: *How can you find the missing parts? What do you notice about the number pairs that add up to 1?*
- Question ❷: *What do the part-whole models and bar model show? What is the same about them? How do they differ?*

IN FOCUS In questions ❶ and ❷, children are presented with a variety of different representations for adding decimals to make 1. They are also introduced to the word 'complement'. Encourage children to try different methods, ensuring they use place value counters next to the diagrams. Discuss their preferred methods and clarify any questions or misconceptions. Question ❸ links the use of concrete resources to the abstract written calculation. Astrid and Ash present the most likely misconception children will encounter. Placing counters in each column will help children visualise the column addition better and understand the value of each digit in the calculation.

STRENGTHEN To support understanding, encourage children to find the answer to any of the questions using more than one method. Children could use the part-whole model or the bar model to find the answer.

DEEPEN After question ❸ b), children could produce their own calculations where two or more decimals are added together to make a whole. They could extend their understanding even further by including both addition and subtraction in each calculation, for example, ☐ − ☐ + ☐ = 1.

ASSESSMENT CHECKPOINT Children should be able to verbalise how they know that their calculations are correct. Children should also be able to explain when and how they can do an exchange when adding decimal numbers that total 1.

ANSWERS

Question ❶ a): 0·6 m
Question ❶ b): 0·51 m
Question ❶ c): 0·32 m
Question ❷ a): 0·8
Question ❷ b): 0·16
Question ❷ c): 0·868
Question ❷ d): 0·479
Question ❸ a): 0·29 + 0·71 = 1
Question ❸ b): 0·724 + 0·276 = 1
Question ❸ c): 0·34 + 0·21 + 0·45 = 1
0·34 − 0·21 + 0·87 = 1
There are many possible solutions.
The sum of the missing digits is 0·766.
For example: 0·234 + 0·383 + 0·383 = 1.

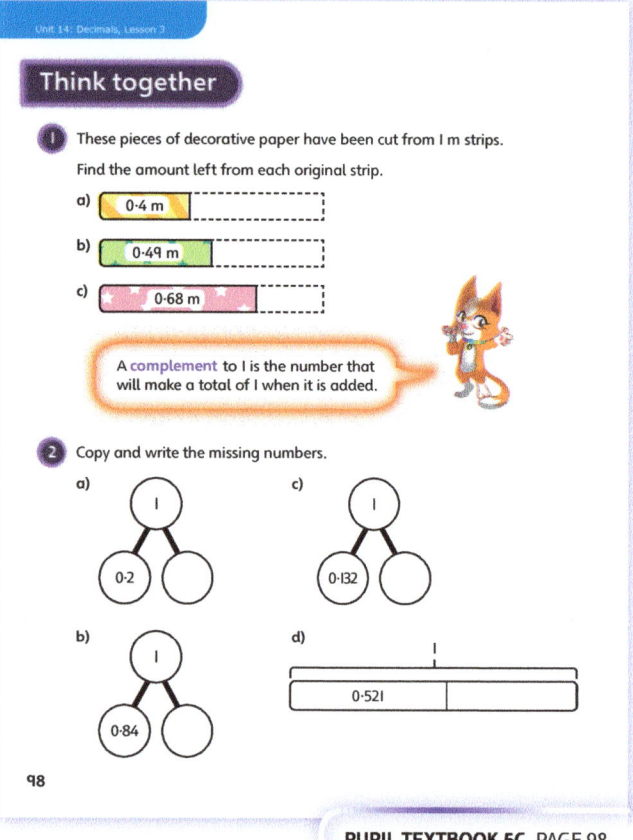

PUPIL TEXTBOOK 5C PAGE 98

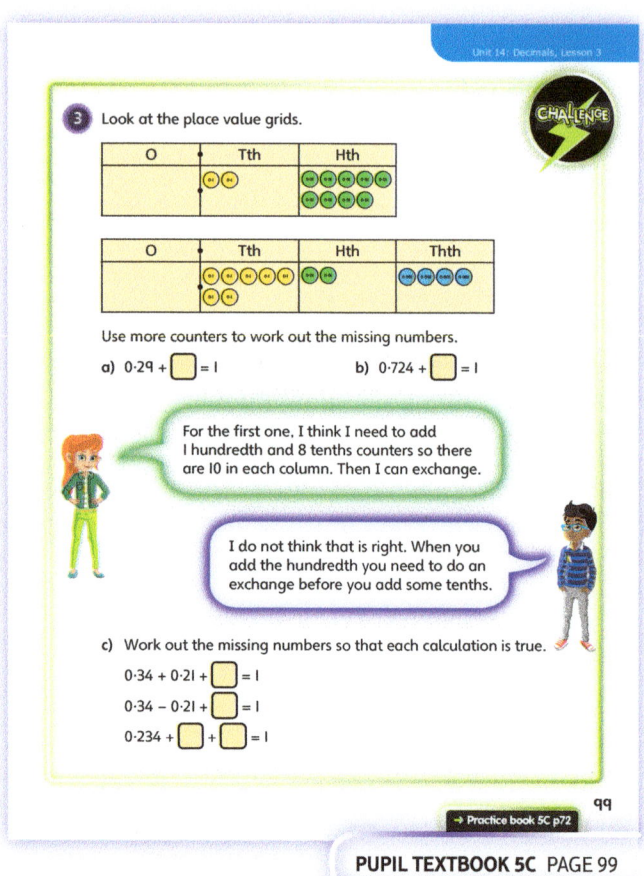

PUPIL TEXTBOOK 5C PAGE 99

Unit 14: Decimals, Lesson 3

Practice

WAYS OF WORKING Independent thinking

IN FOCUS Questions ① and ② aim to consolidate children's understanding of adding decimals that make 1.

In question ③, children need to explain Lexi's mistake. This will demonstrate their understanding of how to display the calculation correctly when using a place value grid.

Question ④ links addition and subtraction. Children should be able to change an addition calculation to a subtraction and vice versa.

STRENGTHEN Strengthen understanding by using concrete manipulatives. Use counters, bar models, number lines or place value grids to help reinforce the concept in a more visual way that children can relate to.

DEEPEN The **Challenge** question provides an opportunity for children to provide multiple solutions to the same question. To extend this, children can be presented with similar questions that vary the number of parts in the whole. Encourage children to find all of the different ways the calculations can be completed.

ASSESSMENT CHECKPOINT Children should be able to use resources to explain the calculations they have completed. Furthermore, they should also be able to draw different representations to show the solutions to the problems.

ANSWERS Answers for the **Practice** part of the lesson can be found in the *Power Maths* online subscription.

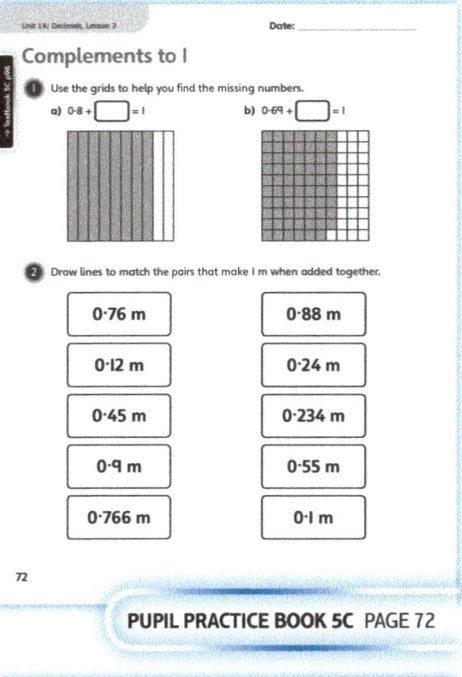

PUPIL PRACTICE BOOK 5C PAGE 72

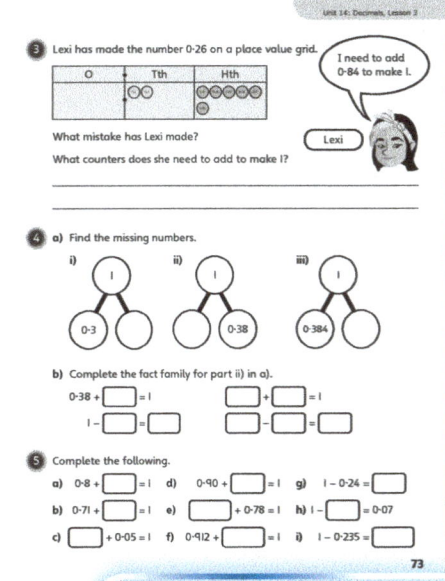

PUPIL PRACTICE BOOK 5C PAGE 73

Reflect

WAYS OF WORKING Independent thinking

IN FOCUS In this **Reflect** activity, children are required to reason whether Andy's calculation is correct or incorrect. They should use accurate vocabulary. They should be encouraged to use diagrams or column addition to support their reasoning.

ASSESSMENT CHECKPOINT The strength of children's written responses should allow assessments to be made about whether they have a secure understanding of the concept of adding decimals, including the addition of decimals that require one or more exchange.

ANSWERS Answers for the **Reflect** part of the lesson can be found in the *Power Maths* online subscription.

After the lesson ⏸

- What percentage of children mastered the lesson?
- Were children able to use a variety of different representations to show their solutions?
- Can children identify calculations that need exchange and complete these calculations correctly?

PUPIL PRACTICE BOOK 5C PAGE 74

135

Unit 14: Decimals, Lesson 4

Add and subtract decimals across 1

Learning focus

In this lesson, children will add numbers less than 1 where the total is greater than 1.

Before you teach

- Can children add fractions with a sum greater than 1 and a denominator of 10?
- Can children make an exchange when using the column addition method?

NATIONAL CURRICULUM LINKS

Year 5 Number – fractions (including decimals and percentages)

Solve problems involving number up to three decimal places.

ASSESSING MASTERY

Children can add decimals less than 1 where the total is greater than 1. For example, when adding 0·5 + 0·7, they are able to explain why the answer is 1·2 and not 0·12. Children can answer abstract questions or those presented with a context and can draw diagrams to explain their thinking.

COMMON MISCONCEPTIONS

Children may not exchange correctly when the sum of the decimals is greater than 1. For example, when adding 0·7 + 0·6, they think that the answer is 0·13 rather than 1·3. Ask:
- *How many tenths are there in 0·7? How many tenths are there in 0·6? How many tenths are there in total? How many tenths are there in one whole? How many tenths are left over?*

STRENGTHENING UNDERSTANDING

Encourage children to clearly describe the place value of each digit and ensure they understand the importance of it, in particular when making an exchange. Some children will benefit from using place value counters. Other children may find it useful to link decimals with fractions out of 10, and to use concrete representations such as a metre stick or a scale that is split into tenths. Seeing what happens when two decimals add to make a number greater than 1 will deepen children's understanding.

GOING DEEPER

Ask children to explore and explain different methods. Encourage children to decide which method is more efficient depending on the context. Children may represent their additions in more than one way: for example, with bar models, number lines and place value counters.

KEY LANGUAGE

In lesson: add, total, column, decimal, whole, tenths, less than (<), greater than (>), equals (=), metres (m), kilograms (kg)

Other language to be used by the teacher: increased by, digit, exchange, ones (1s), hundredths, place value

STRUCTURES AND REPRESENTATIONS

Place value grid, column addition, number line, bar model

RESOURCES

Optional: metre stick, weighing scales

 In the eTextbook of this lesson, you will find interactive links to a selection of teaching tools.

Quick recap

Ask children to mentally add two single-digit numbers that cross 10: for example, 5 + 7. Ask children how they can use this answer to help them with another calculation: for this example, 50 + 70 = ?.

Unit 14: Decimals, Lesson 4

Discover

WAYS OF WORKING Pair work

ASK

- Question 1 a): *What information does the table show?*
- Question 1 a): *Do all plants grow in the same away? What does 'amount grown' mean?*
- Question 1 b): *How tall was the sunflower at the start of the month? How tall was it at the end of the month? What do you need to do to work out how much it has grown?*

IN FOCUS Question 1 a) helps to ensure that children can use a table to derive the relevant information and then interpret how to use it. Children should realise that they must add the two values together to find the answer. Question 1 b) requires children to find the amount by which the sunflower has grown. Children should realise they need to work out the missing number to answer the question. Discuss possible methods of addition and subtraction that can be used to find the solutions. Children can use part-whole models or bar models to represent the problem. Encourage them to use mental calculations and their knowledge of bonds rather than column methods.

PRACTICAL TIPS Show children pictures of fast-growing and slow-growing plants. Investigate and discuss their growth rates and the maximum height they can grow to. Discuss the height that a bamboo can reach each year and compare that to a sunflower or a slower-growing plant, such as a cactus. The more curious children become, the more invested they will be in the task ahead.

ANSWERS

Question 1 a): At the end of the month the height of the bamboo tree is 1·6 m.

Question 1 b): The sunflower has grown by 0·6 m.

PUPIL TEXTBOOK 5C PAGE 100

Share

WAYS OF WORKING Whole class teacher led

ASK

- Question 1 a): *How tall is the bamboo tree at the start? How much has it grown by? How can you find its current height?*
- Question 1 b): *How tall was the sunflower at the start? What height is it now? How can you work out how much it has grown?*

IN FOCUS Question 1 a) requires children to add two decimal numbers where the total is greater than 1. These calculations necessitate one exchange when using the column addition method. Explore the different models and methods that can be used to find the total. Use the number line to further develop understanding of each step. Discuss the number of tenths of each number and the number of tenths that make 1 whole. Then consider how many more you need to add on.

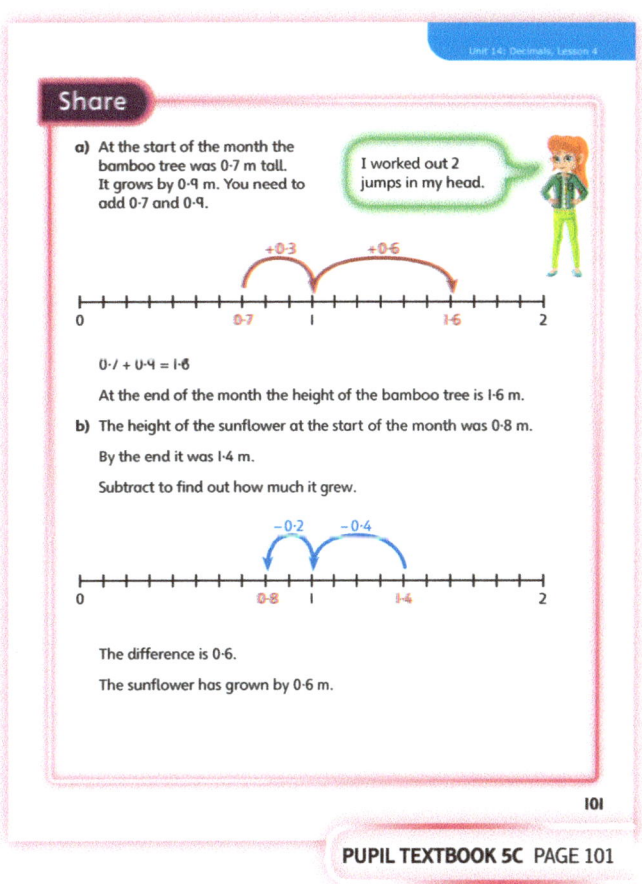

PUPIL TEXTBOOK 5C PAGE 101

Think together

WAYS OF WORKING Whole class teacher led (I do, We do, You do)

ASK

- Question ❶: *What information do you need to find the height of the sunflower at the end of the month?*
- Question ❶ b): *How many tenths are there in 0·5? And in 0·7? How many tenths are there altogether? How many tenths make one whole? What mistake has Emma made?*
- Question ❷: *How can you use the answer to each part to find the answers to the other parts?*

IN FOCUS Question ❶ b) focuses on a common misconception. If children use the column method, make sure that they lay out the columns correctly. Children should be able to explain which place value column they should start with when adding numbers together and why. Use this question to reinforce the place value of each digit and to discuss the exchange of 10 tenths for 1 one that takes place. Question ❷ encourages children to spot patterns and mentally work out the answer to additions rather than resorting to column methods.

STRENGTHEN To support understanding in question ❶ b), show children a weighing scale or a picture of scales showing 0·5 kg. Show children what happens when 0·7 kg is added. A number line can be used alongside the abstract calculation in order to link the concrete, pictorial and abstract representations together.

DEEPEN Take children further in question ❸. Ask them to write number stories or provide a context for each of the calculations. Children can share their stories and discuss how they can be solved. Encourage them to use diagrams in their explanations and to answer any questions their peers may have. As an extension, children could create their own missing number questions.

ASSESSMENT CHECKPOINT Can children add numbers less than 1 with a sum greater than 1? In question ❶ b), can children explain why 0·5 + 0·7 is not 0·12? Can children explain how they have found missing numbers to their partners?

ANSWERS

Question ❶ a): 2·5 m

Question ❶ b): Emma has collected 1·2 kg.
She has forgotten that 10 tenths = 1.
5 tenths + 7 tenths = 12 tenths = 1·2

Question ❷ a): 7 + 8 = 15
0·7 + 0·8 = 1·5
0·07 + 0·08 = 0·15
1·7 + 0·8 = 2·5

Question ❷ b): 12 − 8 = 4
1·2 − 0·8 = 0·4
0·12 − 0·08 = 0·04

Question ❸: 0·7 + 0·8 + 0·3 = 1·8
10 − 1·5 = 8·5
0·99 + 0·99 = 1·98
0·36 + 0·25 = 0·61

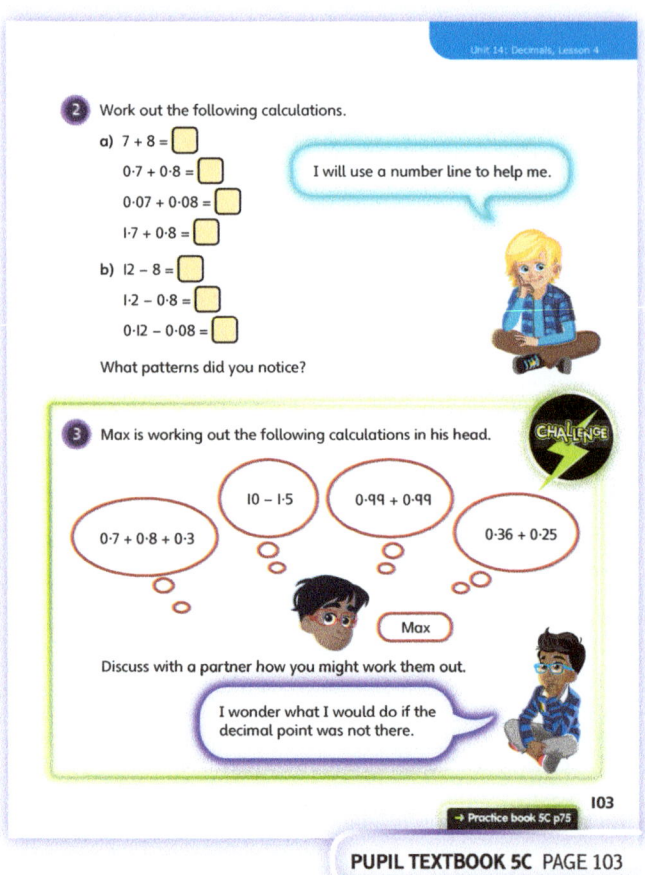

PUPIL TEXTBOOK 5C PAGE 102

PUPIL TEXTBOOK 5C PAGE 103

Unit 14: Decimals, Lesson 4

Practice

WAYS OF WORKING Independent thinking

IN FOCUS Questions 1 a) to c) aim to consolidate children's understanding of how to find the total of two decimal numbers where the answer exceeds 1. Question 1 d) focuses on using a number line to find the answer to a subtraction.

Throughout this section, number lines are a helpful tool. For example, in question 1 a) they allow children to start at 0·7 and count on in five jumps of 0·1 without having to think too much about how the numbers change when they cross 1. This will give them more confidence at counting on through 1, before attempting the same technique without a number line. Question 2 encourages children to use the answer to the first part of each question to work out the other answers mentally. Question 3 applies the same rules as question 2, but uses larger numbers.

STRENGTHEN Encourage children to use counters on a place value grid to support their understanding and to make it clearer when to exchange 10 tenths for 1 one. Children can draw their own diagrams to check their answers are correct.

DEEPEN Question 4 encourages children to problem solve and work out missing digits in addition calculations. This can be explored further by giving children more complex missing number problems. For example, children could find the calculation for two decimal numbers with a total of 1·23. Ask children to find more than one solution.

ASSESSMENT CHECKPOINT Use questions 1 to 3 to check that children can work out the answers to simple decimal additions and subtractions.

ANSWERS Answers for the **Practice** part of the lesson can be found in the *Power Maths* online subscription.

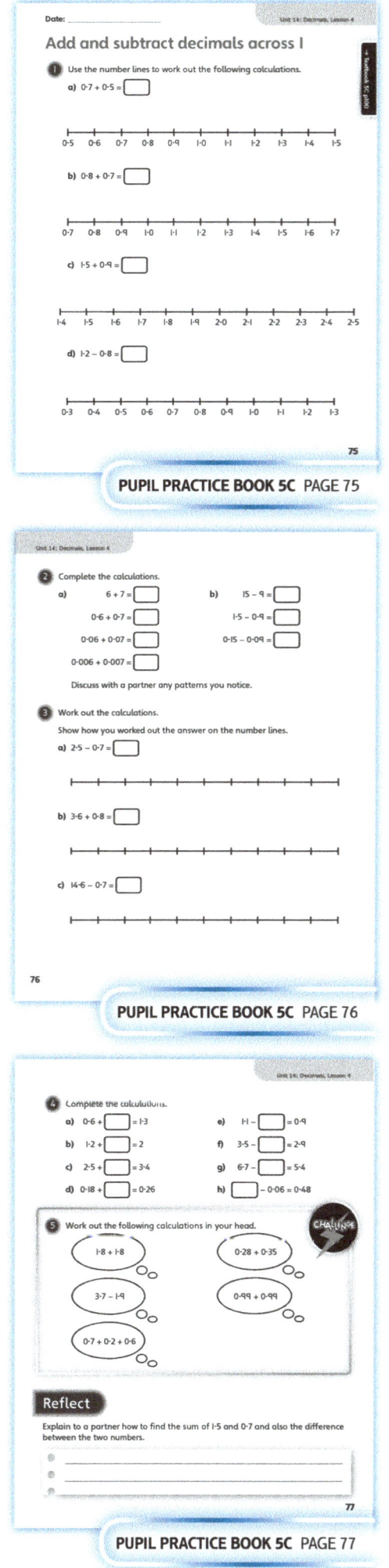

PUPIL PRACTICE BOOK 5C PAGE 75

PUPIL PRACTICE BOOK 5C PAGE 76

PUPIL PRACTICE BOOK 5C PAGE 77

Reflect

WAYS OF WORKING Pair work

IN FOCUS This **Reflect** activity checks children's understanding of what they have learnt in the lesson. They need to complete an addition and subtraction. Keep an eye out for children making the common misconception discussed in the lesson. Encourage children to explain their reasoning as well as to calculate the answer.

ASSESSMENT CHECKPOINT This activity will assess children's ability to add and subtract two decimals where the answer crosses the whole. Look for children who can clearly explain the exchange that takes place and the reason for it. Identify children who still need support to do this.

ANSWERS Answers for the **Reflect** part of the lesson can be found in the *Power Maths* online subscription.

After the lesson

- Can children explain how to add decimal numbers that exceed 1?
- Can children identify and solve calculations that require exchanges?
- Can children explain the importance of place value when adding?
- Are they aware of common mistakes they should try to avoid?

139

Unit 14: Decimals, Lesson 5

Add decimals with the same number of decimal places

Learning focus
In this lesson, children will add two numbers that have the same number of decimal places, such as 2·56 + 7·75.

Before you teach
- Children focused on decimal numbers less than 1 in the previous lesson. Do you need to revisit decimals greater than 1 prior to this lesson?
- Do children use efficient methods to add decimals?

NATIONAL CURRICULUM LINKS

Year 5 Number – fractions (including decimals and percentages)
Solve problems involving number up to three decimal places.

ASSESSING MASTERY

Children can use and explain different methods of adding two decimal numbers that have the same number of decimal places. Within a context, they can decide which method is the most efficient to use.

COMMON MISCONCEPTIONS

Children may not realise that zero is an important placeholder. For instance, when adding 8·02 and 5·01, they may write 8·02 as 8·2 and 5·01 as 5·1, and calculate 8·02 + 5·01 = 13·3. Ask:
- Are 8·2 and 8·02 the same value? What does the '2' in the number represent? Are £8·20 and £8·02 the same amount?

STRENGTHENING UNDERSTANDING

To strengthen learning in this lesson, ask children to place decimal numbers in a place value grid. When they have done this successfully, ask them what the value of each digit is. Use the context of money to explain the differences between £8 and 20p and £8 and 2p.

GOING DEEPER

Deepen learning in this lesson by counting on or counting back in tenths and hundredths from a given decimal number. Can children cross whole number boundaries? For example, if counting in hundredths across 9, children would say: 8·99, 9, 9·01.

KEY LANGUAGE

In lesson: add, total, digit, column, whole, part, place value, ones, tenths, hundredths, less than (<), greater than (>), equals (=), pounds (£)

Other language to be used by the teacher: decimal, decimal places, exchange, pence (p)

STRUCTURES AND REPRESENTATIONS

Number line, place value grid, column addition, part-whole model, bar model

RESOURCES

Mandatory: place value counters

Optional: plastic or paper money, picture of items for sale in the shop

 In the eTextbook of this lesson, you will find interactive links to a selection of teaching tools.

Quick recap
Ask children to add two 2- or 3-digit numbers together. Check they can set out the calculation correctly in columns and that they don't make any common mistakes. Ensure examples require exchanges.

Unit 14: Decimals, Lesson 5

Discover

WAYS OF WORKING Pair work

ASK

- Question 1 a): *What is Max buying? How much money does he have? What do you need to know to calculate whether he has enough money to buy the items he wants?*
- Question 1 b): *Is Jamie paying more or less than Max? How do you know? What method can you use to find out how much Jamie's meal costs?*

IN FOCUS Question 1 a) is important because children have to work with decimal numbers above 1. This calculation requires children to make two exchanges when using the column addition method. Question 1 b) requires children to find the sum of two totals.

PRACTICAL TIPS Give children an opportunity to visualise the problem by using concrete representations. Each table could have its own shop or canteen to represent how children might come across this problem in their everyday lives. The items in the shop can be labelled. Give children a certain amount to spend, for instance £4, and ask them to list the items that they can buy and the ones they cannot. Ask: *How much do these items cost? How much more money do you need to buy this item?*

ANSWERS

Question 1 a): Max's meal costs £4 in total, so he has enough money.

Question 1 b): The total cost of Jamie's meal is £5·35.

PUPIL TEXTBOOK 5C PAGE 104

Share

WAYS OF WORKING Whole class teacher led

ASK

- Question 1 a): *What method can you use to add the two numbers together? Would the answer be different if you chose a different method?*
- Question 1 a): *Which place value column do you need to start with? Will you need to make an exchange?*
- Question 1 b): *Do you need to know the cost of each item Jamie buys to find the total cost of her meal? What is the key word that Jamie says that tells you to add £1.35 to the cost of Max's meal?*

IN FOCUS Question 1 a) is important because children must represent the decimal numbers on a place value grid using place value counters. The use of place value grids in question 1 a) reinforces the place value of each digit in the calculation. Demonstrate why this is important when children are required to carry out the exchange of 10 hundredths for 1 tenth and 10 tenths for 1 one. Question 1 b) is important because children must represent the addition of a whole number and a decimal number with two decimal places on a number line. Doing this makes the learning both visual and practical. Discuss the use of mental addition. Ask: *What would you do if you were in the canteen? How would you calculate the total cost to work out if you had enough money?*

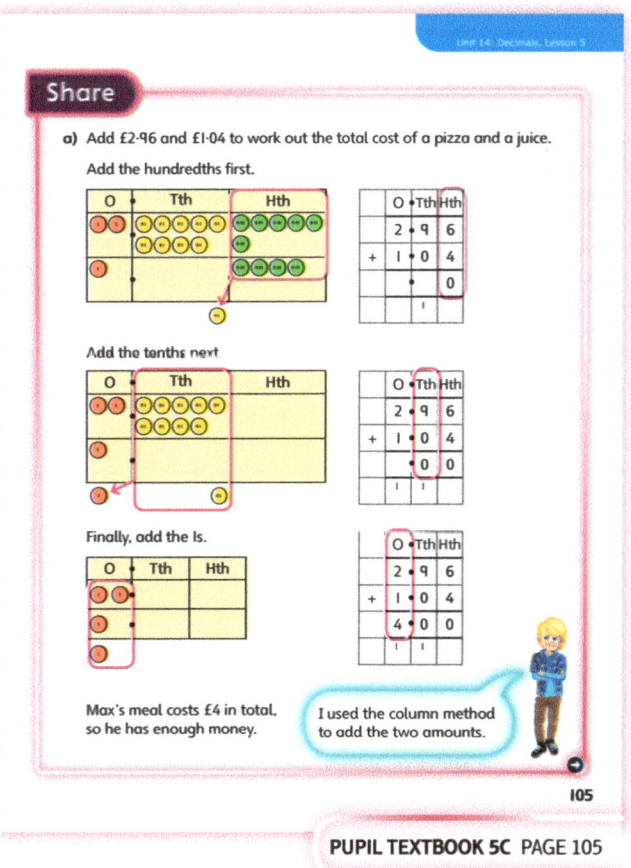

PUPIL TEXTBOOK 5C PAGE 105

141

Think together

WAYS OF WORKING Whole class teacher led (I do, We do, You do)

ASK
- Question 3 a): *How can the last digits help you determine whether you are correct?*
- Question 3 b): *What method can you use to check the answers?*

IN FOCUS Question 2 is important because children have an opportunity to use a part-whole model, a bar model and an abstract calculation to add decimal numbers. This will require a good understanding of what tenths and hundredths mean and how each of the models can be used to find the missing numbers. In question 3 a), when working out which two numbers make a given total, encourage children to look at the last digit in each number instead of carrying out the full calculation.

STRENGTHEN For questions 2 a) and b), provide children with place value counters and a place value grid to use alongside the part-whole model and the bar model. This will help them to find the whole more readily. The place value grid and counters are particularly useful for question 2 c), allowing children to visualise each decimal number.

DEEPEN For question 3, ask children to work out the total for other combinations of numbers or to imagine they bought four items instead of three items. Ask children to make their own questions based on the picture. What key words do they use?

ASSESSMENT CHECKPOINT Can children add decimal numbers with the same number of decimal places? Are children able to use the part-whole model and the bar model to find solutions?

ANSWERS

Question 1 a): £6·09

Question 1 b): £5·27

Question 2 a): 6·6

Question 2 b): 7·685

Question 2 c): 13·63

Question 3 a): ruler, pencils and marbles
£2·38 + £6·47 + £3·15 = £12

Question 3 b): eraser + pencils = £0·94 + £6·47 = £7·41
marbles + notebook = £3·15 + £4·26 = £7·41

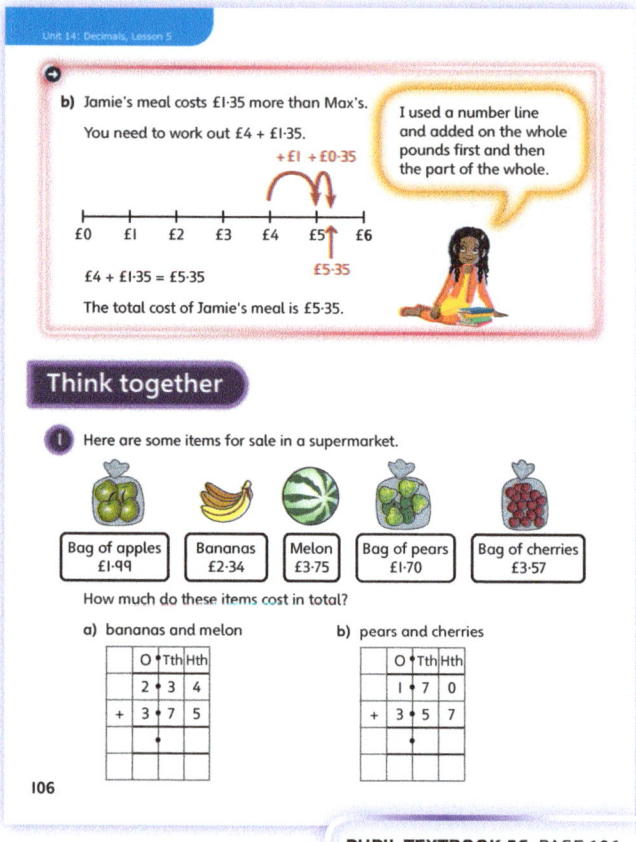

PUPIL TEXTBOOK 5C PAGE 106

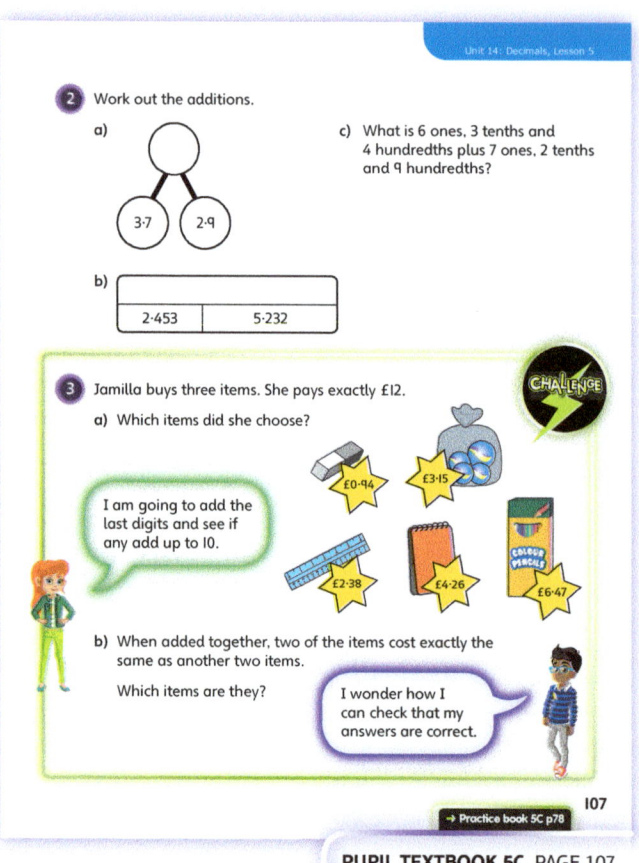

PUPIL TEXTBOOK 5C PAGE 107

Practice

WAYS OF WORKING Independent thinking

IN FOCUS Question ❶ aims to consolidate children's understanding of adding two decimal numbers through the use of the column method. The calculations are already set in the column addition grid and children have to work out the answers, with many of the problems involving exchanges. In question ❷, children solve a problem involving money and, in question ❸, children have to set out their calculations to find the sum.

In question ❹, children will need to use their preferred method to add numbers with two or three decimal places and then use appropriate signs to compare the amounts.

STRENGTHEN To support children who are struggling with the column additions, provide them with concrete materials such as counters and place value grids.

DEEPEN The **Challenge** question provides an opportunity for children to provide multiple solutions to the same question. Children need to read and understand the table of prices to solve this question. Explore this further by asking children to round the numbers to the nearest whole number before adding them mentally. Furthermore, ask children to make their own table of activities and prices. Ask children to work in pairs, and tell them that they each have a given amount of money (which you should specify). Ask children to work out what activities they would like to do, given the money that they have.

THINK DIFFERENTLY Question ❺ will help children understand a common mistake that can be made when one of the numbers has 0 as a hundredths digit. Emphasise the importance of understanding the place value of each column.

ASSESSMENT CHECKPOINT Children are confident in adding two decimal numbers that have the same number of decimal places.

ANSWERS Answers for the **Practice** part of the lesson can be found in the *Power Maths* online subscription.

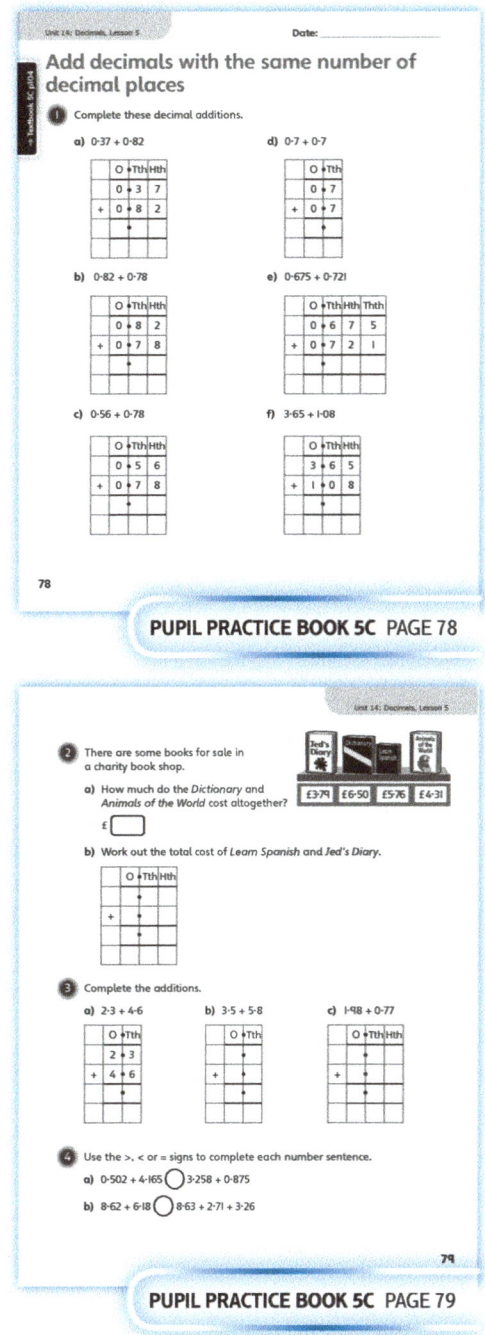

PUPIL PRACTICE BOOK 5C PAGE 78

PUPIL PRACTICE BOOK 5C PAGE 79

Reflect

WAYS OF WORKING Independent thinking

IN FOCUS This activity is a great way for children to reflect on the learning of this lesson. Encourage children to explain how they would carry out the calculation as well as actually answering it. Look for children who are able to add the numbers without support. Offer support or concrete apparatus to children who lack confidence when adding the numbers.

ASSESSMENT CHECKPOINT Assess if children can correctly explain how to find the total of two decimal numbers, emphasising the importance of the place value of each column and identifying the exchanges they need to make.

ANSWERS Answers for the **Reflect** part of the lesson can be found in the *Power Maths* online subscription.

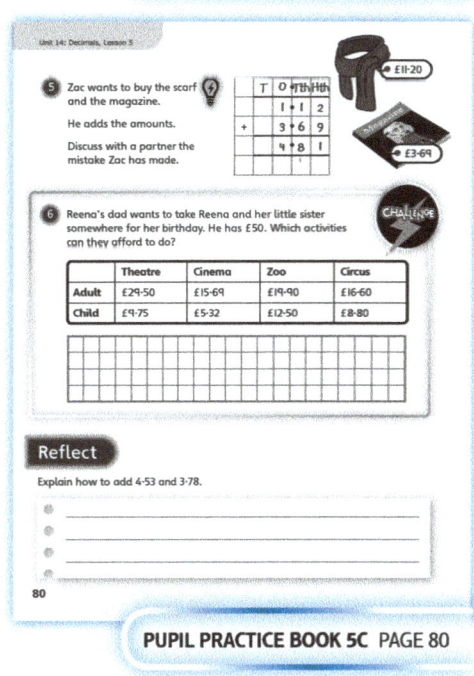

PUPIL PRACTICE BOOK 5C PAGE 80

After the lesson

- Can all children add decimals with two decimal places?
- Can they represent decimal numbers on a place value grid and a number line?
- Are children confident with the key vocabulary of the lesson?

143

Unit 14: Decimals, Lesson 6

Subtract decimals with the same number of decimal places

Learning focus

In this lesson, children will use the column method to subtract decimals in the context of taking away or finding the difference. This will include examples where an exchange is required.

Before you teach

- Can children subtract 2- and 3-digit numbers?
- Do children understand key vocabulary such as 'greater', 'inverse', 'cost' and 'change'?

NATIONAL CURRICULUM LINKS

Year 5 Number – fractions (including decimals and percentages)

Solve problems involving number up to three decimal places.

ASSESSING MASTERY

Children can subtract any two numbers that have the same number of decimal places, such as 12·56 – 7·75. They can recognise why exchanges are needed when subtracting decimal numbers.

COMMON MISCONCEPTIONS

Children may, when subtracting two decimal numbers, subtract the smaller digit from the larger digit. For example, when working out 3·82 – 1·25 = 2·63, they may write 3 in the hundredths column as they have worked out 5 – 2 and not made the exchange. Ask:
- *Which digit do you need to subtract? What will you do if the digit you are subtracting is bigger than the digit you are subtracting from? How do you set out the column method when subtracting numbers?*

STRENGTHENING UNDERSTANDING

Children should first spend time practising subtracting whole numbers with 2 or 3 digits before moving on to subtracting decimals.

GOING DEEPER

Give children missing number problems such as ☐ – 0·78 = 1·65 or ask them to complete ☐ – ☐ = 2·34 in as many different ways as possible.

KEY LANGUAGE

In lesson: subtract, difference, efficient, exchange, greater, change, cheaper, ones (1s), tenths, hundredths, column, inverse, predict, pounds (£), pence (p), kilometres (km)

Other language to be used by the teacher: place value, fewer, more, less

STRUCTURES AND REPRESENTATIONS

Place value grid, column method subtraction, bar model, number line

RESOURCES

Mandatory: place value counters
Optional: paper money, toy items (fruit) to buy

 In the eTextbook of this lesson, you will find interactive links to a selection of teaching tools.

Quick recap

Give children practice using column methods to find the answer to different subtractions. Subtractions should include examples where an exchange is needed.

144

Unit 14: Decimals, Lesson 6

Discover

WAYS OF WORKING Pair work

ASK

- Question 1 a): *How much do a pineapple and a watermelon cost altogether? How much does the pineapple cost? How could you work out how much the watermelon costs?*
- Question 1 b): *When would a shopkeeper give you change? What information do you need to know to calculate the change Amelia will get?*

IN FOCUS Question 1 a) requires children to calculate one part when the other part and the whole are given. It requires children to subtract with two exchanges. Question 1 b) introduces another subtraction. If calculated as a column subtraction, this also requires two exchanges where two of the digits are 0.

PRACTICAL TIPS To assist children in visualising the scenario and in determining what calculations are required, give them paper money and ask them to role-play. Ask how much change they would receive if they bought different items.

ANSWERS

Question 1 a): The watermelon costs £3·49.

Question 1 b): Amelia gets 26p change.

PUPIL TEXTBOOK 5C PAGE 108

Share

WAYS OF WORKING Whole class teacher led

ASK

- Question 1 a): *In which column(s) do you need to make an exchange? How do you know?*
- Question 1 b): *Could you use either an addition or a subtraction on this number line and find the same answer? What would be a different method of presenting this information?*

IN FOCUS For question 1 a), discuss the column subtraction and ensure children are confident in using this layout when subtracting. Check that children can explain how they know that they need to make an exchange. Use the column method to reinforce the place value of each digit when carrying out the calculation and explain the exchanges that take place. For question 1 b), show children the number line and ask: *Why might Flo have counted on to find the difference? How can this be presented as a subtraction?* Encourage children to try to show this as a column subtraction. Reinforce the exchange that happens when the hundredths and tenths digits are 0.

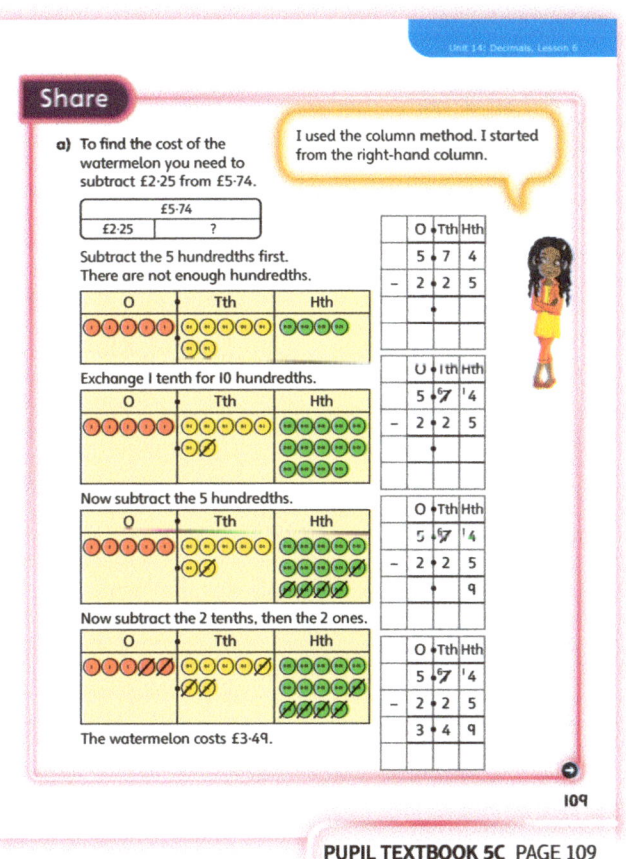

PUPIL TEXTBOOK 5C PAGE 109

145

Unit 14: Decimals, Lesson 6

Think together

WAYS OF WORKING Whole class teacher led (I do, We do, You do)

ASK

- Question ❶: *What calculation do you need to do to calculate 'how much cheaper?' Which place value column should you start with? What exchange do you need to make?*
- Question ❷: *Would it help to predict the number of exchanges? What information would you need to find to do this? Which digits should you look at first?*
- Question ❸ a): *When do you have to exchange to complete the subtraction?*
- Question ❸ b): *Can you find Reena's number in two different ways?*

IN FOCUS Question ❶ requires children to subtract two decimal numbers. The use of place value counters and the column method supports the calculation. Question ❷ encourages children to look at the digits within each number and compare them. Children need to have a clear understanding of when an exchange happens to answer this question. Invite children to share their views. Write their answers on the board, including any mistakes, and use this opportunity to clarify any misconceptions children may still have. Question ❸ encourages children to analyse the methods used and reason why Ebo is not correct. Encourage children to draw their own diagrams to visualise the question better and to identify the subtractions they need to complete.

STRENGTHEN To help children visualise the calculations Reena must make, give children a similar version of the question using whole numbers. For instance, ask: *I am thinking of a number. I add 1, then add 2. My answer is 7. What number was I thinking of?* Once children are clear on the strategy to use, they can apply this to find Reena's number.

DEEPEN Further develop understanding by presenting children with questions similar to question ❷, and ask them to predict the number of exchanges. Extend this further by asking children to write their own 'predict' questions and ask them to think of how they can make their questions easier or harder.

ASSESSMENT CHECKPOINT Can children compare different methods to carry out a subtraction calculation? Are they able to explain their steps and the reasoning they used to find their answers?

ANSWERS

Question ❶: £1·63

Question ❷ a): 0·82 − 0·38 = 0·44

Question ❷ b): 3·25 − 1·73 = 1·52

Question ❷ c): 37·5 − 13·9 = 23·6

Question ❷ d): 2·054 − 1·375 = 0·679

Question ❸ a): Ebo has the correct calculation but he has subtracted 3 from 7 instead of exchanging and subtracting 7 from 13. Lexi's number is 4·6.

Question ❸ b): Ebo's number is 8·73. Reena's number is 7·24.

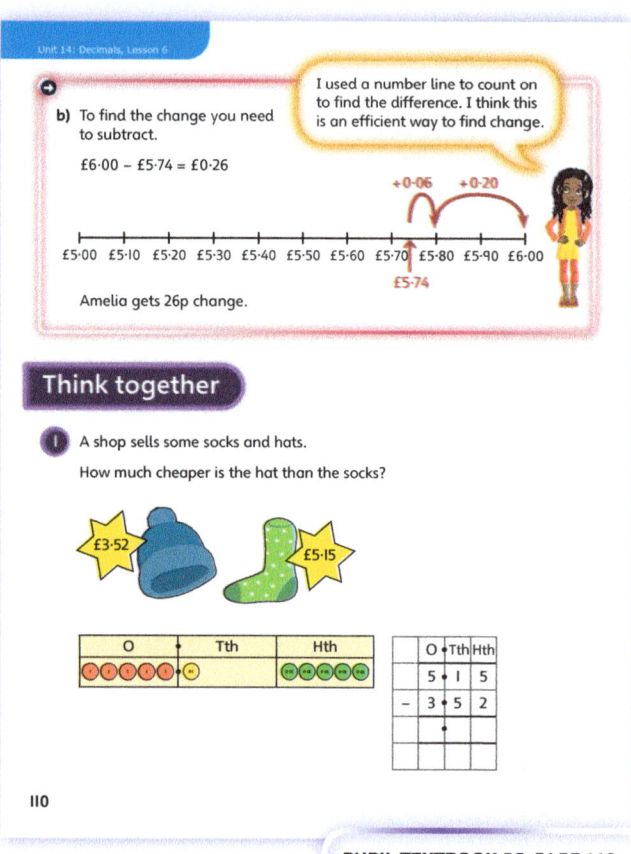

PUPIL TEXTBOOK 5C PAGE 110

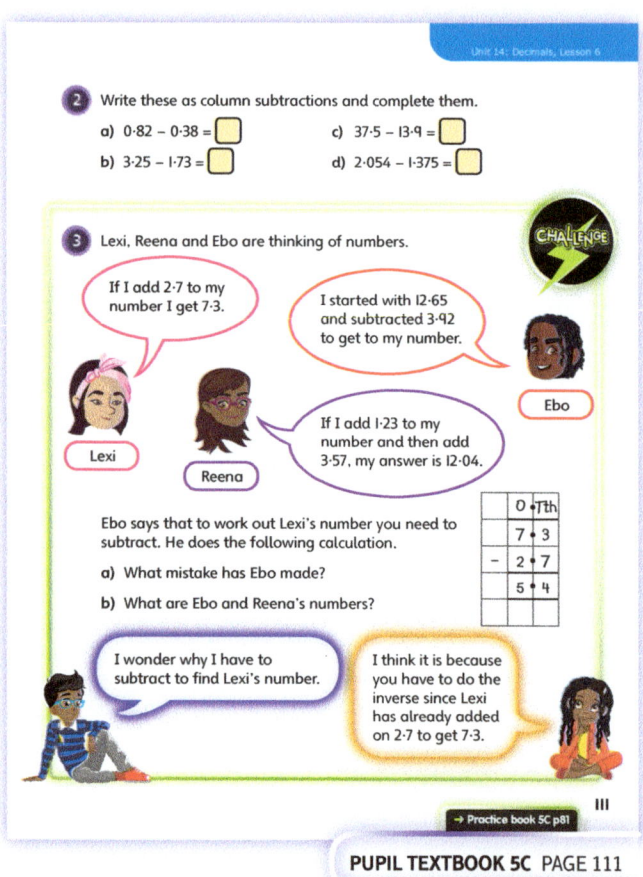

PUPIL TEXTBOOK 5C PAGE 111

146

Unit 14: Decimals, Lesson 6

Practice

WAYS OF WORKING Independent thinking

IN FOCUS Questions ❶ and ❷ aim to consolidate children's understanding of subtracting decimal numbers. Provide time for discussion of the different methods children used to find their answers.

Question ❸ presents another misconception that occurs when subtracting from a number that has 0 as a hundredths digit. Use this opportunity to clarify the exchange of 1 tenth for 10 hundredths to make this calculation work.

Question ❹ develops children's understanding of decimals as part of a whole in a problem-solving context. Encourage children to use a bar model to explain their answers and use the correct mathematical vocabulary when explaining their reasoning.

STRENGTHEN Ask children to think of the possible mistakes that could happen when subtracting decimals. Write their ideas on the board. Invite children to think of ways and strategies to avoid these mistakes. Then write their suggestions on the board next to the mistakes using a different coloured pen. Children may find it easier to visualise question ❸ if it is presented in a real-life scenario. Ask: *You had £6·20, you spent 59p. How much money do you have left?*

DEEPEN Provide children with missing number questions, similar in format to question ❺. These types of questions involve a problem-solving element and allow children to explore different paths to achieve the answer. Allow them to work through and discuss the additional problems. Children should then pose questions of their own for a partner to answer.

THINK DIFFERENTLY Although the question text refers to additions, question ❺ requires children to use the inverse operation in order to find the missing numbers. Look for children setting out the calculation correctly, and encourage them to use place value counters if they require extra support.

ASSESSMENT CHECKPOINT Can children subtract decimals? Are children's diagrams accurate and their written processes fluent? Can they provide solutions by subtracting decimals?

ANSWERS Answers for the **Practice** part of the lesson can be found in the *Power Maths* online subscription.

Reflect

WAYS OF WORKING Independent thinking

IN FOCUS This question highlights children's understanding of the difference between subtractions that involve exchange and those that do not. Children need to look at the calculations and notice how the digits in each of the numbers differ. Encourage children to use key vocabulary and the correct place value terms when explaining their reasoning.

ASSESSMENT CHECKPOINT Can children explain how to subtract one decimal number from another? Can children explain how they know whether the subtraction calculation will necessitate an exchange?

ANSWERS Answers for the **Reflect** part of the lesson can be found in the *Power Maths* online subscription.

After the lesson

- Can children use the column method to subtract two decimal numbers?
- Can children identify where exchanges will occur in subtractions?
- Can children predict what the last digit of the answer will be when subtracting one decimal number from another?

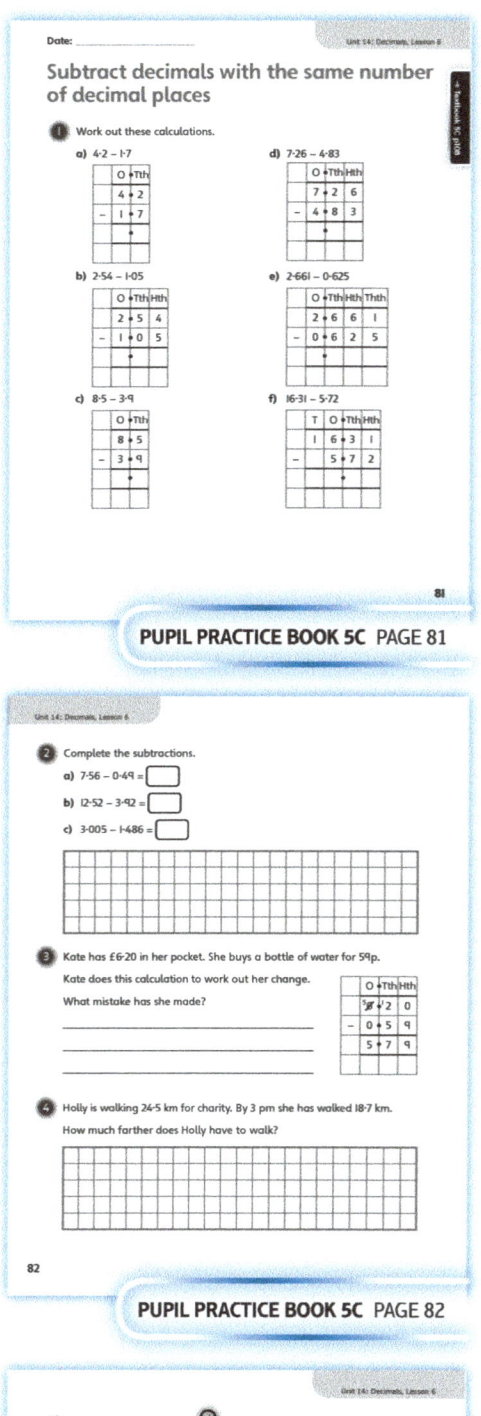

PUPIL PRACTICE BOOK 5C PAGE 81

PUPIL PRACTICE BOOK 5C PAGE 82

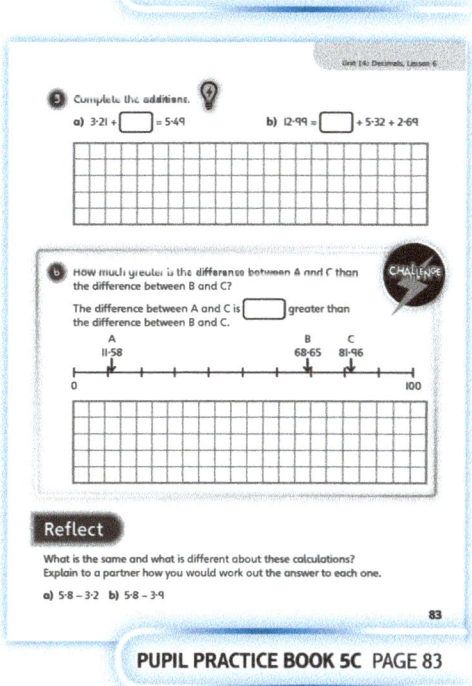

PUPIL PRACTICE BOOK 5C PAGE 83

147

Unit 14: Decimals, Lesson 7

Add decimals with a different number of decimal places

Learning focus

In this lesson, children will add and subtract decimals with a different number of decimal places. This includes examples where an exchange is required or where children must identify the mistake in a calculation.

Before you teach

- Do children know how to accurately lay out a column addition and subtraction?
- Do children know how to exchange when adding and subtracting using the column method?

NATIONAL CURRICULUM LINKS

Year 5 Number – fractions (including decimals and percentages)

Solve problems involving number up to three decimal places.

ASSESSING MASTERY

Children can add and subtract decimals that have a different number of decimal places. They can use addition and subtraction to check the answers to their calculations.

COMMON MISCONCEPTIONS

Children may make mistakes when using column addition or subtraction. They may align the numbers from right to left rather than according to their place value, in the way shown in the example. Ask:
- *Does the answer make sense? Can 2·3 and 4·61 add together to make 4·84? Which digits represent the ones, the tenths, the hundredths? Can 2 ones + 4 ones = 4 ones?*

	O	Tth	Hth	Thth
	0	2	3	0
+	4	6	1	0
	4	8	4	0

STRENGTHENING UNDERSTANDING

Encourage children to use a place value grid to help them identify the calculations more readily. Do this alongside the column addition or subtraction so that children can see how the concrete and abstract representations link together.

GOING DEEPER

Give children six number cards, such as 0, 1, 2, 3, 4 and 5. Ask them to use each digit once only to make two decimal numbers – for example, 2·304 and 1·5. Ask children to add the numbers. How many different decimal numbers can they make, and add, in 2 minutes?

KEY LANGUAGE

In lesson: add, subtract, check, digit, decimal point, hundredths, column, difference, sum, shortest, less, further, addition pyramid

Other language to be used by the teacher: inverse, operation, ones, tenths, equal, exchange, place value, greatest

STRUCTURES AND REPRESENTATIONS

Column addition, column subtraction, addition pyramid, place value grid, number line

RESOURCES

Optional: place value counters, 0–9 number cards, metre stick

 In the eTextbook of this lesson, you will find interactive links to a selection of teaching tools.

Quick recap

Children practise adding integers with different numbers of digits using column addition. Look for children lining up the digits correctly from the right. Explain that in reality we are lining up the integers by *place value* – each 1s digit is vertically aligned, as is each 10s digit, etc. Discuss possible misconceptions.

Unit 14: Decimals, Lesson 7

Discover

WAYS OF WORKING Pair work

ASK

- Question 1 a): *What calculation do you need to do to calculate how far Ambika's paper plane flew? What operation will you use? What method will you use to complete the addition?*
- Question 1 b): *What calculation do you need to do to calculate the distance Lee's paper plane flew? What is the key word in the question? Will you add or subtract? Why?*

IN FOCUS Question 1 a) is used to find the sum of two decimals with different numbers of decimal places. Question 1 b) is used to find the difference between two decimals with the same number of decimal places.

PRACTICAL TIPS Ask children to make paper planes. Invite three children to throw their paper planes and ask children to use a metre ruler to measure how far each plane flew. Write the three numbers on the board. Ask children to think of questions that you could ask them to solve using these numbers.

ANSWERS

Question 1 a): Ambika's paper plane flew 5·83 m.

Question 1 b): Lee's paper plane flies 3·81 m.

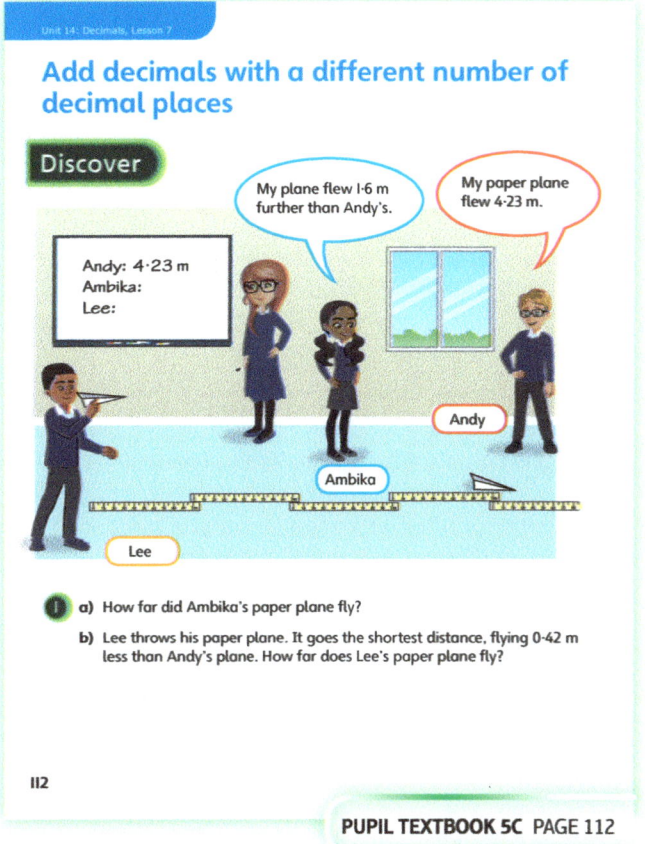

PUPIL TEXTBOOK 5C PAGE 112

Share

WAYS OF WORKING Whole class teacher led

ASK

- Question 1 a): *Why do we need to add? What numbers should we add together?*
- Question 1 a): *Why is it important to line the numbers up at the decimal point?*
- Question 1 a): *Why has Dexter added a 0 in the hundredths column? Will this change the answer? Are 1·6 and 1·60 different? Why do we use 1·60?*
- Question 1 b): *Look at the methods used. How many cm are there in 1 m?*

IN FOCUS For question 1 a), it is vital that children fully understand that 1·6 and 1·60 are equal. Invite one child to use a metre ruler and measure 1·6 m from a fixed point, such as a door. Ask the child to stand at that point. Ask a second child to measure 160 cm or 1·60 m using a metre ruler, from the same starting point. Both children will be standing at the same point. This should enable them to recognise for themselves that 1·6 m and 1·60 m are the same distance. Now ask each child to walk 42 cm back towards the starting point. Again, both children should notice that they have reached the same point. Ask children to calculate how far from the door they are. Write 1·6 − 0·42 = 1·18 m and 1·60 − 0·42 = 1·18 m. Ensure children understand why both calculations have the same answer.

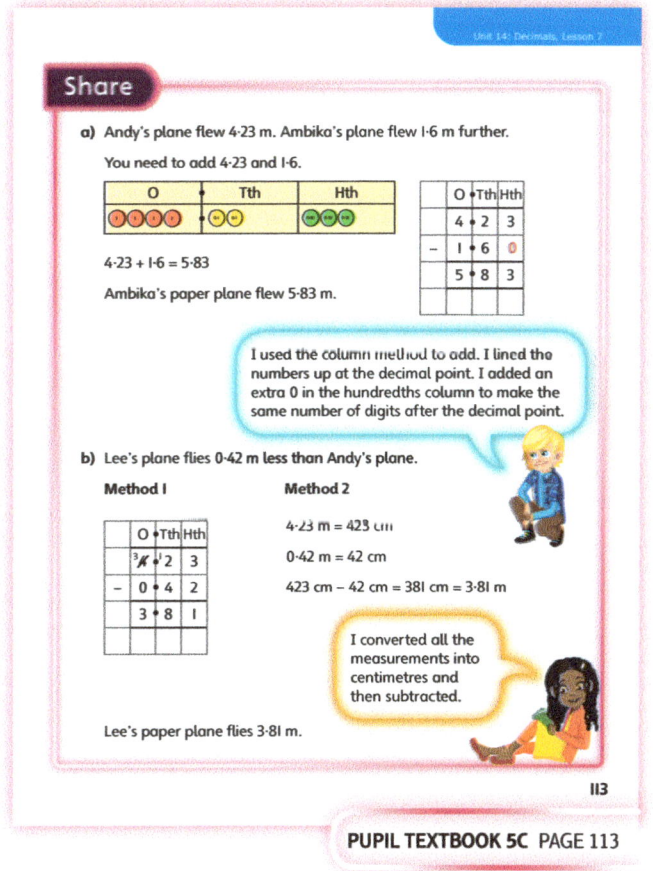

PUPIL TEXTBOOK 5C PAGE 113

149

Think together

WAYS OF WORKING Whole class teacher led (I do, We do, You do)

ASK

- Question ❶: *What is the same in this question and the Discover question? What is different?*
- Question ❷: *How does an addition pyramid work? What do you have to do to calculate each missing step? Should you add or subtract?*
- Question ❸ a): *Can you spot the mistakes? How do you know if an answer is wrong? Is each answer too big or too small?*

IN FOCUS Question ❶ asks children to perform simple additions of decimals with differing numbers of decimal places. In the examples the digits have already been lined up. Question ❷ features an addition pyramid to provide children with lots of addition practice. Question ❸ looks at different mistakes that children may make. Discuss whether 1·79 is a sensible answer to 4·5 + 1·34. Encourage children to round each decimal amount to the nearest whole to find an approximate total: 4·5 rounds up to 5, and 1·34 rounds down to 1, so you would expect the sum to be close to 5 + 1 = 6. Ask children to describe the mistakes that have been made. Discuss why you do not align the numbers from the right this time, like you do with integers. Instead you align decimal additions at the decimal point.

STRENGTHEN Encourage children to show their workings clearly for question ❸. If children are using the column method, they need to ensure that the digits are being lined up at the decimal point; children also need to ensure that they add an extra 0 in the hundredth column so that there are the same number of digits after the decimal point.

DEEPEN Ask children to add the numbers in question ❸ mentally. What method did they use? Did they round the numbers up or down? Discuss the methods with the whole class and encourage children to use estimation to check their answers.

ASSESSMENT CHECKPOINT Are children confident with laying the column method out neatly and accurately? Are they confident in using a variety of strategies when adding, subtracting or identifying mistakes that may happen?

ANSWERS

Question ❶ a): 4·05 m

Question ❶ b): 5·521 m

Question ❷: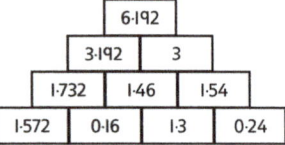

Question ❸ a): 4·5 + 1·34
The digits are not lined up correctly. There are 4 ones and 5 tenths, not 4 tenths and 5 hundredths. 82·43 + 1·89 The numbers in the top row have not been adjusted after the exchanging.

Question ❸ b): 4·5 + 1·34 = 5·84
82·43 + 1·89 = 84·32

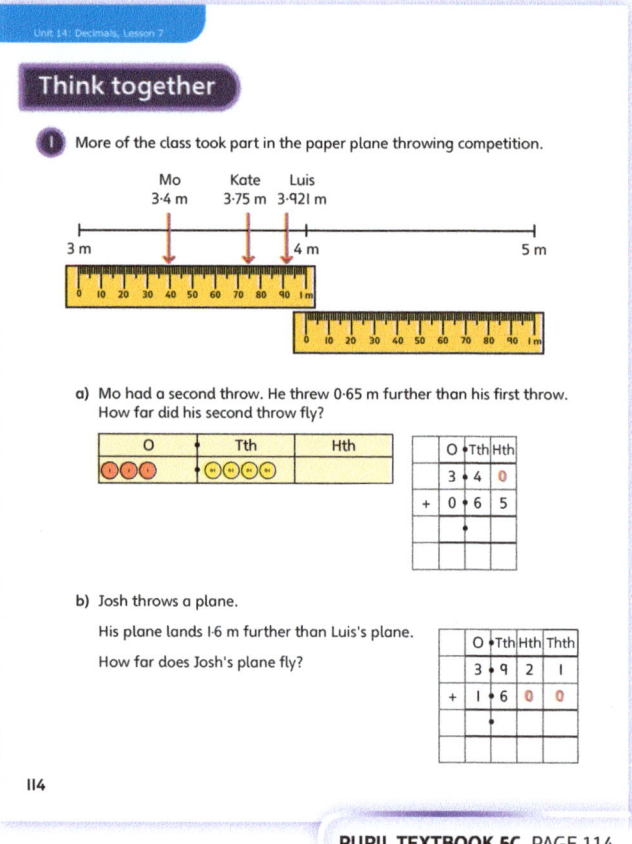

PUPIL TEXTBOOK 5C PAGE 114

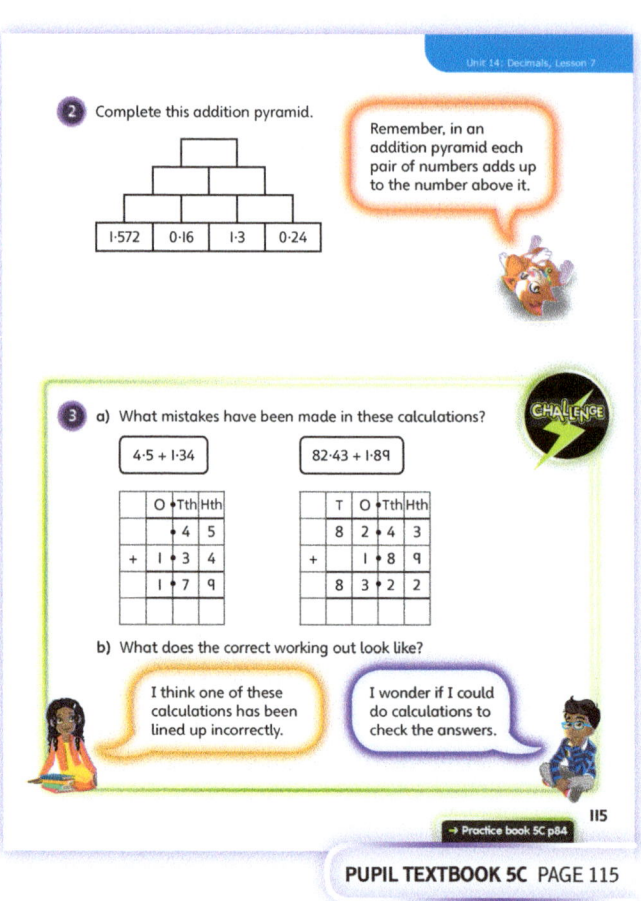

PUPIL TEXTBOOK 5C PAGE 115

Unit 14: Decimals, Lesson 7

Practice

WAYS OF WORKING Independent thinking

IN FOCUS Questions ① to ③ provide practice in adding and subtracting decimals with a different number of decimal places. Encourage children to set out their calculations in columns in questions ③ and ④. In question ⑤, focus the learning on identifying Zac's mistake rather than simply subtracting the numbers. Ask children to double check that they have aligned the digits with the decimal point. Encourage children to check their answers are reasonable by approximating the decimals and doing a quick mental calculation.

STRENGTHEN Support children by providing place value counters and place value grids so that they can see how digits line up. For example, when adding 1·7 and 2·52, children should put 7 counters in the tenths column and 5 counters underneath, because, in each number, the 7 and 5 represent tenths. At each stage ask: *What is the value of each digit in the number?*

DEEPEN Question ⑥ can be explored further by asking children to make their own 'always, sometimes, never' questions based on what they have learnt so far in the lesson. Encourage children to support their answers with examples and diagrams.

THINK DIFFERENTLY Question ⑦ asks children to find the sum of A and B. Children should use their knowledge of number lines to find the intervals and then work out what each arrow is pointing to. Ask: *Look at this question. What is different about this question compared to other ones you have solved so far? What does each point represent? What do you need to do to calculate the sum of the two numbers?*

ASSESSMENT CHECKPOINT Children are confident in using the column layout to add and subtract decimals with different numbers of decimal places.

ANSWERS Answers for the **Practice** part of the lesson can be found in the *Power Maths* online subscription.

Reflect

WAYS OF WORKING Independent thinking

IN FOCUS This **Reflect** activity checks children's understanding of how to add or subtract decimals with a different number of decimal places. Encourage children to explain their ideas clearly, paying attention to mistakes that might happen and the two main misconceptions:
1. not lining numbers up by the decimal place
2. forgetting to add 0 placeholders when two decimals do not have the same number of decimal places.

ASSESSMENT CHECKPOINT Children should be able to articulate the methods they use to add or subtract decimals with a different number of decimal places. They are able to explain their reasoning.

ANSWERS Answers for the **Reflect** part of the lesson can be found in the *Power Maths* online subscription.

After the lesson

- Do children understand the importance of setting up the column method correctly?
- Can children employ a variety of strategies to solve problems?
- Can children describe the mistakes that may happen when adding or subtracting decimals?

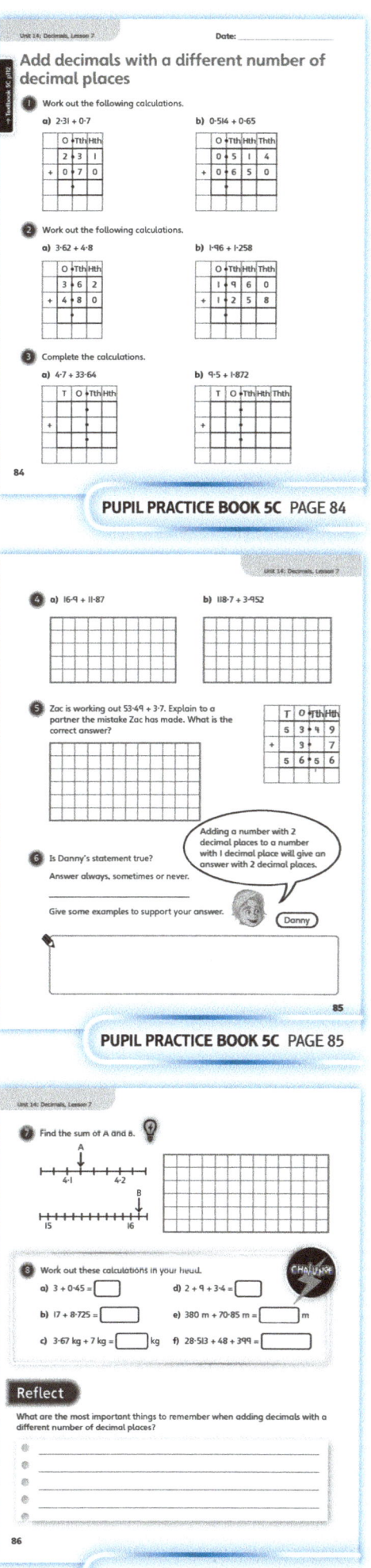

PUPIL PRACTICE BOOK 5C PAGE 84

PUPIL PRACTICE BOOK 5C PAGE 85

PUPIL PRACTICE BOOK 5C PAGE 86

Unit 14: Decimals, Lesson 8

Subtract decimals with a different number of decimal places

Learning focus
In this lesson, children will add and subtract decimal numbers with up to 4 digits to and from whole numbers. They will perform exchanges when there are zeros in the columns.

Before you teach
- How will you explain what to do if there is a 0 in a column required for an exchange?
- Do children know how to lay out a column subtraction accurately?

NATIONAL CURRICULUM LINKS

Year 5 Number – fractions (including decimals and percentages)

Solve problems involving number up to three decimal places.

ASSESSING MASTERY

Children can add or subtract a whole number and a decimal. They can talk through their methods, demonstrating a clear understanding of place value when a zero is in the column in which an exchange is required. Finally, children can show their addition or subtraction calculations using place value counters and the column method of addition or subtraction.

COMMON MISCONCEPTIONS

Children may try to make an exchange but see that there is a 0 in the next column. They may exchange the hundredths but not change the tenths or ones. Ask:
- *What should you do if there is a 0 in a column you need for an exchange?*

Children may also think that when you subtract from 0, the answer is 0 or the number itself. For instance, 0 – 2 = 2. Ask:
- *Can you show me 0 – 2? Can the answer be 2? Can you buy a notebook for £2 if you have no money?*

STRENGTHENING UNDERSTANDING

Children should gain confidence by subtracting whole numbers with 1, 2 or 3 digits from multiples of 100 and 1,000 before moving on to subtracting decimals from whole numbers. Children should be able to work methodically and solve questions such as 100 – 235 or 1,000 – 342. Revisit subtraction of decimals from 1. Ask: *What is 1 – 0·2? 1 – 0·23? 1 – 0·456?*

GOING DEEPER

Deepen learning in this lesson by providing children with subtractions that have exchange mistakes in them. Can children spot the mistakes and reason why the mistakes may have been made?

KEY LANGUAGE

In lesson: place value, digits, mass, decimal, take away, minus, add, subtract, column addition, column subtraction, whole, part

Other language to be used by the teacher: difference, fewer, exchange, less than, greater than, thousandths, hundredths, tenths, ones

STRUCTURES AND REPRESENTATIONS

Place value grid, column addition, column subtraction, number line

RESOURCES

Mandatory: place value counters, weighing scales, measuring jug

 In the eTextbook of this lesson, you will find interactive links to a selection of teaching tools.

Quick recap

Provide children with practice subtracting decimals using the column method. Then ask children to add numbers such as 12,714 + 729. Check that children line the digits up correctly.

Unit 14: Decimals, Lesson 8

Discover

WAYS OF WORKING Pair work

ASK

- Question ① a): *How much juice is in each bottle? Why would you need to know the total amount? What calculation do you need to do?*
- Question ① b): *How much flour was there in the bag to start with? How much flour is on the scales?*

IN FOCUS Question ① a) requires children to add a whole number and a decimal. Question ① b) introduces the subtraction of a decimal from a whole number. Children are focusing on exploring what happens when an exchange is needed and there is a 0 in the column they need to exchange from.

PRACTICAL TIPS You may want to use digital weighing scales or a 2 or 3 litre measuring jug. Ask children to first measure 2 litres of water or weigh 2 kg. Allow children to be curious and explore what happens when the weight or the capacity change. Write the number statements on the board and discuss what happens in each instance. Are the numbers getting bigger or smaller?

ANSWERS

Question ① a): There is 6·25 l of juice in the two bottles in total.

Question ① b): There is 1·704 kg of flour left in the bag.

PUPIL TEXTBOOK 5C PAGE 116

Share

WAYS OF WORKING Whole class teacher led

ASK

- Question ① a): *Why are there red zeros after the 5 in 5·00 + 1·25? What is their purpose?*
- Question ① b): *Where do you start the column subtraction?*
- Question ① b): *What steps do you need to go through when there is a 0 in the column you need to use for an exchange? How many exchanges do we have to do? What does the small '1' mean? Where does the '9' come from?*

IN FOCUS For question ① b), model the subtraction using the column method format. Show children the importance of the layout, how to strike through the numbers being exchanged, and how to put the small 1 for the exchanged number. Ensure that children understand every step of the calculation and reinforce the place value of each digit. Use place value counters to reinforce that 1 = 10 tenths, 1 tenth = 10 hundredths and 1 hundredth = 10 thousandths.

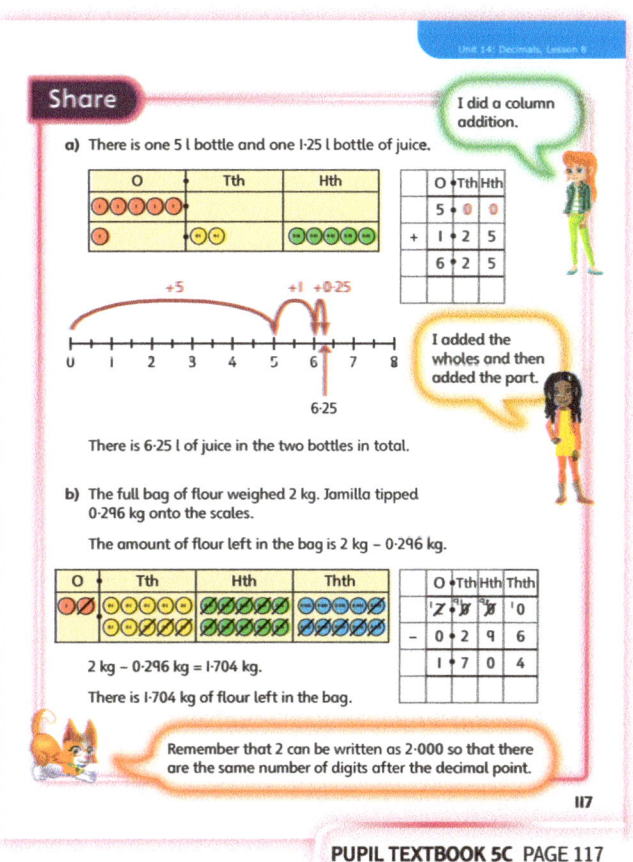

PUPIL TEXTBOOK 5C PAGE 117

153

Unit 14: Decimals, Lesson 8

Think together

WAYS OF WORKING Whole class teacher led (I do, We do, You do)

ASK

- Questions ❶ a) and b): *Why do you need to add the zeros in? Why do you need to add a different number of zeros each time?*
- Question ❶ c): *How many zeros do you need to add?*
- Question ❷: *What do you need to make sure happens with the numbers? What mistakes do you need to avoid?*
- Question ❸: *What do both methods have in common? How do you decide what you should change each number to? Which method will you use?*

IN FOCUS In question ❶, children are subtracting decimals with different numbers of decimal places from a whole number. With scaffolding, they are supported to add in zeros to make the numbers of digits after the decimal place the same. This allows them to subtract correctly. In question ❷, children are again presented with a series of calculations with differing numbers of decimal places but without the scaffolding. Children should set out the calculations as a column subtraction. Look for children making sure they add in extra zeros to make the numbers the same length.

STRENGTHEN Provide children with place value counters in question ❸ and ask them to set out each of Jamilla's calculations. Ask them to first set out 2 – 0·296, then demonstrate what they need to do to change to 1·999 – 0·295. If necessary, simplify to whole numbers such as 20 – 3 and ask them to think why that is the same as 19 – 2 or 19 – 3 + 1. Allow time for children to explore these methods before progressing to question ❸ b).

DEEPEN Question ❸ offers a great opportunity to link back to previous work on adding and subtracting whole numbers using knowledge of place value. Children should be able to spot the links between all the calculations. Challenge them to think of other methods that they can use to complete each calculation. Ask: *What method would you use if you subtracted 296 from 2,000? Can you use the same method to work out 2 – 0·296?*

ASSESSMENT CHECKPOINT Use questions ❶ and ❷ to check that children can subtract decimals with differing numbers of decimal places. Check that children line up the digits in the column subtractions correctly.

ANSWERS

Question ❶ a): 2·2 kg more

Question ❶ b): 1·35 kg more

Question ❶ c): 2·642 litres

Question ❷: a) 7·6 – 3·52 = 4·08   d) 17·68 – 3·9 = 13·78
b) 7·68 – 3·5 = 4·18   e) 4·2 – 1·79 = 2·41
c) 7·68 – 3·9 = 3·78   f) 4·25 – 1·795 = 2·455

Question ❸ a): Children choose the method they prefer.

Question ❸ b): 6 – 3·45 = 2·55   26 – 2·8 = 23·2
3 – 0·914 = 2·086

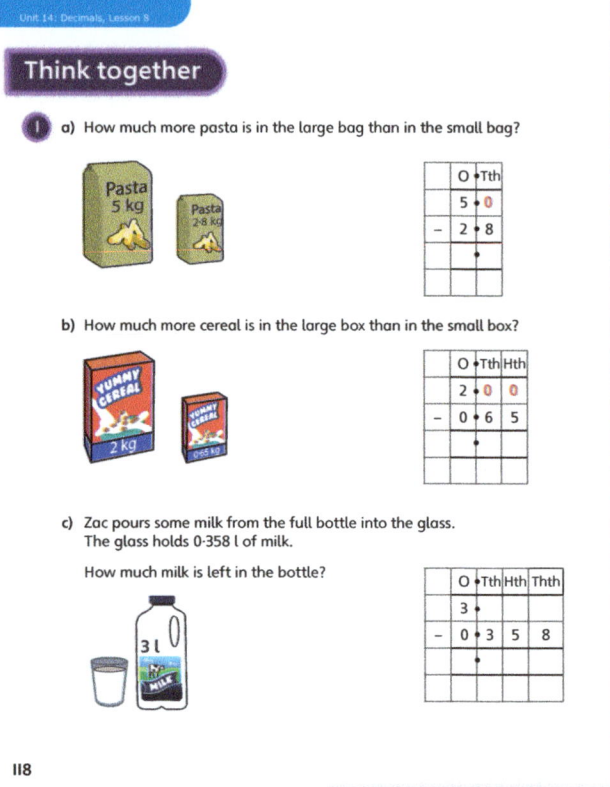

PUPIL TEXTBOOK 5C PAGE 118

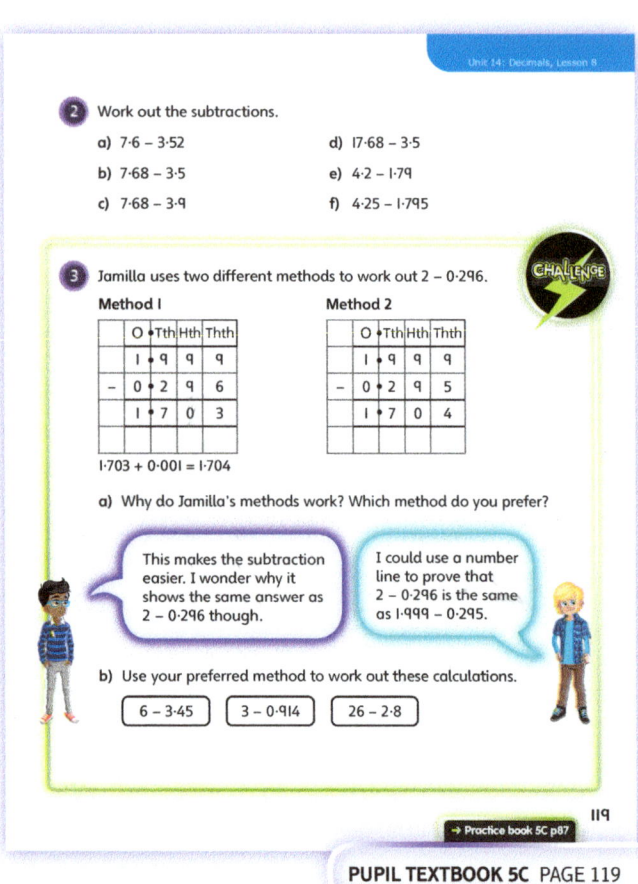

PUPIL TEXTBOOK 5C PAGE 119

Unit 14: Decimals, Lesson 8

Practice

WAYS OF WORKING Independent thinking

IN FOCUS Question ❶ provides fluency practice of the methods used in the Textbook. Children should be adding extra zeros to make the numbers the same length.

Question ❷ asks children to find solutions and work out the missing numbers whilst linking addition and subtraction. This will focus children's thinking and push them to work backwards. Support them by asking them to talk you through where they should start and how they can check that their answer is correct. Question ❷ also offers another opportunity to subtract a decimal from a whole number by manipulating the numbers to simplify the calculation.

Questions ❸ and ❹ introduce a context for adding or subtracting a whole number with a decimal. Ask children to write the appropriate units of measurement with their answers.

STRENGTHEN Encourage children to use counters on a place value grid and number lines to support their learning. When the calculation is not given in a column layout, encourage them to write it in columns and offer support and feedback on their efforts.

DEEPEN Explore the **Challenge** question in more depth by providing similar problems or other missing number questions.

ASSESSMENT CHECKPOINT Check that children are confident in subtracting decimals with different numbers of decimal places. They should be able to subtract a decimal from a whole number using different methods with confidence.

ANSWERS Answers for the **Practice** part of the lesson can be found in the *Power Maths* online subscription.

Reflect

WAYS OF WORKING Independent thinking

IN FOCUS This **Reflect** activity checks children's understanding of applying the method where they manipulate the numbers to calculate the difference. Encourage children to explain by demonstrating the calculation on a number line. Focus their attention not only on the answer but on noticing what is the same and what is different between the calculations. Children should explain that both numbers (6·9 and 2·3) are 0·1 less than the original numbers, hence the difference between the numbers will be the same.

ASSESSMENT CHECKPOINT Can children explain how a subtraction between a whole number and a decimal can be manipulated in order to make it easier to solve? Can children use a number line to support their answer?

ANSWERS Answers for the **Reflect** part of the lesson can be found in the *Power Maths* online subscription.

After the lesson

- Can children explain how to exchange when there is a 0 in the column required for an exchange?
- Do children employ a variety of strategies to complete the calculations?
- Are children able to use a variety of different representations to show their solutions?
- Can children accurately manipulate numbers in order to make a calculation more efficient to solve?

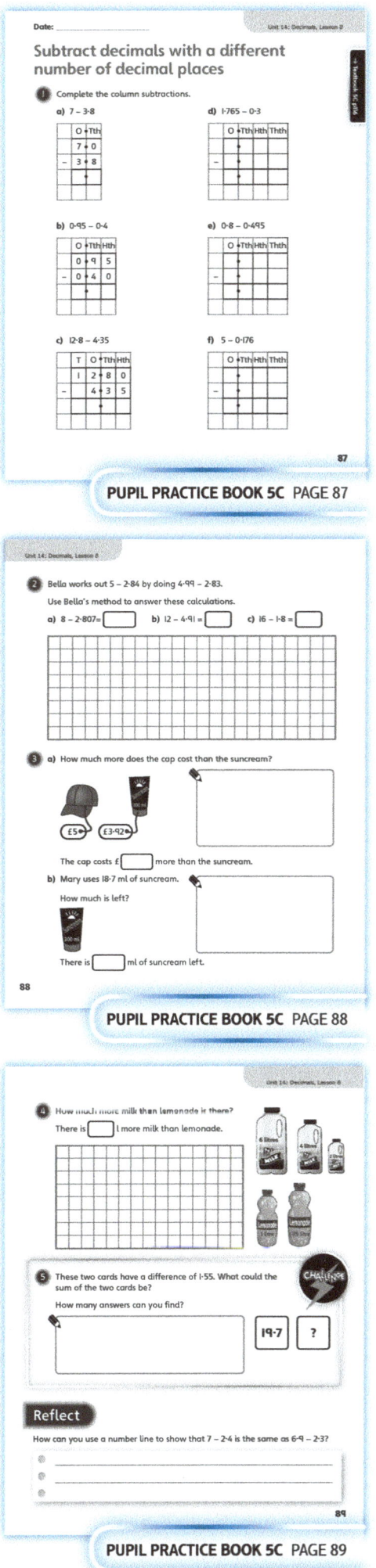

PUPIL PRACTICE BOOK 5C PAGE 87

PUPIL PRACTICE BOOK 5C PAGE 88

PUPIL PRACTICE BOOK 5C PAGE 89

Unit 14: Decimals, Lesson 9

Problem solving with decimals 1

Learning focus
In this lesson, children will learn strategies for solving problems involving adding and subtracting numbers with up to three decimal places.

Before you teach
- Do children know how to lay out column additions and subtractions and make multiple exchanges?
- Do children understand key vocabulary such as 'more than' and 'less than'?
- Do they consider the context in which the vocabulary has been used?

NATIONAL CURRICULUM LINKS

Year 5 Number – fractions (including decimals and percentages)
Solve problems involving number up to three decimal places.

ASSESSING MASTERY
Children can solve problems that involve a combination of adding and subtracting numbers with up to three decimal places and making multiple exchanges.

COMMON MISCONCEPTIONS
Sometimes, children assume that certain words always indicate a particular calculation without considering the context. For example, they may think that 'more' always means adding and 'less' always means subtracting. Ask:
- What are you trying to find out? What kind of answer are you expecting to get? Have you used all the information from the question? How can you check your answer is correct? If you substitute the answer back into the question, does it make sense?

STRENGTHENING UNDERSTANDING
Children should focus on finding solutions to problems with either addition or subtraction first, before attempting questions that require both operations. Encourage children to use a bar model or number line to aid their understanding or to check their answer. Use different contexts with smaller numbers. Ask: *What operation should you use?*

GOING DEEPER
Give children multi-step problems that can be solved in two or more different ways. Ask them to draw diagrams and bar models and generalise where possible. Challenge them to find all of the different ways of finding solutions to problems.

KEY LANGUAGE
In lesson: more than, difference, total, method, efficient, mass, weigh, heavier, kilogram (kg)

Other language to be used by the teacher: altogether, addition, subtraction, combine, compare

STRUCTURES AND REPRESENTATIONS
Bar model, column addition, column subtraction

RESOURCES
Optional: weighing scales

 In the eTextbook of this lesson, you will find interactive links to a selection of teaching tools.

Quick recap
Write four decimals on the board: 1·7, 3·95, 0·68 and 2·905. Ask children to pick two numbers and add them together. They should then tell a partner their total. Can their partner work out which numbers they chose?

Unit 14: Decimals, Lesson 9

Discover

WAYS OF WORKING Pair work

ASK

- Question 1 a): *How much does the astronaut weigh on Earth? And on the moon? What does 'how much greater' mean?*
- Question 1 b): *What do you need to know to find the mass of the life support? What do you know about the mass of the life support?*

IN FOCUS Question 1 a) requires children to subtract to calculate the difference between the mass of the astronaut on Earth and on the moon. Question 1 b) is a two-step addition problem that involves working out the mass of the life support and then the total combined mass of the spacesuit and life support.

PRACTICAL TIPS Discuss the changes in the astronaut's weight on Earth and on the moon. Show examples of a variety of objects and their weights on Earth: for example, a car, TV and a melon. Ask children to predict what their weight might be on the moon. Will it be more than or less than their weight on Earth? Why is it important for the astronauts to know what the weights of different objects will be? Pay attention to vocabulary used, such as 'more than', 'less than', 'difference', 'how much more?' and 'how much less?'. Discuss the calculations that children need to do in each instance.

ANSWERS

Question 1 a): The weight of the astronaut on Earth is 64·05 kg more than on the moon.

Question 1 b): The total mass is 189·98 kg.

Share

WAYS OF WORKING Whole class teacher led

ASK

- Question 1 a): *What calculation do you need to do if you are finding 'how much greater'? Are you adding or subtracting?*
- Question 1 b): *If the mass of the life support is 90·2 kg greater than the mass of the spacesuit, how could you work out the life support's mass?*
- Question 1 b): *How can you work out the total mass of the spacesuit and the life support?*

IN FOCUS For question 1 a), show children the bar model and ask how they know that finding the difference will lead them to performing a subtraction. Ensure children are confident with the correct column layout for a decimal subtraction. For question 1 b), look for children who just add the numbers together, instead of finding the mass of the life support first. Highlighting the bar model should help children realise why this will not work. Discuss the key words in the question and how children could work out the mass of the life support. Emphasise the importance of reading the question carefully. Reinforce the place value of each digit when carrying out the calculations and address the need for an exchange.

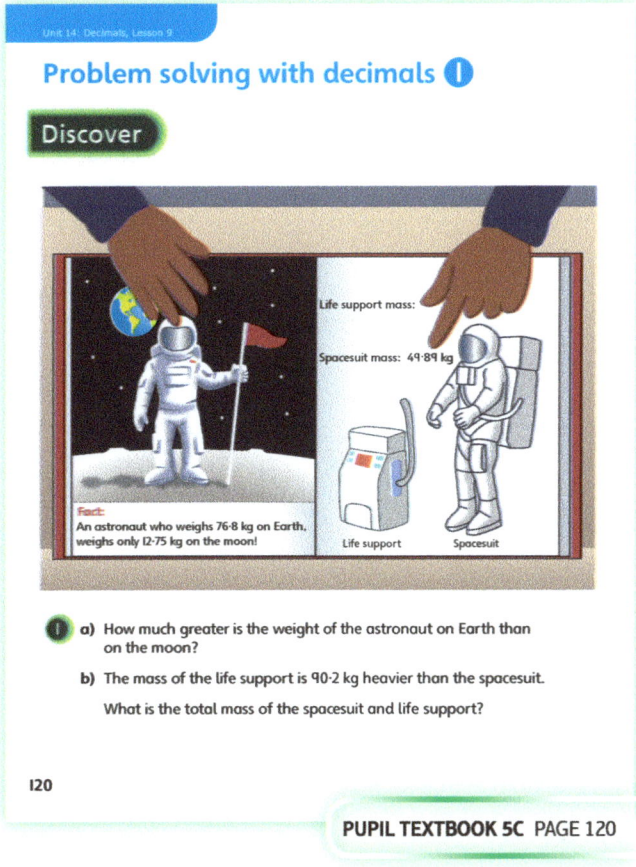

PUPIL TEXTBOOK 5C PAGE 120

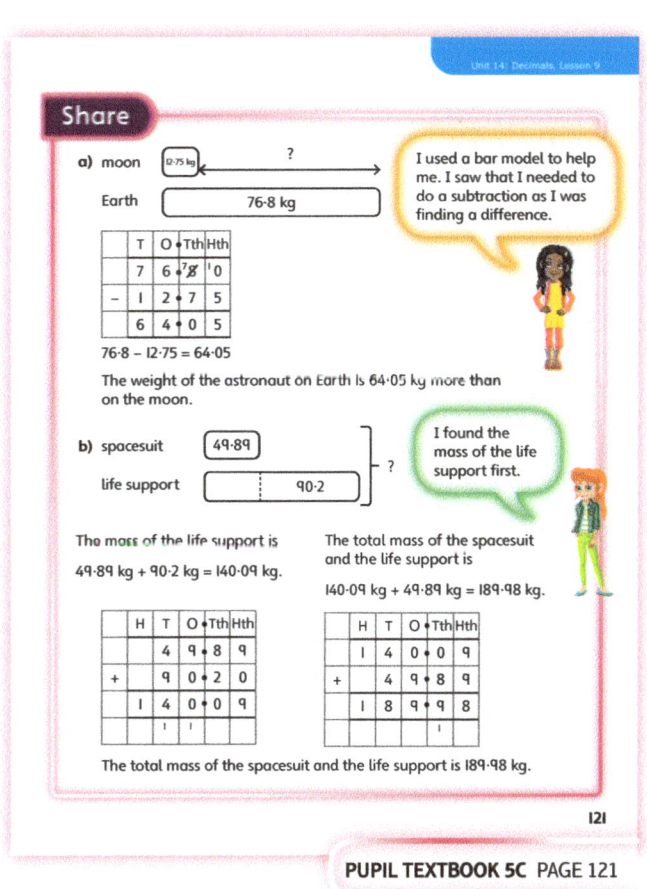

PUPIL TEXTBOOK 5C PAGE 121

157

Think together

WAYS OF WORKING Whole class teacher led (I do, We do, You do)

ASK

- Question ❶: *What calculation do you need to do? What word(s) tell you this?*
- Question ❷: *How much does each magazine cost? Look at the bar model. What do you need to do to find the total?*
- Question ❸ a): *How can you use the mass of Rock B to find the mass of Rock A or Rock C? What do you then need to do to find the total mass? Are there two different ways to find the total?*

IN FOCUS In question ❶, children may not understand why 'how much further?' is being associated with a subtraction. Use the bar model to show why they need to subtract. Encourage children to use the column subtraction method. Question ❷ requires children to add three decimal numbers together. Show the bar model and discuss why they need to carry out an addition calculation. Discuss the different methods that could be used: adding all three numbers at once in a column addition; or adding two numbers in a column addition then adding the other number to this answer; or carrying out the addition calculations mentally. Could they manipulate any of the numbers to make the calculation simpler? In question ❸ a), children should first write down the masses of the three rocks as A: (12 – 3·6) kg, B: 12 kg, C: (12 + 4·75) kg. Then, they could either: work out the individual masses of A and C and add all three together; or they could multiply 12 by 3, then subtract 3·6 and add 4·75.

STRENGTHEN To support understanding, children should be able to describe a question and represent it using a bar model or diagram. They should write the information they know on the bar model and use a question mark to show what they are trying to find out.

DEEPEN Question ❸ can be explored further by changing the mass of rock B and asking children to find the new masses of rocks A and C. Alternatively, tell children that the difference in mass between two new rocks D and E is 5·25 kg (for example). Can children think of any two values that the masses of rocks D and E could be?

ASSESSMENT CHECKPOINT Can children solve problems that use decimal numbers and involve addition, subtraction and a combination of the two? Confident answers to question ❸ will indicate that children have a firm grasp of the learning in this section.

ANSWERS

Question ❶: 7·644 m

Question ❷: £13·86
Methods could include column addition or rounding to the nearest pound and adjusting.

Question ❸ a): The total mass is 37·15 kg
A = 12 – 3·6 = 8·4 kg
C = 12 + 4·75 = 16·75 kg
A + B + C = 8·4 kg + 12 kg + 16·75 kg
Or 3 × 12 kg – 3·6 kg + 4·75 kg

Question ❸ b): Rock C weighs 8·35 kg more than rock A.
16·75 – 8·4 or 3·6 + 4·75

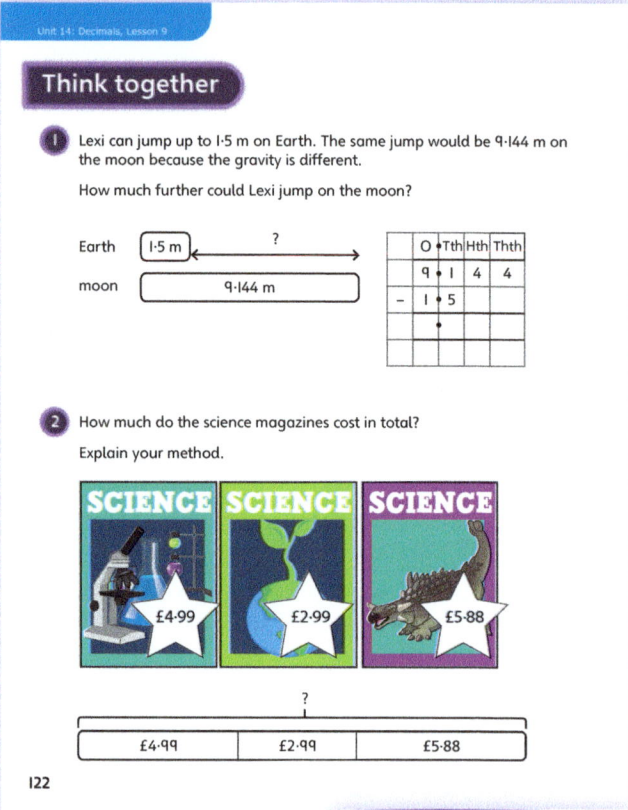

PUPIL TEXTBOOK 5C PAGE 122

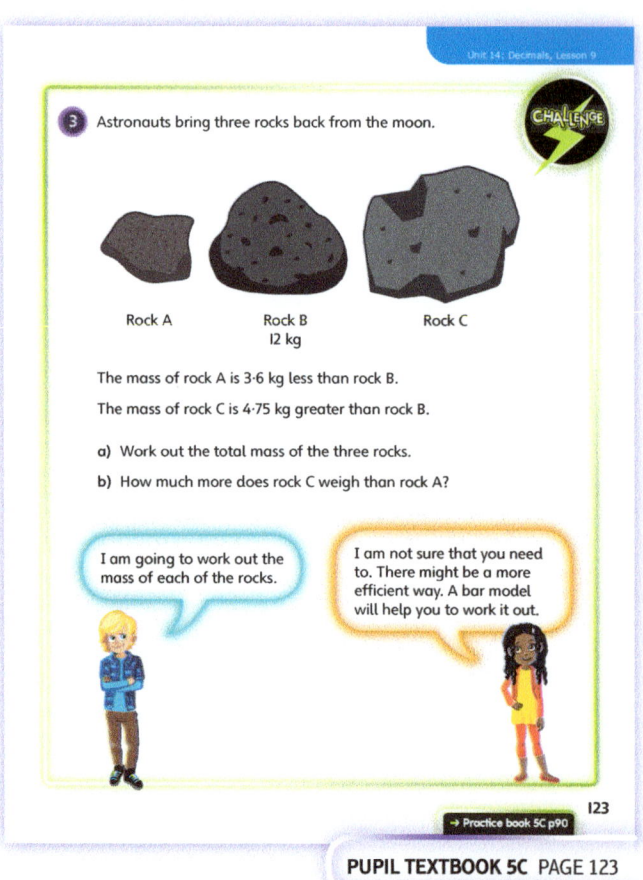

PUPIL TEXTBOOK 5C PAGE 123

Unit 14: Decimals, Lesson 9

Practice

WAYS OF WORKING Independent thinking

IN FOCUS For question ❶, ask children to focus on the calculations that arise from each bar model, and to show their method for solving them. Encourage children to discuss the different ways they could find a solution to question ❶ c). Questions ❷ and ❸ aim to consolidate children's understanding of problem solving within a context. Urge children to draw a bar model and to add in any information as they complete the calculation.

STRENGTHEN Before children answer question ❹, substitute the decimals for whole numbers. For example, find two numbers that add up to 5 but have a difference of 1. Ask children to use a bar model. From this, lead on to identifying what children need to do to answer the actual question.

DEEPEN Question ❷ can be explored in greater depth by asking children to create their own word problem based on the numbers used in question ❷. For example: Bella throws a javelin 26·3 m. Her second throw travels 6·85 m less than the first throw. How far did Bella throw the second javelin? Ask: *What will the key words in your word problem be? Can you put the key words into a context? Are you going to create an addition or subtraction problem?*

THINK DIFFERENTLY In question ❺, children are asked to subtract four numbers with varying numbers of decimal places. Support children by asking them to pay attention to the place value of each digit in the calculation. Encourage children to use counters on a place value grid to support their understanding. Challenge children to work out the calculation mentally and share their strategy with the whole class.

ASSESSMENT CHECKPOINT Can children confidently add and subtract decimals and solve problems that involve combinations of these calculations?

ANSWERS Answers for the **Practice** part of the lesson can be found in the *Power Maths* online subscription.

Reflect

WAYS OF WORKING Independent thinking

IN FOCUS This **Reflect** activity checks children's understanding of using bar models to solve addition and subtraction problems. Encourage children to write a two-step problem. Pay close attention to the vocabulary they use. Ask children to find solutions to their problems. As an extension, are they able to find solutions in a different way?

ASSESSMENT CHECKPOINT Can children use the appropriate vocabulary to write a two-step problem that requires the use of both addition and subtraction to solve it?

ANSWERS Answers for the **Reflect** part of the lesson can be found in the *Power Maths* online subscription.

After the lesson

- Can children apply what they know about the addition and subtraction of decimal numbers in problem solving contexts?
- Are children confident identifying each step that is required to solve a two-step problem?

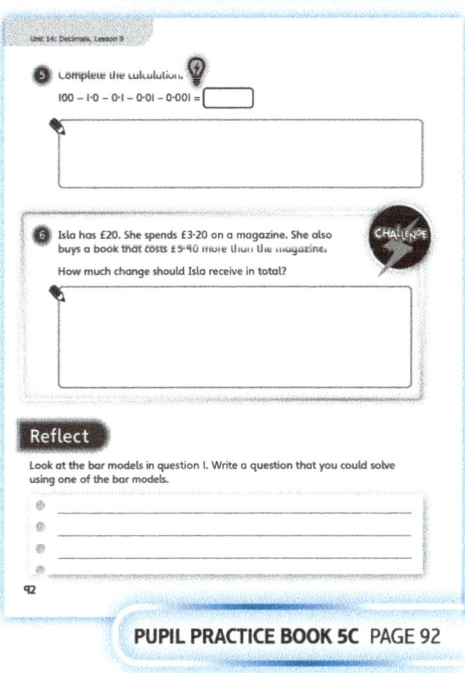

PUPIL PRACTICE BOOK 5C PAGE 90

PUPIL PRACTICE BOOK 5C PAGE 91

PUPIL PRACTICE BOOK 5C PAGE 92

Unit 14: Decimals, Lesson 10

Problem solving with decimals 2

Learning focus
In this lesson, children will learn how to solve more complex addition and subtraction multi-step problems. They will interpret and identify the information necessary to solve the problems.

Before you teach
- Are children confident with breaking a problem down into smaller parts?
- Can children select the information they know, and decide what is unknown or needs to be discovered?

NATIONAL CURRICULUM LINKS

Year 5 Number – fractions (including decimals and percentages)
Solve problems involving number up to three decimal places.

ASSESSING MASTERY
Children can solve more complex multi-step problems that involve adding and subtracting decimals and where the information is represented in tables or needs to be extracted from sentences.

COMMON MISCONCEPTIONS
Children may not understand what calculation a problem requires them to undertake and so will just guess whether they need to add or subtract. This is particularly the case where the question is long. Ask:
- *What is the question about? Can you describe it? Why do you think this information is provided? What can you find out from it? What model could you draw to help you?*

STRENGTHENING UNDERSTANDING
Encourage children to sketch, act out situations or use concrete materials – whatever helps solidify the concept for them. Remind them to read the problem carefully a number of times until they fully understand what is needed. Encourage them to discuss the problem with a partner or rewrite the question in their own words.

GOING DEEPER
Ask children to consider solving the problem in a different way. Can they simplify it at all? Encourage them to think 'what if?' and see if they can draw on the strategies used to solve one problem to help them with another. Encourage the use of logical thought processes and generalisation where possible.

KEY LANGUAGE
In lesson: how much?, balance, distance, decimal, multiply, more than (>), less than (<), total, mass, weight, add, remove, reduce, difference, kilogram (kg)

Other language to be used by the teacher: greatest, fewer, compare, increase, identical

STRUCTURES AND REPRESENTATIONS
Bar model, number line, column subtraction

RESOURCES
Mandatory: weighing scales, blank comparison bar models

 In the eTextbook of this lesson, you will find interactive links to a selection of teaching tools.

Quick recap
Ask children to complete different decimal addition and subtraction questions.

Unit 14: Decimals, Lesson 10

Discover

WAYS OF WORKING Pair work

ASK

- Question 1 a): *What can you see in the picture? What do all the scales have in common? How do they differ? Which one is the balance scale?*
- Question 1 a) *What is the mass of the oats? What is the mass of the pears? What is their combined mass?*
- Question 1 b): *When does a scale balance? What happens if the left-hand side is heavier than the right-hand side? What do Emma and Ebo need to do to get the scales to balance?*

IN FOCUS Question 1 a) requires children to identify what will happen to the scale when the mass at either side changes. In question 1 b), children can use two different methods to balance the scales. They could either increase the mass of sugar (on one side of the balance) or reduce the mass of oats (on the other side of the balance).

PRACTICAL TIPS Use a scale to reinforce the concept of mass measured in kilograms. Gradually change the mass of one of the items you are weighing and note what happens to the scales when you do so. Allow children to explore what would happen if they increased both sides by the same amount, or reduced both sides by different amounts.

ANSWERS

Question 1 a): The oats and pears will still be heavier as 3·49 kg > 3 kg.

Question 1 b): 3·49 kg – 3 kg = 0·49 kg
Emma and Ebo can add 0·49 kg of sugar to the bag of sugar.
Or they can remove 0·49 kg of oats from the bag of oats.

Share

WAYS OF WORKING Whole class teacher led

ASK

- Question 1 a): *What increments does the scale go up in?*
- Question 1 b): *How could you balance the scales? What calculation do you need to do so that both sides have an equal mass?*

IN FOCUS For question 1 a), show all the scales. Discuss how the children know the arrow is pointing to 2·6 kg on the analogue scales. Discuss with children the methods that can be used to find the combined mass of 2·6 + 0·89 (the oats plus the pears). Children could count on from 2·6 or use column addition. The sugar weighs 3 kg. Discuss whether the combined oats and pears weigh more than 3 kg. Ask: *Which side of the scales will be heavier?* Question 1 b) could be further explained using a comparison bar model. Ask children what calculation they need to do so that the bars are of equal length. The use of the bar model will help children visualise why they need to either add 0·49 kg of sugar or subtract 0·49 kg of oats to balance the scales and do the calculation.

PUPIL TEXTBOOK 5C PAGE 124

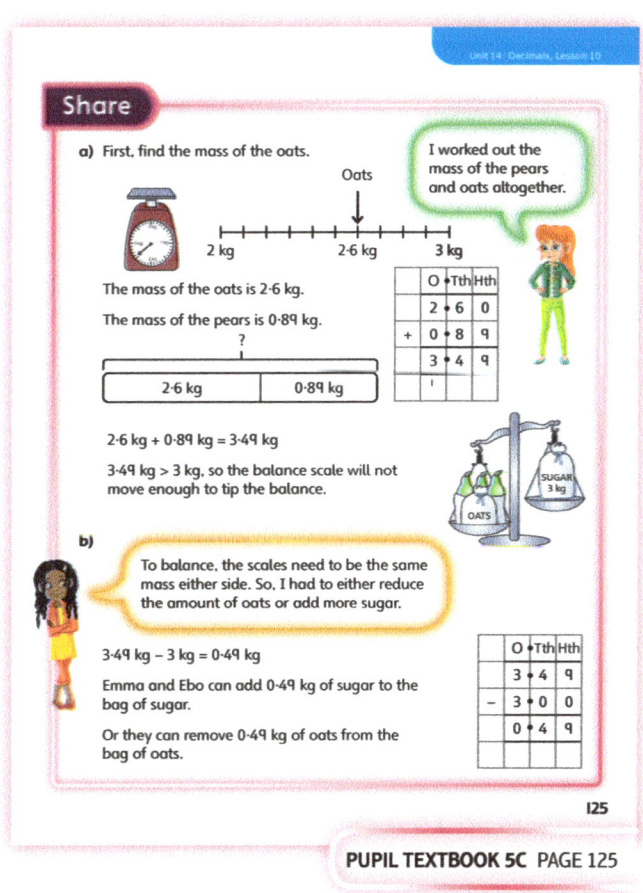

PUPIL TEXTBOOK 5C PAGE 125

Unit 14: Decimals, Lesson 10

Think together

WAYS OF WORKING Whole class teacher led (I do, We do, You do)

ASK

- Question ❶: *How much sugar was in the bag to start with? How much sugar has been used?*
- Question ❷: *What do you need to know to find the total mass? What calculation could you do to work out the mass of the flour on each spoon?*
- Question ❸: *What is the distance between the 1st and 2nd lamp post? How can you use this information?*

IN FOCUS Question ❶ requires children to find the solution using addition and comparison. Ask them to find a different way to find the solution and share their ideas with the class. To find the solution to question ❷, children will need to both add and subtract decimals. Discuss what 'less' means and encourage children to draw a comparison bar model (like the example) to identify what calculations are needed. They can use the column method for their calculations. For question ❸, children may find it useful to draw a line and mark each of the lamp posts on the line and label the distances between them. They should use a bar model and write the numbers on each of the bars. Discuss the need to add 5·85 m three times and the different strategies that children could use.

STRENGTHEN Provide a blank comparison bar model in question ❷ and ask children to fill in the information they know. Reinforce the link between the information in the word problem and how it can be represented on a model. Look for children who add or subtract the numbers in the question but are unsure of why they need to do so. Break the question into smaller parts. Ask: *What does that mean? What can you find from this information?*

DEEPEN Question ❸ can be explored further by asking children to find the total distance between all the posts or to compare the distance between each pair of lamp posts. Take the opportunity to challenge children to practise all the different strategies they have learned so far to add and subtract decimals.

ASSESSMENT CHECKPOINT Can children recognise mathematical language within problems? Can they correctly identify which operations needs to be used? Are children able to use the addition and subtraction of decimals to find the solutions to multi-step problems, or do they need further practice?

ANSWERS

Question ❶: 1·61 kg

Question ❷: 21·3 g
  18·6 − 15·9 = 2·7   2·7 + 18·9 = 21·3
  Or 18·6 + 18·6 − 15·9 = 21·3

Question ❸: The distance between the 3rd and 4th lamp post is 5·511 m.
  3 × 5·85 = 5·85 + 5·85 + 5·85 = 17·55
  17·55 − 5·85 − 6·189 = 5·511

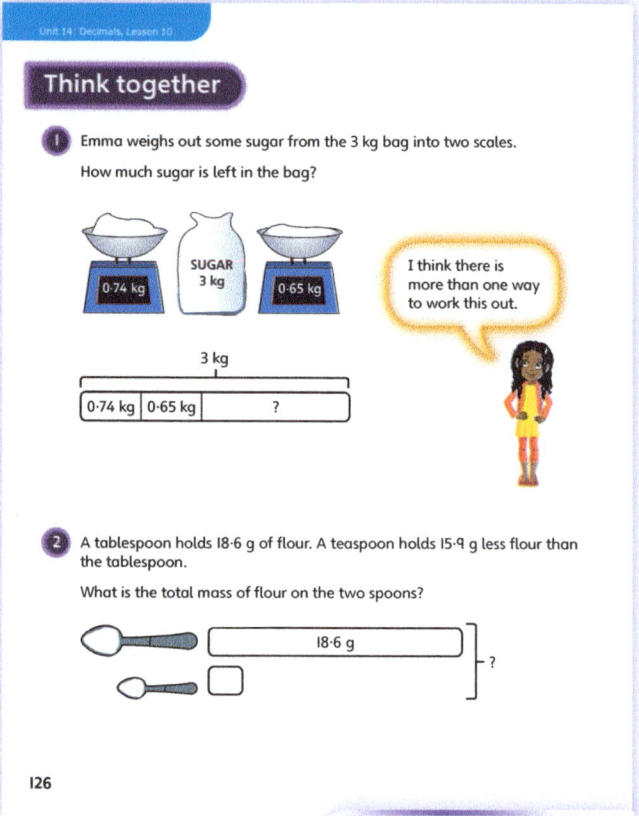

PUPIL TEXTBOOK 5C PAGE 126

PUPIL TEXTBOOK 5C PAGE 127

Unit 14: Decimals, Lesson 10

Practice

WAYS OF WORKING Independent thinking

IN FOCUS Questions 2 and 3 consolidate children's understanding of problem solving within a context. Encourage children to use a comparison bar model, asking questions to help them do this correctly. Ask: *Which bar will be longest? How do you know? Which bar will be shortest?* Ensure children are sure which operation is required and that they do not just add or subtract the numbers given in the question.

STRENGTHEN Suggest that children draw a bar model to help them grasp which calculations are needed in question 6. Alternatively, provide blank bar models so they can fill the numbers in themselves.

DEEPEN Ask children to find the solution to question 7 using number lines and a bar model to support their answers. Ask children to compare both models and strategies and discuss which other types of question both models could be used for.

THINK DIFFERENTLY Children have come across questions similar to question 5 when calculating the perimeter of shapes. Some children may find it challenging to link the shapes and decimals together. Ask children to consider what they would do if they were working with whole numbers instead of decimals. For example, ask: *What if the length of the rectangle was 10 cm and the width was 4 cm?*

ASSESSMENT CHECKPOINT Look at children's responses and working out for question 3. Can children identify with confidence the relevant information needed to solve a multi-step problem involving the addition and subtraction of decimals? Do children follow a logical and systematic approach to their problem solving?

ANSWERS Answers for the **Practice** part of the lesson can be found in the *Power Maths* online subscription.

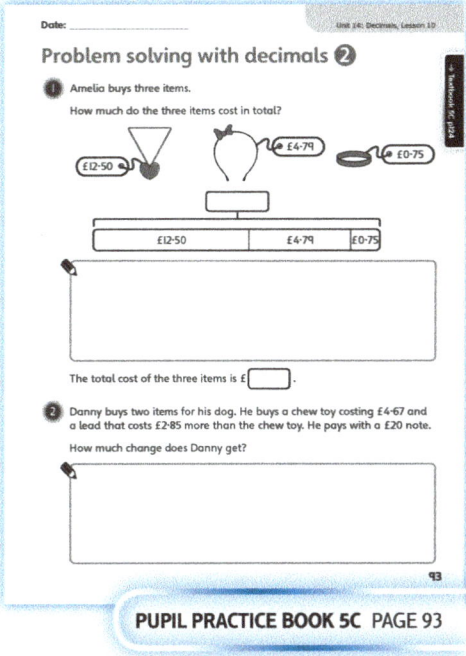

PUPIL PRACTICE BOOK 5C PAGE 93

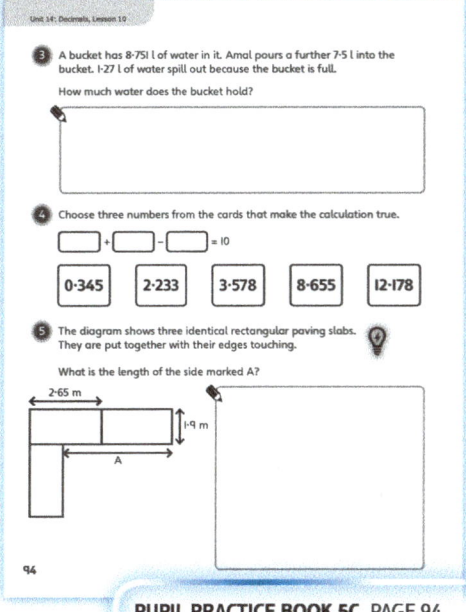

PUPIL PRACTICE BOOK 5C PAGE 94

Reflect

WAYS OF WORKING Pair work

IN FOCUS This question allows children to self-assess their learning. Children need to note the language and numbers used in the question. They need to think of the operations and strategy they will use to reach the solution 3·21 kg. This is a good opportunity to identify children that need extra support and to clarify any misconceptions that they may have.

ASSESSMENT CHECKPOINT Can children create their own addition and subtraction problem-solving questions? What calculations will they provide that will solve their questions?

ANSWERS Answers for the **Reflect** part of the lesson can be found in the *Power Maths* online subscription.

After the lesson

- Can children identify the important mathematical language within a problem?
- Can children identify the correct operation and employ a suitable, efficient method?

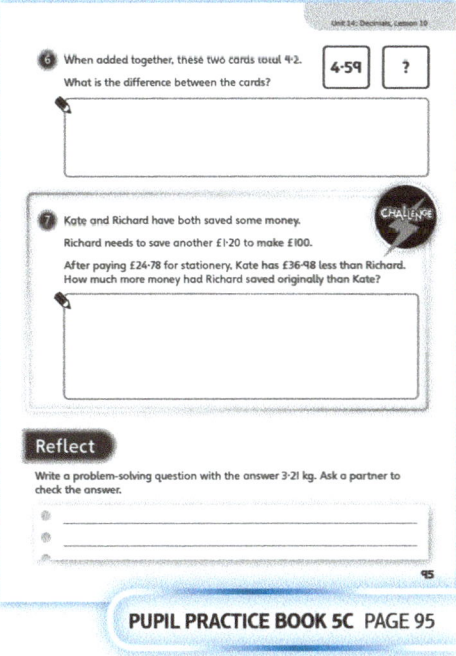

PUPIL PRACTICE BOOK 5C PAGE 95

163

Unit 14: Decimals, Lesson 11

Decimal sequences

Learning focus
In this lesson, children will use their understanding of decimal numbers to count and complete decimal sequences. They will describe the rule that decimal sequences follow and use it to calculate missing terms.

Before you teach
- How can you link sequences to children's real-life experiences or other areas of the curriculum?
- What concrete resources could you use to demonstrate the order of the numbers children will be studying?

NATIONAL CURRICULUM LINKS

Year 5 Number – fractions (including decimals and percentages)
Read, write, order and compare numbers with up to three decimal places.

ASSESSING MASTERY
Children can reliably count in decimals and continue the count accurately. They can order and complete number sequences and write down the rule that a decimal sequence follows.

COMMON MISCONCEPTIONS
Children may make mistakes when ordering the numbers in a sequence if they neglect to consider the place value of each digit. For example, they may say that the next number in the sequence 0, 0·4, 0·8 is 0·12 rather than 1·2. Ask:
- *What is the rule that the sequence follows? How many tenths do you have? How many tenths are you adding? What happens if you add another 4 tenths?*

STRENGTHENING UNDERSTANDING
To support children as they complete the number sequences in this lesson, it may help to revisit what they have learnt so far about the addition and subtraction of decimals. Show children a concrete representation of a decimal number sequence using place value counters. Ask: *What do you notice about how the number of counters changes?* Choose one of the numbers at random and hide it and its concrete representation. Ask: *Which number have I hidden? How do you know?*

GOING DEEPER
Give children two decimal numbers and ask them to create number sequences that include the two numbers without them being next to each other in the sequence. Children can challenge a partner to complete their number sequence. Ask: *Can you write your sequence so that the numbers increase? How would the sequence change if the numbers decreased instead?*

KEY LANGUAGE
In lesson: decimal, add, subtract, sequence, order, rule, amount, count up, pattern

Other language to be used by the teacher: count on, count back, jumps (or steps)

STRUCTURES AND REPRESENTATIONS
Number line, tables

RESOURCES
Optional: place value counters, plant growth chart, number lines, decimal number cards

 In the eTextbook of this lesson, you will find interactive links to a selection of teaching tools.

Quick recap
Give children a starting number, say 6. Ask them to keep adding 15 to the number. What is the first number they say above 100? Vary the activity by giving children a different starting number and a different amount by which to increase it each time.

Unit 14: Decimals, Lesson 11

Discover

WAYS OF WORKING Pair work

ASK

- Question 1 a): *How tall is the rose bush to start with? How can you find out how tall the rose bush will be next month?*
- Question 1 a): *What pattern or rule do the numbers follow?*
- Question 1 b): *What is the height of the rose bush? What operation do you need to do to find its height last month?*
- Question 1 b): *How could you predict the number of months it took to grow from 60 cm to 87·2 cm?*

IN FOCUS Question 1 a) invites children to write down a sequence of decimal numbers when the rule is given. Children should notice that the numbers follow a rule. Ensure they recognise that the difference between any two consecutive heights is 2·5 cm. Questions 1 a) and b) link to previous learning. Children should be able to use their understanding of adding and subtracting decimals to work out the heights of the plants over the next few months or over previous months.

PRACTICAL TIPS Give children a growth chart of different plants showing how much each plant grows every month or year. Ask them to investigate what will happen to the height of each plant. They could draw a plant for each month and label each picture with its increasing height month-on-month. Discuss which plants grow faster or slower.

ANSWERS

Question 1 a):

Month	April	May	June	July	Aug	Sept	Oct
Height (cm)	15·4	17·9	20·4	22·9	25·4	27·9	30·4

The rule is to add 2·5 each time.

Question 1 b):

May	June	July	Aug	Sept	Oct	Nov	Dec	Jan	Feb	Mar	April
59·7	62·2	64·7	67·2	69·7	72·2	74·7	77·2	79·7	82·2	84·7	87·2

11 months ago, the rose bush was shorter than 60 cm. So, the rose bush has been over 60 cm tall for the last 10 months.

Share

WAYS OF WORKING Whole class teacher led

ASK

- Question 1 a): *What do the jumps show? How are the number line, table and Discover picture linked?*
- Question 1 a): *What other sequences have you encountered?*
- Question 1 b): *Which way do you count on the number line? Does the size of the jumps change when you count backwards?*

IN FOCUS For questions 1 a) and b), it will be important for children to recognise how adding and subtracting decimals is essential when completing the number sequences. The importance of using a number line and being systematic by organising the data in a table is made explicit. The use of jumps forwards and backwards highlights the fact that the inverse of addition is subtraction and vice versa.

165

PUPIL TEXTBOOK 5C PAGE 128

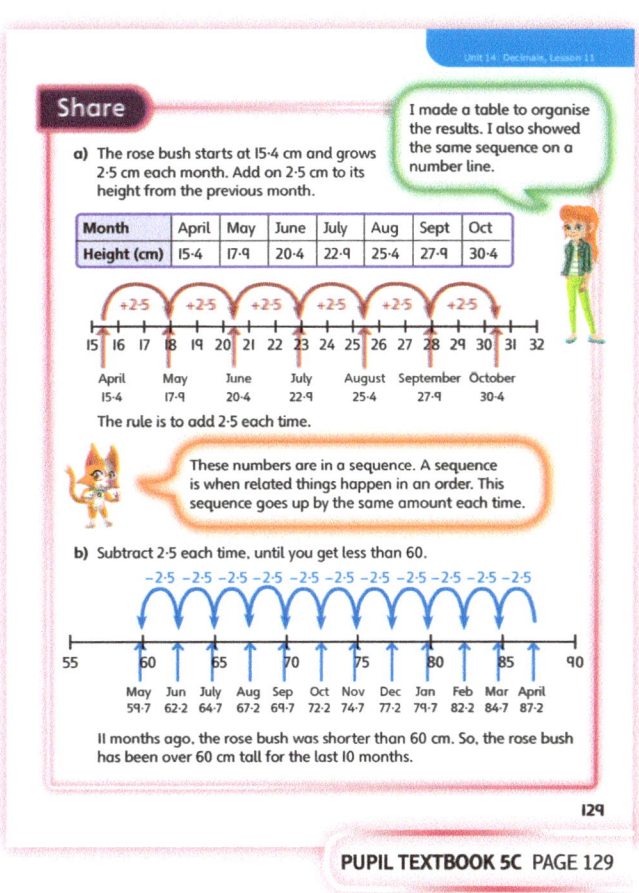

PUPIL TEXTBOOK 5C PAGE 129

Think together

WAYS OF WORKING Whole class teacher led (I do, We do, You do)

ASK

- Question ❶: *How does the table help show the sequence clearly?*
- Question ❶: *What do you need to know to find the missing values for each sequence?*
- Question ❷: *How will you know what number the sequences count in?*
- Question ❷: *Will you add or subtract to find the missing numbers?*
- Question ❸: *Why do you need to put the decimal cards in order?*

IN FOCUS Question ❶ requires children to extract the necessary information from the table to find the next term. Children need to use the clues given in the partially completed sequences to find the rules. Discuss with children how to go about finding each rule first, using the correct language such as 'the tenths increase by' to describe it.

STRENGTHEN To help children to complete the sequences in question ❷, ask: *How many equal intervals are there between one number and the next? If there are five equal intervals, what decimal would each interval represent?*

DEEPEN Question ❷ c) deepens children's fluency in recognising sequences of decimals on a number line. To extend further, ask children to reverse the numbers, so the first number in the sequence is 8 and the last is 7. Ask: *Will the rule change now? Will the size of the jump be different?*

ASSESSMENT CHECKPOINT Children are able to use tables and number lines to represent a number sequence. They can describe the rule that decimals follow, predict the next term and complete the missing terms in a given sequence.

ANSWERS

Question ❶:

	April	May	June	July	Aug	Sept	Oct
White rose	15·1	15·2	15·3	15·4	15·5	15·6	15·7
Climbing rose	10·0	12·6	15·2	17·8	20·4	23	25·6
Wild rose	12·429	12·43	12·431	12·432	12·433	12·434	12·435

Question ❷ a): 20·5, 20·75, 21, 21·25, 21·5, 21·75, 22

Question ❷ b): 0·65, 0·68, 0·71, 0·74, 0·77, 0·8, 0·83, 0·86

Question ❷ c): 7·0, 7·2, 7·4, 7·6, 7·8, 8·0

Question ❸ a): The sequence has a difference of 0·1 between each number. The cards that are covered up are 3·6 and 3·8.

Question ❸ b): The sequence increases by 3·1 each time. The given numbers, in order, are 29·4, 32·5, 35·6, 38·7 and 41·8. The missing numbers could be 23·2 and 26·3, 26·3 and 44·9, or 44·9 and 48·0.
51·1 is the first number in the sequence above 50.

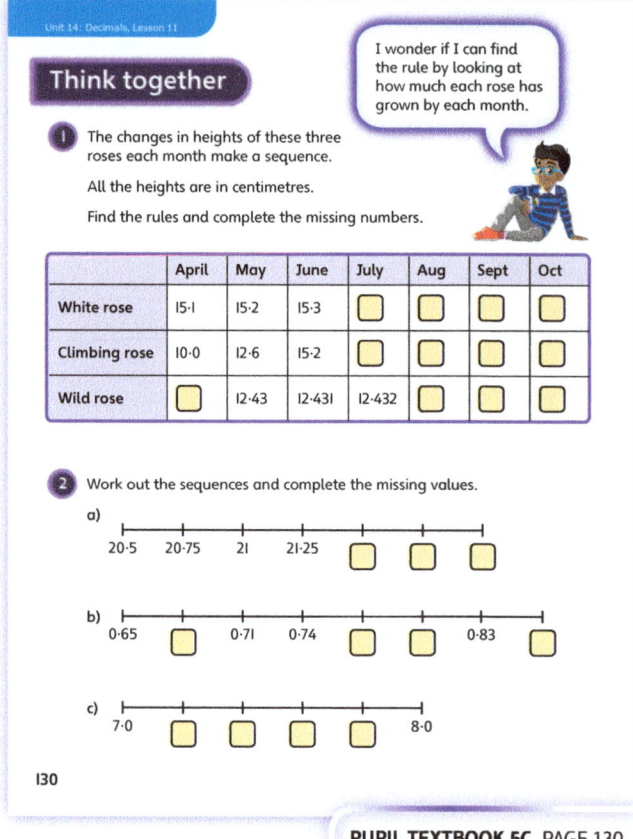

PUPIL TEXTBOOK 5C PAGE 130

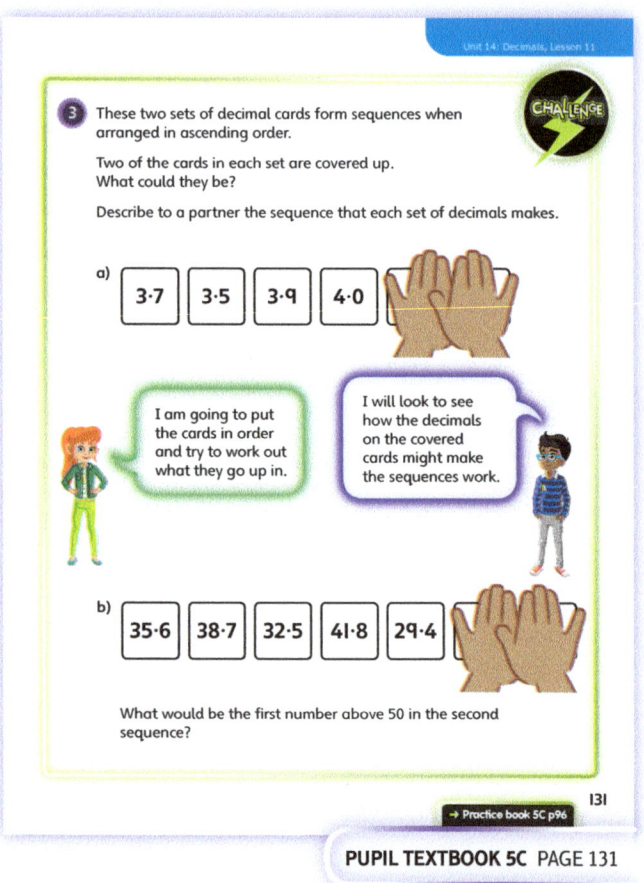

PUPIL TEXTBOOK 5C PAGE 131

Practice

WAYS OF WORKING Independent thinking

IN FOCUS Questions ❶ and ❷ present number sequences with missing numbers in different positions. This is important as it not only develops children's fluency in calculating the term-to-term rule and finding the next or previous term, but also makes them consider which operation they need to use. Children should be able to decide whether they add or subtract in each instance.

STRENGTHEN For question ❺, children may find it difficult to add multiples of 0·2, particularly when crossing 0: for example, 12·85 + 0·2. Ensure that children have access to number lines and place value counters. Look for children who say 12·87 and ask them to use a concrete resource to model the sequence. By this point, children should be able to independently choose a number line to demonstrate a sequence. Write examples of mistakes on the white board and invite the class to discuss the calculations and correct the mistakes.

DEEPEN Extend the context of question ❼ by asking children to explore how many rounds it would take Toshi to run a total of 30 m. Alternatively, change the distance of 0·4 m to 0·48 m and ask children to describe how the total distance travelled changes from one round to the next.

ASSESSMENT CHECKPOINT Are children confident in recognising and understanding how a number sequence can be represented? Are children able to complete the missing numbers in the sequences? Can they recognise the rule a number sequence is following?

ANSWERS Answers for the **Practice** part of the lesson can be found in the *Power Maths* online subscription.

Reflect

WAYS OF WORKING Pair work

IN FOCUS This question offers a good opportunity to assess children's fluency with the concepts covered in both this and previous lessons. Encourage children to consider using multiple ways of representing the number sequence they have made, including a number line and a table. Ask children to use the correct mathematical vocabulary when describing the rule their sequence follows.

ASSESSMENT CHECKPOINT Are there children who use mixed whole and decimal numbers, numbers with more than one decimal place or pictures to represent their number sequence? Do they simplify the decimal numbers in their sequences, for instance 0·6, 0·8, 1, 1·2? Children should begin to recognise that there is a limitless number of ways to represent their number sequence.

ANSWERS Answers for the **Reflect** part of the lesson can be found in the *Power Maths* online subscription.

After the lesson

- Do children understand what a decimal sequence is?
- What concrete resources would have supported the learning in this lesson?

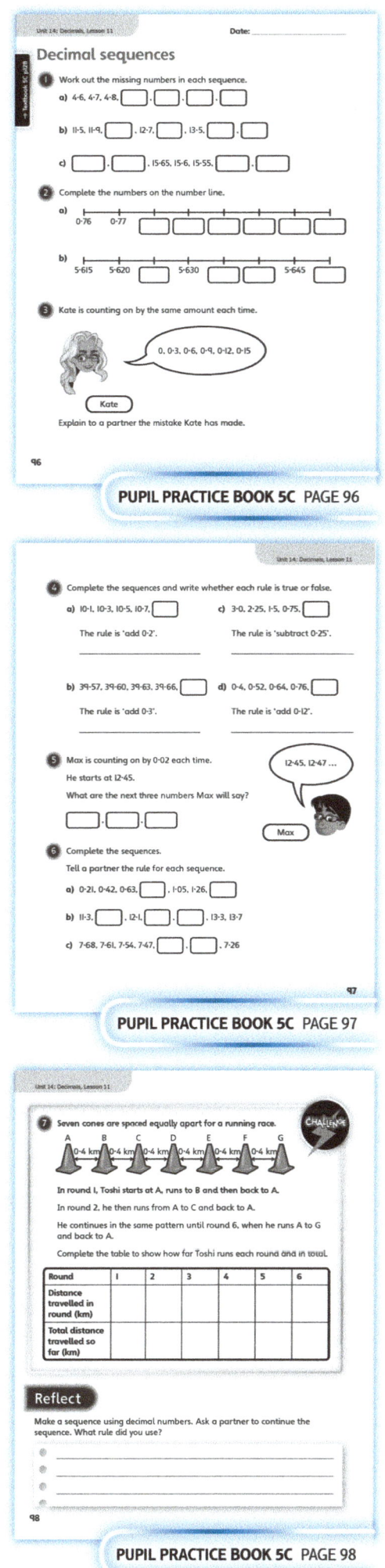

167

Unit 14: Decimals, Lesson 12

Multiply by 10

Learning focus
In this lesson, children will use their understanding of place value to develop fluency in multiplying decimals by 10.

Before you teach
- Check that children understand the concept of multiplication.
- Check that children can show a multiplication on a place value grid.

NATIONAL CURRICULUM LINKS

Year 5 Number – fractions (including decimals and percentages)

Recognise and use thousandths and relate them to tenths, hundredths and decimal equivalents.

Solve problems involving number up to three decimal places.

ASSESSING MASTERY

Children can confidently and fluently multiply any integer or decimal (to three decimal places) by 10. They can explain what happens to the digits in the numbers.

COMMON MISCONCEPTIONS

Children may think that to multiply by 10, they just add a zero to the end of the number. This is a particularly common misconception when multiplying decimals. For example, they may have noticed that when they multiply integers by 10, it appears a 0 is added to the end of a number, so they think that 1·7 multiplied by 10 is 1·70. Ask:

- *When you multiply by 10, what happens to the digits in the number?*

Use a place value grid to explain that they shift to the left by 1 digit. Explain that this is why it looks like you put a zero on the end as the zero is a placeholder. Show that when you multiply 1·7 by 10, it becomes 17.

STRENGTHENING UNDERSTANDING

In order to support children's understanding, use counters (or base 10 equipment) on a place value grid. Ask children to make a decimal number: for example, 15·6. Discuss the values of each digit and support children to help them realise that they have 10 tens, 50 ones and 60 tenths. This should be done step by step. Encourage children to notice that when multiplying 15·6 by 10, they can exchange this for 1 hundred, 5 tens and 6 ones, making the number 156.

GOING DEEPER

Ask children to investigate a Gattegno chart and its use. Ask them to use it to multiply 15, 178, 1·8 and 0·59 by 10. What do they notice?

KEY LANGUAGE

In lesson: multiply, place value, digits, 100, 10, 1, tenths, hundredths, thousandths

STRUCTURES AND REPRESENTATIONS

Place value grids

RESOURCES

Mandatory: place value grids, counters

Optional: base 10 equipment, Gattegno charts

 In the eTextbook of this lesson, you will find interactive links to a selection of teaching tools.

Quick recap

Ask children what they remember about multiplying by 10. Ask them to multiply integers by 10. Start with numbers such as 7, 9, 15, 178, etc.; and then move on to numbers such as 120, 300, 1,005, etc. Ask: *What patterns do you notice?*

Discover

WAYS OF WORKING Pair work

ASK

- Question 1 a): *Do you think Aki is right? Why might he think the way he does? What mistake has he made?*
- Question 1 b): *What do 0·1 counters look like? What can you do with these counters?*

IN FOCUS In question 1 a), ask children to think about the mistake that Aki has made. Encourage children to think about why Aki has made his mistake by asking them to multiply simple integers by 10 and to see what happens. In question 1 b), children use a place value grid and show how they can exchange ten 0·1 counters for one 1s counter.

PRACTICAL TIPS Ensure that place value counters and grids are available for children to use.

ANSWERS

Question 1 a): Aki is using a method that works for whole numbers (placing a zero) but it does not work for decimals.

Question 1 b): 10 × 0·1 = 1

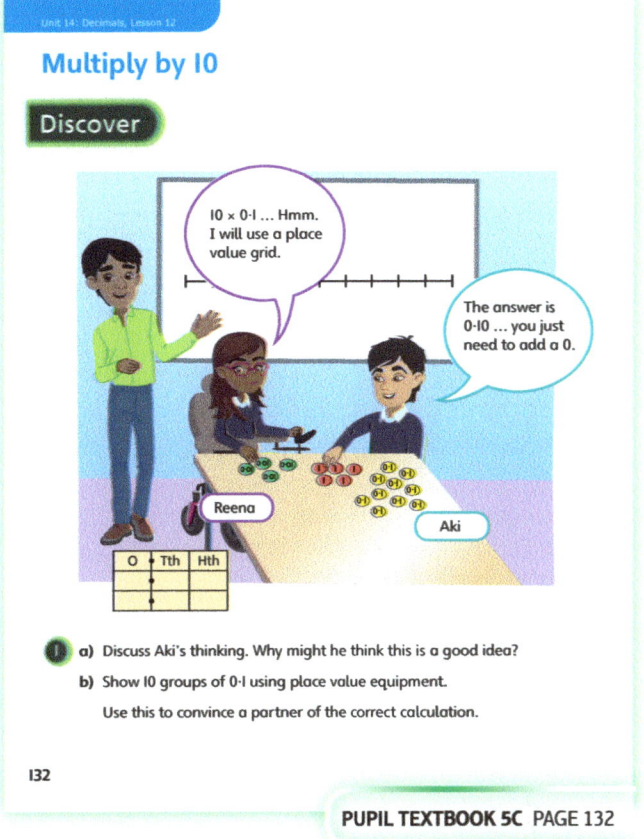

PUPIL TEXTBOOK 5C PAGE 132

Share

WAYS OF WORKING Whole class teacher led

ASK

- Question 1 a): *What do you notice when you multiply a positive whole number or integer by 10? Why might Aki think the answer is 0·10? Why is this wrong?*
- Question 1 b): *Which column do the 0·1 counters go in? What happens when you have 10 of these counters? What can you exchange them for?*

IN FOCUS In question 1 a), discuss the examples in which you multiply a whole number or integer by 10. Children may notice the pattern of the zero being 'added'. Discuss why this happens and how the digits are essentially moving to the left by a place. You may want to show this on a place value grid or Gattegno chart, depending on children's prior confidence with multiplying by 10. For question 1 b), ask children to work through the example, where 10 counters are placed in the tenths columns and can be exchanged for one 1. Use Dexter's comment to discuss with children how 1 is ten times the size of 0·1, linking this to the counters in the place value chart.

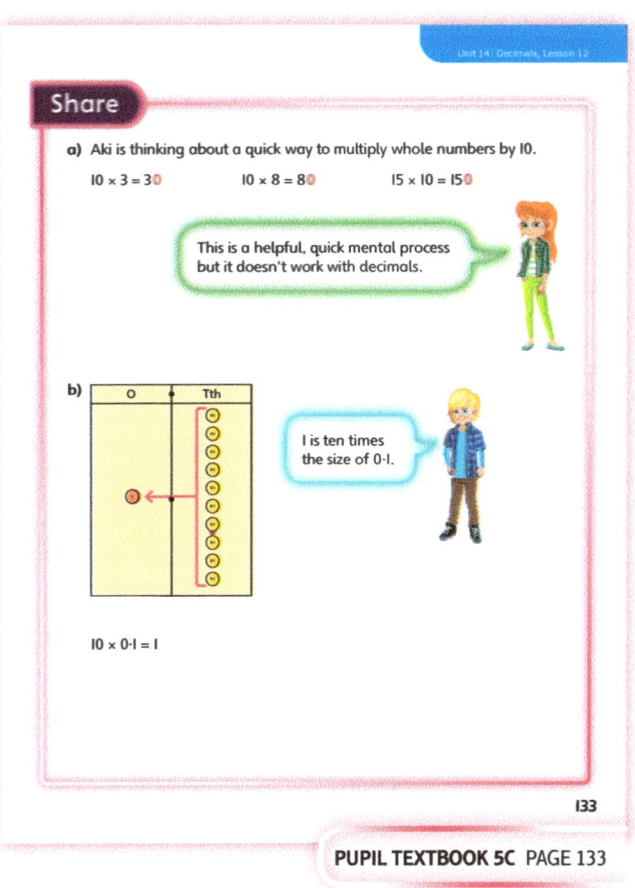

PUPIL TEXTBOOK 5C PAGE 133

Think together

WAYS OF WORKING Whole class teacher led (I do, We do, You do)

ASK

- Question ① a): *How do the place value grids show the multiplication? How can you work out the answers by exchanging?*
- Question ① b): *What do you notice happens to the digits? Is this the same as what happened before?*
- Question ②: *What patterns do you notice?*
- Question ③: *How can you work out the missing numbers? What do you do differently this time?*

IN FOCUS In question ① a), children use place value grids to help support their understanding of multiplying decimals by 10. Encourage them to use counters to make the numbers themselves and physically make the exchange. Ask them to discuss what happens to the digits of the numbers. In question ① b), children use their knowledge of part a) to work out the answers by considering what happens to the digits. The numbers are presented in a place value grid for support. In question ②, children should be more confident with finding the answer mentally. Some of the multiplications in question ③ require children to think about what number must have been multiplied by 10 to give the answer, prompting them to think about doing the opposite.

STRENGTHEN For questions ② and ③, support children by putting numbers into place value grids or using counters. Remind children what happens to the digits when you multiply a whole number by 10 and explain the same happens when you multiply a decimal by 10.

DEEPEN Ask children questions such as 12·6 × 10 × 10. Discuss what happens when they repeatedly multiply decimals by 10. Can they link this to multiplying by 10, 100 and 1,000?

ASSESSMENT CHECKPOINT Use questions ① and ② to assess whether children can multiply decimals by 10 confidently and whether they can explain the process.

ANSWERS

Question ① a): i) 0·14 × 10 = 1·4   ii) 2·3 × 10 = 23

Question ① b): i) 3·7 × 10 = 37   iii) 2·39 × 10 = 23·9
               ii) 4·5 × 10 = 45   iv) 0·196 × 10 = 1·96

Question ② a): 0·1 × 10 = 1
              1·2 × 10 = 12
              5·7 × 10 = 57
              19·1 × 10 = 191

Question ② b): 0·72 × 10 = 7·2   c): 0·256 × 10 = 2·56
              1·25 × 10 = 12·5   1·256 × 10 = 12·56
              5·71 × 10 = 57·1   31·126 × 10 = 311·26
              19·16 × 10 = 191·6

Question ② d): The digits stay the same and in the same order but their place value increases 10 times.

Question ③ a): 10 × 3·9 = 39   d): 1·262 × 10 = 12·62
             b): 10 × 11·6 = 116   e): 0·32 × 10 = 3·2
             c): 0·456 × 10 = 4·56   f): 1·586 × 10 = 15·86

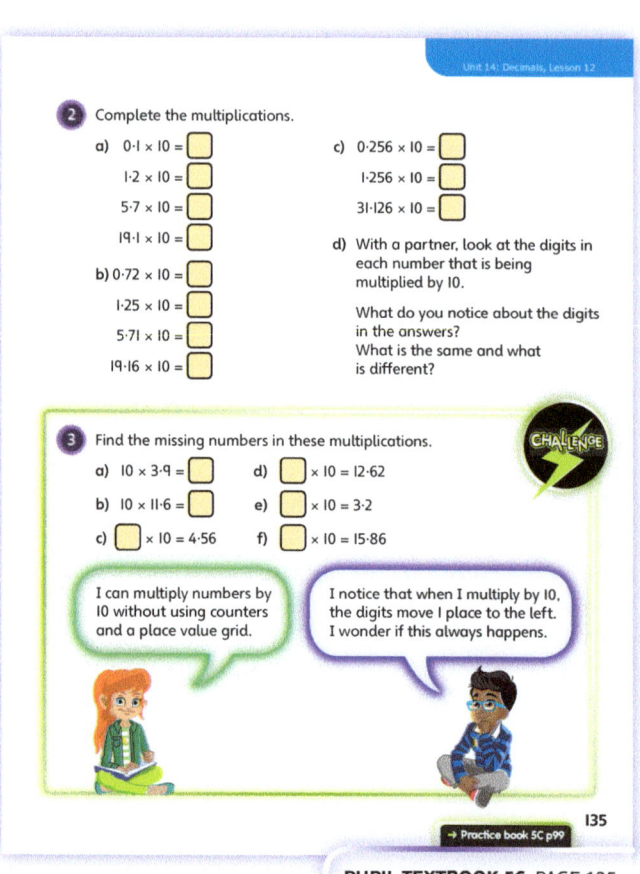

PUPIL TEXTBOOK 5C PAGE 134

PUPIL TEXTBOOK 5C PAGE 135

Unit 14: Decimals, Lesson 12

Practice

WAYS OF WORKING Independent thinking

IN FOCUS This **Practice** follows a very similar structure to the main lesson. In question ①, children are presented with counters on a place value grid to help them multiply by 10. The structure will help them see the exchanges. By this stage, some children may be confident with multiplying by 10 and don't need the supporting structures to help them. In question ②, children are given place value grids to help them understand that the digits shift one place to the left. Children write their answers in the place value grid. Question ③ addresses a common misconception where some children think they just need to append a zero to a decimal when they multiply by 10. Questions ④ and ⑤ provide practice at multiplying decimals by 10. Encourage children to take their time and think about each answer before writing it down.

STRENGTHEN For questions ②, ④ and ⑤, support children by putting numbers into place value grids or using counters. Remind children what happens to the digits when you multiply a whole number by 10 and explain that the same happens when you multiply a decimal by 10.

DEEPEN Challenge children to solve similar questions to question ⑦, where they have to multiply by 10, as well as apply other operations, to solve a problem in a context.

ASSESSMENT CHECKPOINT Use questions ②, ④ and ⑤ to assess whether children can multiply decimals by 10 confidently and whether they can explain what happens.

ANSWERS Answers for the **Practice** part of the lesson can be found in the *Power Maths* online subscription.

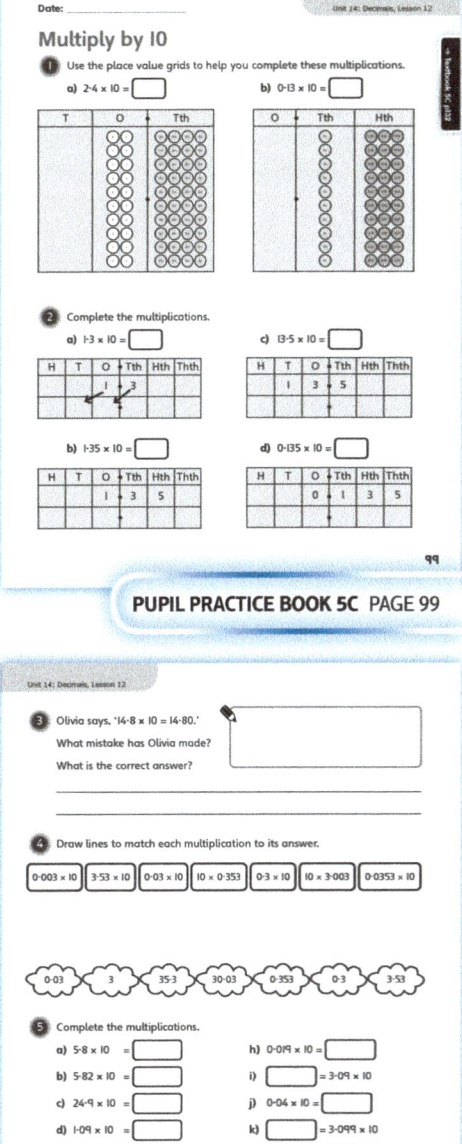

PUPIL PRACTICE BOOK 5C PAGE 99

PUPIL PRACTICE BOOK 5C PAGE 100

Reflect

WAYS OF WORKING Independent thinking

IN FOCUS Children discuss among themselves what happens when they multiply a number by 10. They should use language used in the lesson in support. Children may also want to use a place value grid to explain their understanding.

ASSESSMENT CHECKPOINT Ensure children use the correct language and are not just saying phrases such as 'add a zero'. They should be able to say that the digits shift one place to the left, and understand that they need to put 0 placeholders in at times.

ANSWERS Answers for the **Reflect** part of the lesson can be found in the *Power Maths* online subscription.

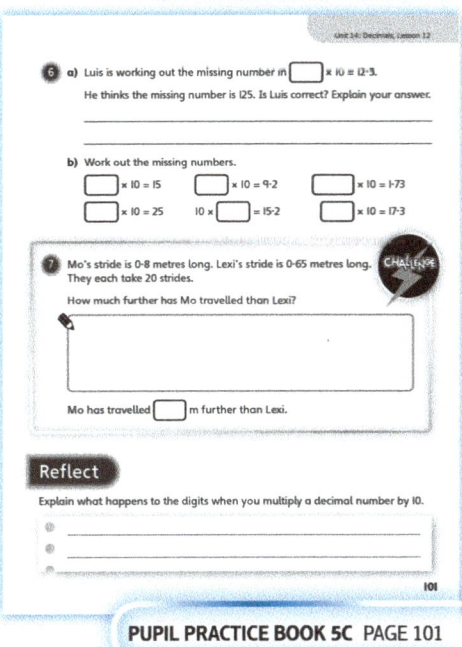

PUPIL PRACTICE BOOK 5C PAGE 101

After the lesson ⏸

- Can children explain what happens to the digits of a number when multiplying by 10?
- Can children multiply a decimal with up to three decimal places by 10?

171

Unit 14: Decimals, Lesson 13

Multiply by 10, 100 and 1,000

Learning focus
In this lesson, children will use their understanding of place value to develop fluency in multiplying decimals by 10, 100 and 1,000.

Before you teach
- How confident are children at multiplying decimals by 10?

NATIONAL CURRICULUM LINKS

Year 5 Number – fractions (including decimals and percentages)

Recognise and use thousandths and relate them to tenths, hundredths and decimal equivalents.

Solve problems involving number up to three decimal places.

ASSESSING MASTERY

Children can reliably and fluently multiply decimal numbers by 10, 100 and 1,000. They can link their understanding of place value to their calculations and confidently represent their thinking using concrete, pictorial and abstract representations.

COMMON MISCONCEPTIONS

Children may get confused about which way the digits 'move' when multiplying by 10, 100 and 1,000 and, instead of multiplying by 10, 100 or 1,000, actually divide by 10, 100 or 1,000. For example, they may say that 4·32 × 100 = 0·0432 rather than 432. Ask:
- *Can you show what you mean by 'moving the digits'? Do numbers get bigger or smaller when multiplied by 100?*

STRENGTHENING UNDERSTANDING

Using a place value grid and place value counters will help children better visualise what happens when a decimal number is multiplied by 10, 100 or 1,000.

GOING DEEPER

Children could create contextual word problems that require someone to multiply by 10, 100 or 1,000. Using real-life situations will help children to see more clearly what happens to each decimal when it becomes 10, 100 or 1,000 times bigger.

KEY LANGUAGE

In lesson: multiply, place value, weight, digit, column, kilogram (kg), mass

Other language to be used by the teacher: tenths, hundredths, thousandths, 1s, 10s, 100s, 1,000s, 10,000s

STRUCTURES AND REPRESENTATIONS

Place value grid

RESOURCES

Mandatory: base 10 equipment, place value counters

Optional: printed place value grids, building blocks, cubes

 In the eTextbook of this lesson, you will find interactive links to a selection of teaching tools.

Quick recap

Check that children can multiply integers by 10. Challenge children to multiply numbers by 20, 30 or another multiple of 10. Keep numbers low as you are checking their understanding of the method rather than focusing on the complexity of the calculation.

Unit 14: Decimals, Lesson 13

Discover

WAYS OF WORKING Pair work

ASK

- Question 1 a): *How many bags are there in each sack?*
- Question 1 a): *What is a pallet? How many sacks are on a pallet?*
- Question 1 a): *How many pallets are on each lorry? How can you then work out how many bags are on a lorry?*
- Question 1 b): *How can you use your answers to 1 a) to work out the different answers here?*

IN FOCUS Questions 1 a) and b) will give children their first opportunity to multiply decimals by 10, 100 and 1,000. Children will use a place value grid and the associative property of multiplication (100 = 10 × 10 and 1,000 = 10 × 10 × 10) to work out each multiplication.

PRACTICAL TIPS Use place value counters, building blocks or cubes to act out the question. For example, to represent 2·5 kg you could make a tower with 2 yellow blocks to show the 1s, and 5 blue blocks to show the tenths. Put 10 of the towers into a box or bag. Show children 10 boxes or bags. By seeing 10 groups of 10 towers, children will visualise the question better and understand that multiplying by 100 is the same as multiplying by 10 and 10.

ANSWERS

Question 1 a): There are 1,000 bags on the lorry.

Question 1 b): The mass of all the potatoes on the lorry is 2,500 kg.

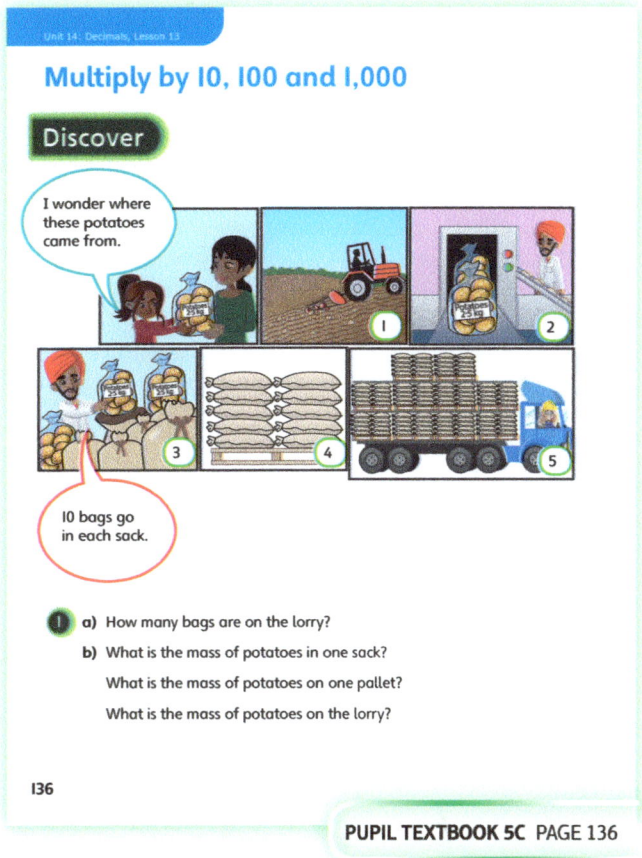

PUPIL TEXTBOOK 5C PAGE 136

Share

WAYS OF WORKING Whole class teacher led

ASK

- Question 1 a): *What do the images show? Can you see where the 100 sacks come from? Why are there 1,000 bags on each pallet?*
- Question 1 b): *What do the arrows show? How do the numbers change and stay the same as you multiply by 10 and 100? Why are the zeros important?*

IN FOCUS It will be important in questions 1 a) and b) to ensure children recognise that, when multiplying, each digit becomes 10, 100 or 1,000 times bigger. This should be supported with the use of concrete resources so children can manipulate and experience the difference between multiplying by 10 once, twice or three times.

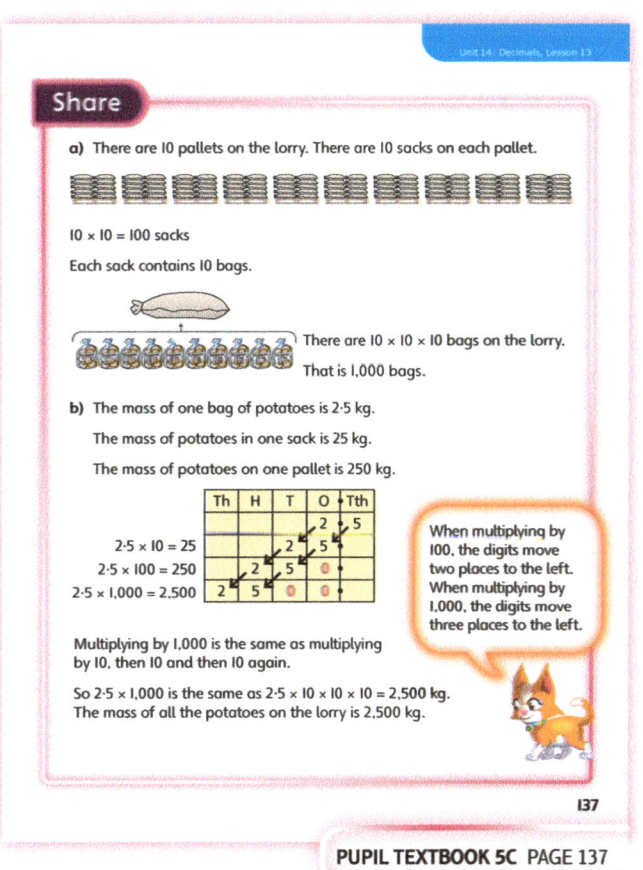

PUPIL TEXTBOOK 5C PAGE 137

173

Think together

WAYS OF WORKING Whole class teacher led (I do, We do, You do)

ASK
- Question ❶: *What do the arrows show?*
- Question ❷: *What happens when you repeatedly multiply by 10? How can you multiply by 100 and 1,000 without using the place value grid? Does this work for all decimal numbers?*
- Question ❸: *What is the chart called? What numbers are shown? What happens when you multiply by 10, 100 and 1,000? Where are the numbers?*

IN FOCUS Question ❶ moves to a more abstract way of recording a multiplication by showing it using a partially completed place value grid. Encourage children to recognise how a product is similar and different when the same number is multiplied by 10, 100 and 1,000. Question ❷ encourages children to generalise. Children should notice the links between multiplying the same decimal number by 100 and 1,000. They will also notice what happens when numbers with one, two and three decimal places are multiplied by 100 and 1,000.

STRENGTHEN If children are finding it difficult to generalise in question ❷, it may help them to build the calculations using concrete representations. Draw a 3 × 2 grid on the board. Write some different lengths in metres: for example, 0·12 m, 1·35 m, 0·45 m. Ask children to copy the grid but write each length in cm instead of m: 12 cm, 135 cm, 45 cm. Give children a metre ruler to check their answers.

DEEPEN Deepen children's fluency with multiplying by 10, 100 and 1,000 by asking them to make up problems with a measurement context for a partner to solve. Ask: *What information is given? How can you answer this question?*

ASSESSMENT CHECKPOINT Can children multiply decimal numbers by 10, 100 and 1,000? Can they use place value grids or Gattegno charts to show this? Evaluate children's answers to question ❸ as an indication of their understanding.

ANSWERS

Question ❶: 3·7 × 10 = 37
3·7 × 100 = 370
3·7 × 1,000 = 3,700

Question ❷ a): 1·72 × 10 = 17·2
1·72 × 100 = 172
1·72 × 1,000 = 1,720

Question ❷ b): 4·13 × 1,000 = 4,130   c) 39·3 × 100 = 3,930
0·413 × 1,000 = 413   3·93 × 100 = 393
0·041 × 1,000 = 41   0·393 × 100 = 39·3

Question ❸ a): 0·8 × 10 = 8
0·4 × 100 = 40
0·2 × 1,000 = 200

Question ❸ b): Move each digit up two rows.
1,000 + 20 + 40 + 7
12·47 × 100 = 124·7

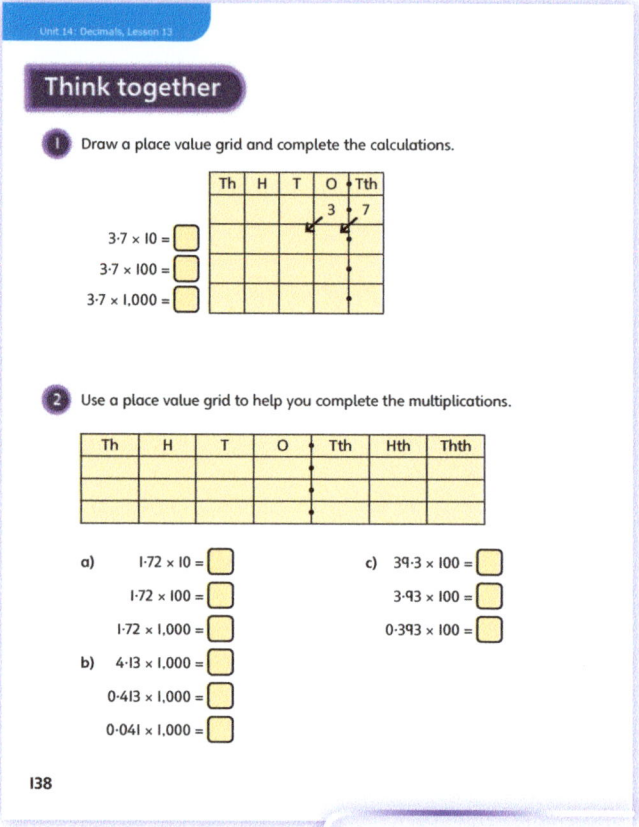

PUPIL TEXTBOOK 5C PAGE 138

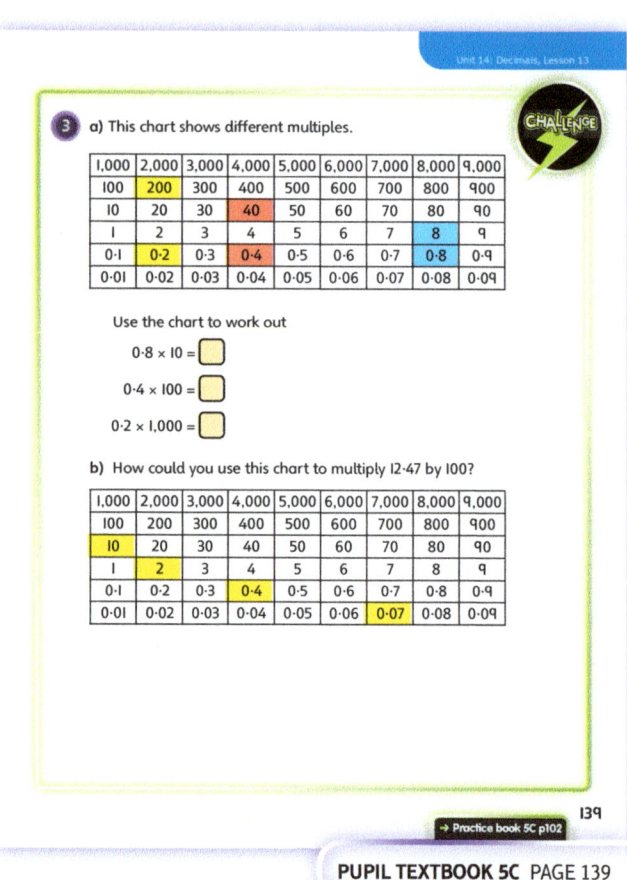

PUPIL TEXTBOOK 5C PAGE 139

Unit 14: Decimals, Lesson 13

Practice

WAYS OF WORKING Independent thinking

IN FOCUS Question ❶ develops children's ability to multiply by 10, 100 and 1,000 while scaffolding their learning with a place value grid. In question ❷, the scaffolding is removed and children are required to find the missing numbers when decimals are multiplied by 100 and 1,000. Question ❸ develops children's fluency and reasoning, by presenting them with a real-life context.

STRENGTHEN For question ❹, encourage children to build the calculations first with concrete materials, and then to look for patterns to help them complete the rest. Ask: *Can you complete the first calculation? How about the second? Can you spot any patterns in the numbers that will help you complete the next two calculations? How can you apply what you have noticed to the next column?*

DEEPEN Deepen the learning from question ❺ by asking children to find three ways to complete ☐ × ☐ > ☐ × ☐. Ask: *Can you convince a partner that you are correct? What resources can you use to support your answer?*

ASSESSMENT CHECKPOINT Children use their understanding of place value and the associative properties of multiplication to multiply decimal numbers by 10, 100 and 1,000. They understand that multiplying by 100 means they need to multiply by 10 and 10 again, whilst multiplying by 1,000 means that they need to multiply by 10, 10 and 10 again.

ANSWERS Answers for the **Practice** part of the lesson can be found in the *Power Maths* online subscription.

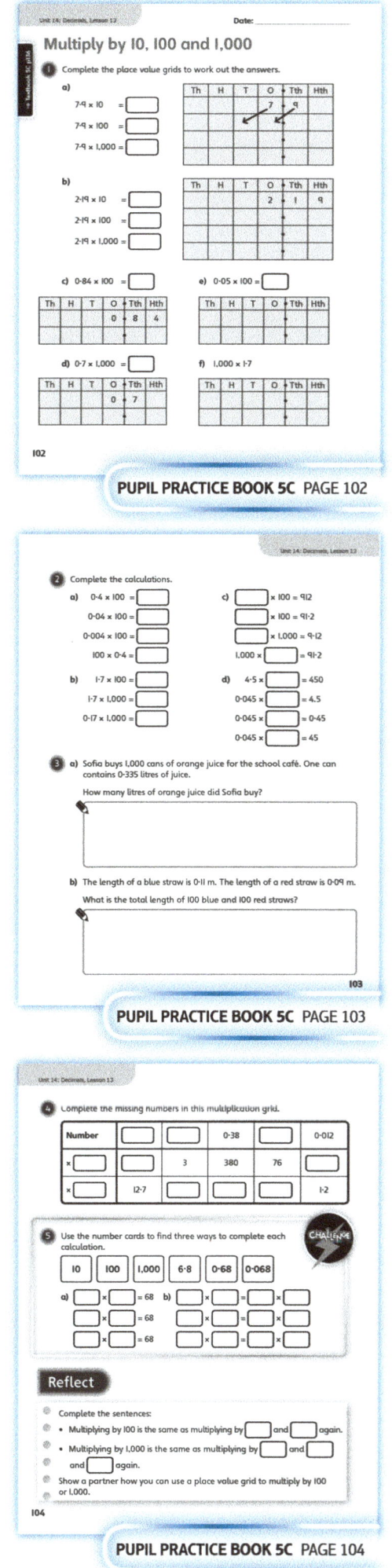

PUPIL PRACTICE BOOK 5C PAGE 102

PUPIL PRACTICE BOOK 5C PAGE 103

PUPIL PRACTICE BOOK 5C PAGE 104

Reflect

WAYS OF WORKING Pair work

IN FOCUS This question provides an opportunity for children to combine their learning from the previous lessons with their learning from this lesson. Children do not simply repeat rules and procedures but are able to use a place value grid to demonstrate what happens when a number is multiplied by 100 and 1,000. Children should recognise with confidence the multiplication facts they have focused on in this lesson and be able to use them to multiply by 10, 100 and 1,000. This also provides an opportunity for children to relate their understanding to a partner in their own words.

ASSESSMENT CHECKPOINT Look for fluency in children when they are recognising and interpreting the concrete representations of a number being multiplied by 10, 100 and 1,000. Children should be able to link the representation to an appropriate calculation confidently.

ANSWERS Answers for the **Reflect** part of the lesson can be found in the *Power Maths* online subscription.

After the lesson

- Did children recognise the usefulness of the properties of multiplication when multiplying by 10, 100 and 1,000?
- Have children mastered the concept of multiplying decimals by 10, 100 and 1,000? Are they able to multiply fluently?

Unit 14: Decimals, Lesson 14

Divide by 10

Learning focus
In this lesson, children will recap how to divide integers and digits by 10.

Before you teach
- Check that children understand the concept of sharing and grouping.
- Check that children know decimal equivalents of tenths and hundredths.
- Check that children can convert between km, m and cm.

NATIONAL CURRICULUM LINKS

Year 5 Number – multiplication and division

Multiply and divide whole numbers and those involving decimals by 10, 100 and 1,000.

Year 5 Number – fractions (including decimals and percentages)

Recognise and use thousandths and relate them to tenths, hundredths and decimal equivalents.

Solve problems involving number up to three decimal places.

ASSESSING MASTERY

Children can confidently and fluently divide any integer or decimal (to two decimal places) by 10. They can explain what happens to the digits in the numbers.

COMMON MISCONCEPTIONS

Children may link dividing by 10 as 'taking off' a zero from the end of the number. For example, they may think that they cannot divide 25 by 10 because 25 does not end in a zero. Use the concept of money to support children in their understanding. Explain you have £25 which you need to share between 10 people. Tell children how you can share the £20 first and each person would get £2. Then ask:
- How can you share the remaining £5 between 10 people? Discuss the fact that £5 = 500p and therefore each person gets 500p ÷ 10 = 50p more. So, in total, each person receives £2·50.

Explain how to introduce the decimal point. You may also wish to use counters to support the division.

STRENGTHENING UNDERSTANDING

In order to support children's understanding, use counters (or base 10 equipment) on a place value grid. Ask children to make a number: for example, 236. Discuss the values of the digits in the number and support children to realise that they can think about the number as 20 tens, 30 ones and 60 tenths. This should be done step by step. Encourage children to then notice, using grouping or sharing, that 236 divided by 10 is equal to 23·6. They can use a place value grid to help them understand what happens with the digits when dividing by 10.

GOING DEEPER

Ask children to try to use the Gattegno chart that was introduced in the previous lesson to work out 236, 12,718 and 23·8 divided by 10. Can they generalise how to use the chart for division?

KEY LANGUAGE

In lesson: divide, metres (m), centimetres (cm), kilograms, litres, digits

Other language to be used by the teacher: place value, 1, grouping, 10s, tenths, hundredths, thousandths

STRUCTURES AND REPRESENTATIONS

Place value charts

RESOURCES

Mandatory: place value grid, counters

Optional: base 10 equipment, place value grids, Gattegno charts

In the eTextbook of this lesson, you will find interactive links to a selection of teaching tools.

Quick recap

Ask children to write decimal equivalents to the fractions $\frac{1}{10}$ and $\frac{2}{10}$. Challenge children by providing them with different fractions and seeing how many decimal equivalents they can write. Can they write any above 1?

Discover

WAYS OF WORKING Pair work

ASK

- Question 1 a): *What is Lexi doing in the picture? How many hand spans has she made? How long are the hand spans in total? What is this in centimetres? How long is each one?*
- Question 1 b): *How can you make 0·9 on a place value grid? What happens when you divide by 10? What happens to the digits?*

IN FOCUS This lesson reminds children of the concept of dividing by 10. In question 1 a), discuss as a class what Lexi is doing and then allow children to discuss the different ways they could get the answer. The route many might follow is to convert to cm and then divide by 10. In question 1 b), encourage children to make the number 0·9 on a place value grid and then divide by 10; by exchanging 9 tenths for 90 hundredths, children get 0·09. By moving the counters, children may start to see what happens to the digits.

PRACTICAL TIPS You can replicate the hand span activity as a whole class.

ANSWERS

Question 1 a): Each hand span is 0·09 m wide.

Question 1 b): The digits move one place to the left when you divide by 10.

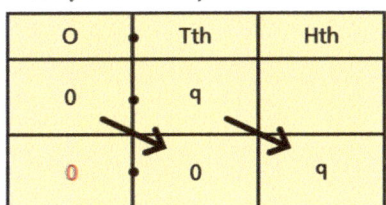

Share

WAYS OF WORKING Whole class teacher led

ASK

- Question 1 a): *How does the bar model represent the situation? How can you use your knowledge of multiplying by 10 to help you? What has Dexter done to work out the answer? Does anyone find this a simpler approach?*
- Question 1 b): *What happens to the digits in the place value chart when you divide by 10? Can you come up with a general rule?*

IN FOCUS This lesson should be a recap for most children. It builds on the previous lesson on multiplication. Children may already know that the digits of the number move one place to the right when you divide by 10. However, they may not have done this confidently with decimals until now. If children have struggled to understand this concept, you may want to use place value counters and grids and have them make 9 tenths and exchange them for 90 hundredths. This will show children why, when you divide 9 tenths by 10, you get 9 hundredths.

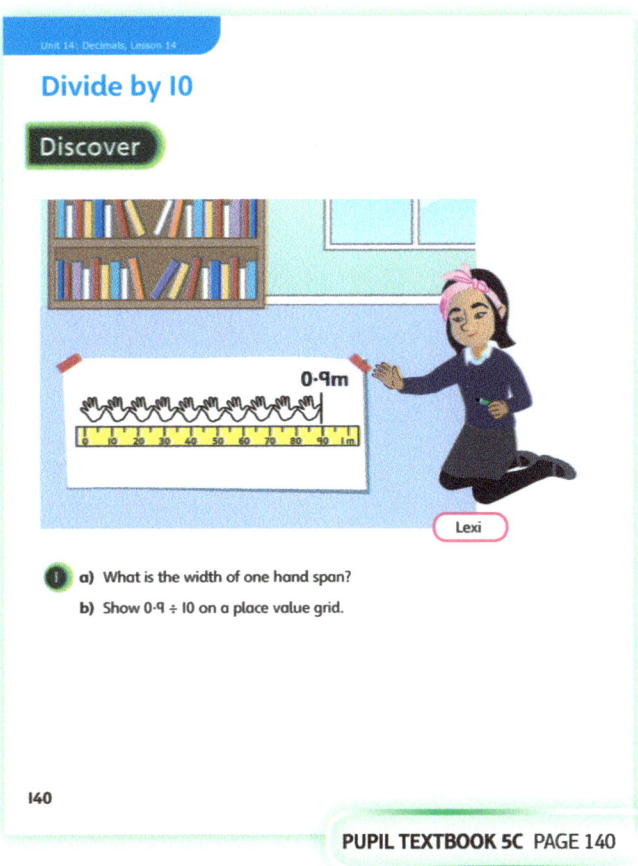

PUPIL TEXTBOOK 5C PAGE 140

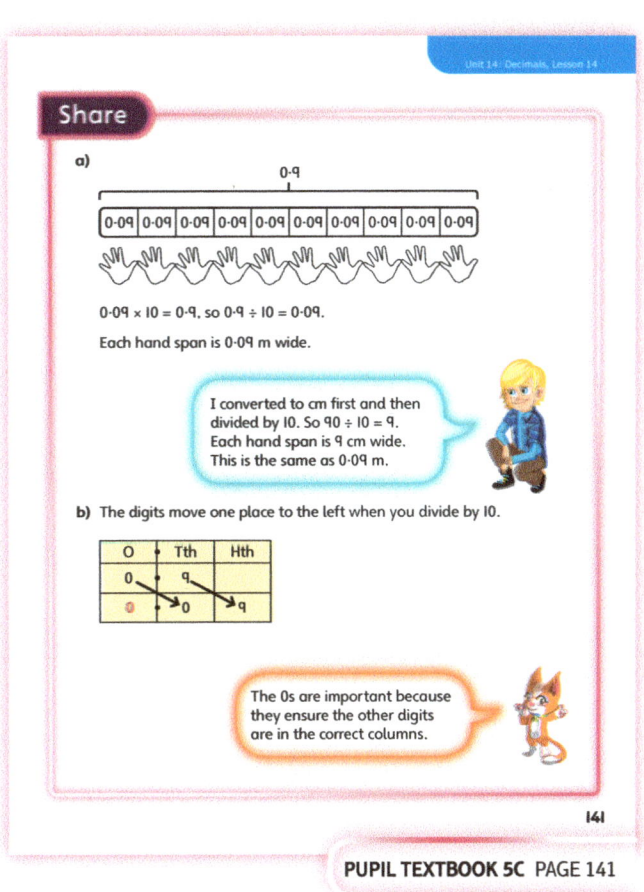

PUPIL TEXTBOOK 5C PAGE 141

Think together

WAYS OF WORKING Whole class teacher led (I do, We do, You do)

ASK

- Question ❶: *What is the same and what is different about this situation? How can you use the place value grid to help you work out the answer?*
- Question ❷: *How can the place value grid help you? What is different about parts d) and f)? How can you use your grid to help you with these parts?*
- Question ❸: *What different methods can you use to find the answers? What problem of your own could you make up?*

IN FOCUS Question ❶ provides an example, in context, of dividing by 10. It is similar to the question in **Discover**. Discuss with children how they can use the place value grid to see how the digits in the number move one place to the right. In question ❷, children are asked to work through six examples of dividing by 10. They may use the place value grid to help, but encourage children to visualise in their heads the movement of the digits, so they don't need to write it out each time. Children need to be careful with parts d) and f) as the answers have been provided and they have to work out the starting numbers. Children should consider how best to do this.

STRENGTHEN If children are struggling to grasp 'why' the digits move, use counters or base 10 equipment and a place value grid to help them. Have children make a number: for example, 2·6. Support them step by step to help them realise that, to divide by 10, they should think about the 2 as 20 tenths, and the 6 tenths as 60 hundredths. They should then notice, using grouping or sharing, that 2·6 divided by 10 is equal to 0·26. The use of the place value grid and counters will help children begin to understand what happens with the digits.

DEEPEN Ask children to extend their understanding by dividing by a multiple of 10. For example, they could work out 2·6 divided by 20 or 3·96 divided by 30.

ASSESSMENT CHECKPOINT Questions ❶ and ❷ will help you determine if children are confident with dividing a number by 10. Ask children more questions like this if they are unsure. Ideally children need to be moving away from using the place value grid as support.

ANSWERS

Question ❶: 0·26 m

Question ❷ a): 0·92 ÷ 10 = 0·092   d) 58·6 ÷ 10 = 5·86

Question ❷ b): 53·6 ÷ 10 = 5·36   e) 89·02 ÷ 10 = 8·902

Question ❷ c): 95 ÷ 10 = 9·5   f) 10·02 ÷ 10 = 1·002

Question ❸ a): 0·295 ml
(2·25 + 0·7) ÷ 10
Or 2·25 ÷ 10 + 0·7 ÷ 10

Question ❸ b): 100 ml costs 12p.
200 ml costs 24p.

Question ❸ c): 0·8 kg

Question ❸ d): Various responses are possible.

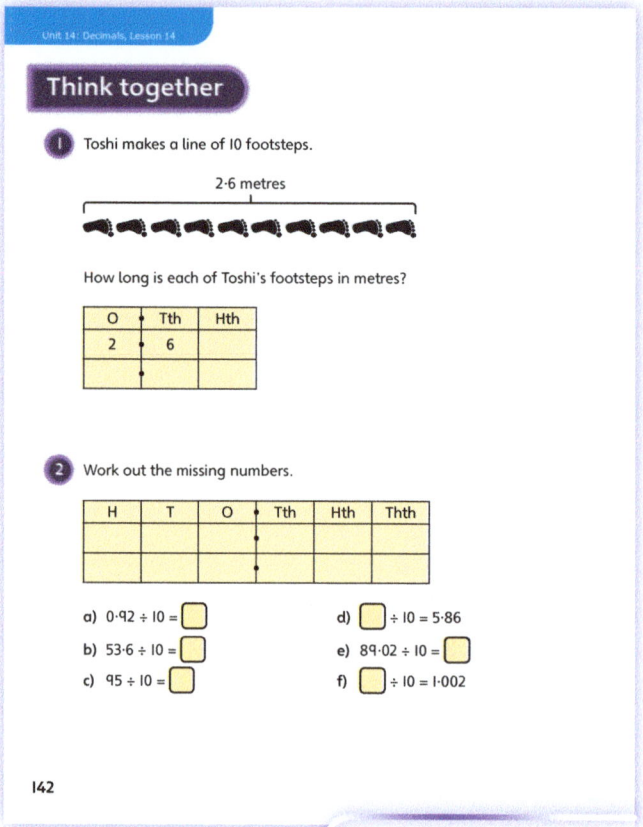

PUPIL TEXTBOOK 5C PAGE 142

PUPIL TEXTBOOK 5C PAGE 143

Unit 14: Decimals, Lesson 14

Practice

WAYS OF WORKING Independent thinking

IN FOCUS Question 1 provides an example of dividing by 10 using counters for support. Question 2 provides place value grids to help children divide by 10. Children should notice that the digits of the number move 1 place to the right. They will also see that they need to introduce 0 at times as a placeholder. Questions 3 to 5 provide more abstract practice, sometimes in context. By this stage, children should be able to attempt the questions without the need of support. If children still need to use a place value grid, that is fine, but encourage them to have a think about their answers first. Question 6 asks children to discuss a common misconception with money.

STRENGTHEN To support children with finding answers to the more abstract questions from question 3 onwards, provide children with a wipe-clean place value grid. They can then draw arrows to show how digits move. If children are relying too heavily on the place value grid, try to encourage them to visualise it in their heads and to have a go without the grid. Then get them to use the grid to check their answers.

DEEPEN Ask children to extend their understanding by dividing by a multiple of 10. For example, they could work out 2·6 divided by 20 or 3·96 divided by 30.

THINK DIFFERENTLY In question 7, children may want to convert units first. For example, children need to know that there are 10 lots of 100 ml in 1 litre.

ASSESSMENT CHECKPOINT Use the first five questions to check children's confidence with dividing by 10. Can they do so fluently and confidently without the support of place value grids?

ANSWERS Answers for the **Practice** part of the lesson can be found in the *Power Maths* online subscription.

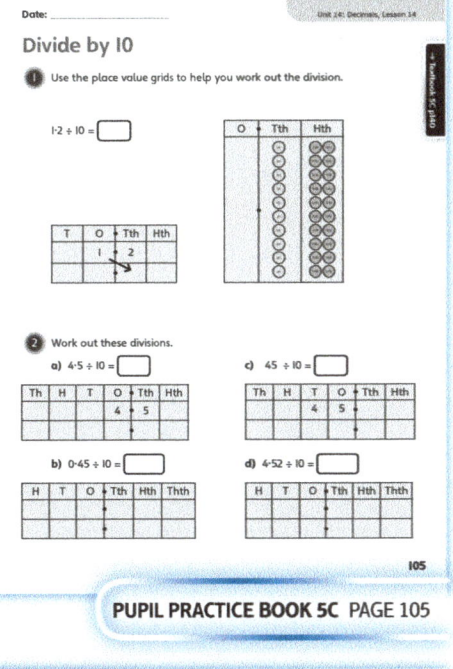

PUPIL PRACTICE BOOK 5C PAGE 105

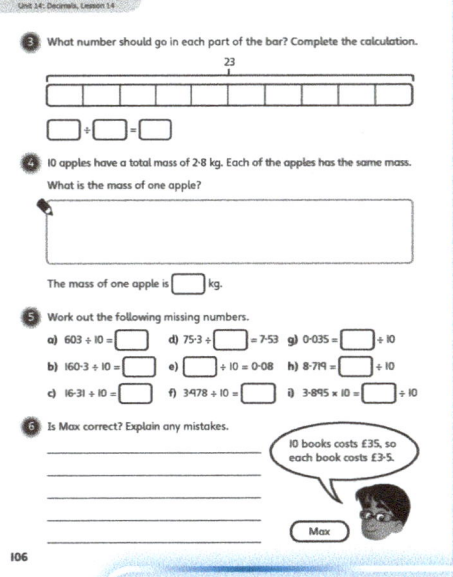

PUPIL PRACTICE BOOK 5C PAGE 106

Reflect

WAYS OF WORKING Pair work

IN FOCUS Children work in pairs to rehearse an explanation for the answer. Children may talk about the different methods that could be used. For some children, you may want them to explain why their method works.

ASSESSMENT CHECKPOINT Use their responses to check that children know that the digits move one place to the right when dividing by 10.

ANSWERS Answers for the **Reflect** part of the lesson can be found in the *Power Maths* online subscription.

After the lesson

- Are children able to divide an integer and decimal (with up to two decimal places) by 10?
- Can children explain that when dividing by 10 the digits move one place to the right?

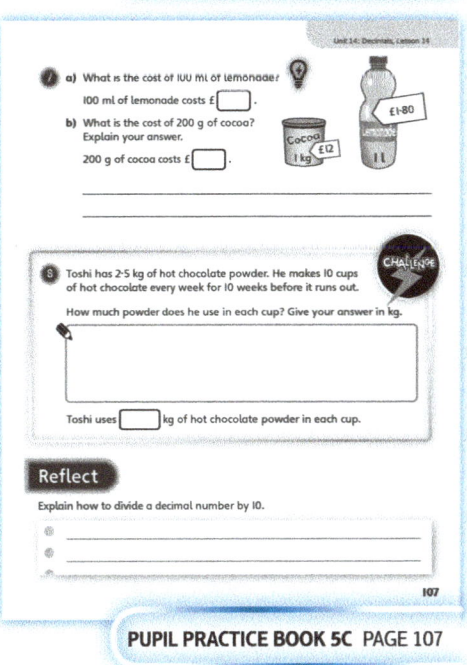

PUPIL PRACTICE BOOK 5C PAGE 107

Unit 14: Decimals, Lesson 15

Divide by 10, 100 and 1,000

Learning focus
In this lesson, children will use their understanding of place value and division of decimals by 10, to develop fluency in dividing decimal numbers by 10, 100 and 1,000.

Before you teach
- How confident were children with multiplying by 10, 100 and 1,000? Are there any misconceptions that need addressing first?
- How will you make clear the link between this lesson and the last?
- How can you strengthen children's understanding of the inverse relationship between multiplication and division?

NATIONAL CURRICULUM LINKS

Year 5 Number – fractions (including decimals and percentages)

Recognise and use thousandths and relate them to tenths, hundredths and decimal equivalents.

Solve problems involving number up to three decimal places.

ASSESSING MASTERY

Children can reliably and fluently use their knowledge of division by 10 to divide decimal numbers by 100 and 1,000. They can represent their thinking confidently using concrete, pictorial and abstract representations.

COMMON MISCONCEPTIONS

Children may link dividing by 10, 100 or 1,000 to always having a set number of zeros after the decimal point. For example, they know that 0·6 ÷ 10 = 0·06, and so assume that 1·6 ÷ 10 = 0·016. Ask:
- Can you show me what dividing by 10 would look like in a place value grid? Why is there a '0' in the answer of 0·6 ÷ 10 = 0·06?

STRENGTHENING UNDERSTANDING

Give children a strip of paper 16 cm long. Ask: *When 16 is divided by 10, is the answer 1·6 cm, 0·16 cm or 0·016 cm?* Discuss where the number 0·016 would be if the strip of paper was a number line. Children should establish that the lengths of 0·16 cm and 0·016 cm are too small to be a tenth of the length of the strip.

GOING DEEPER

Give children these statements to complete: 5 ÷ ☐ = 0·5 ÷ 10, 0·4 ÷ 100 = ☐ ÷ 1,000. Ask: *Is there only one answer?*

KEY LANGUAGE

In lesson: divide, place, share, mass, rule, equal, represent, kilograms (kg), litres (l), kilometres (km)

Other language to be used by the teacher: place value, 1s, grouping, 10s, tenth, hundredth, thousandth

STRUCTURES AND REPRESENTATIONS

Place value grid, bar model

RESOURCES

Mandatory: place value counters

Optional: printed place value grids, one large and ten small boxes

 In the eTextbook of this lesson, you will find interactive links to a selection of teaching tools.

Quick recap

Ask children to divide numbers by 10, 100 and 1,000 that give a whole number answer (for example, 90 ÷ 10, 1,200 ÷ 100). Then move on to more challenging calculations.

Discover

WAYS OF WORKING Pair work

ASK

- Question 1 a): *What is a sachet? What is a carton? How many sachets are there in a carton?*
- Question 1 a): *How many cartons are there in a large box? What calculation do you need to do to work out how many sachets are in a large box?*
- Question 1 b): *How can you work out the mass of each sachet?*

IN FOCUS For question 1 a), discuss the sizes and names of the containers in the picture (namely 'sachet', 'carton' and 'large box') so that children are comfortable with what each container is and what it is called. Some children may assume that the answer is 10 or be unsure what to do, as they simply look at the picture and not at the information it provides. Highlight the importance of looking carefully at the information and deciding how to use it. Question 1 b) gives children the first opportunity to understand the effect of dividing a decimal by 100, by dividing by 10 and 10 again.

PRACTICAL TIPS To help engage children, create a 'magic box'. Put ten smaller boxes inside a large 'magic box'. Each of the smaller boxes should hold ten counters (or sweets). Ask: *If the magic box has a mass of 1·2 kg, what is the mass of each small box?* This activity will give children a hands-on, contextualised experience to relate to the problem in the lesson and should help them to identify a strategy to find the solution.

ANSWERS

Question 1 a): There are 100 sachets of curry powder in the large box.

Question 1 b): There is 0·085 kg of curry powder in each sachet.

Share

WAYS OF WORKING Whole class teacher led

ASK

- Question 1 a): *What information do you need to know to find the number of sachets in the large box? What calculation do you need to do?*
- Question 1 b): *What calculation does the first bar model show? Explain why.*
- Question 1 b): *What calculation does the second bar model show? What does the place value grid show?*
- Question 1 b): *How is Flo's method of dividing by 100 the same or different to Astrid's? Whose is more efficient?*

IN FOCUS Make sure children understand the place value grids in question 1 b) and that they are confident in how to use them. Draw children's attention to the calculation of dividing by 10, then dividing by 10 again, and ask them to think of a single calculation that they could use instead. Show them Flo's method of dividing by 100. If children are unsure, provide examples with whole numbers, such as 700 ÷ 10 = 70 and 70 ÷ 10 = 7. Ask: *What number can you divide 700 by that gives the answer 7?*

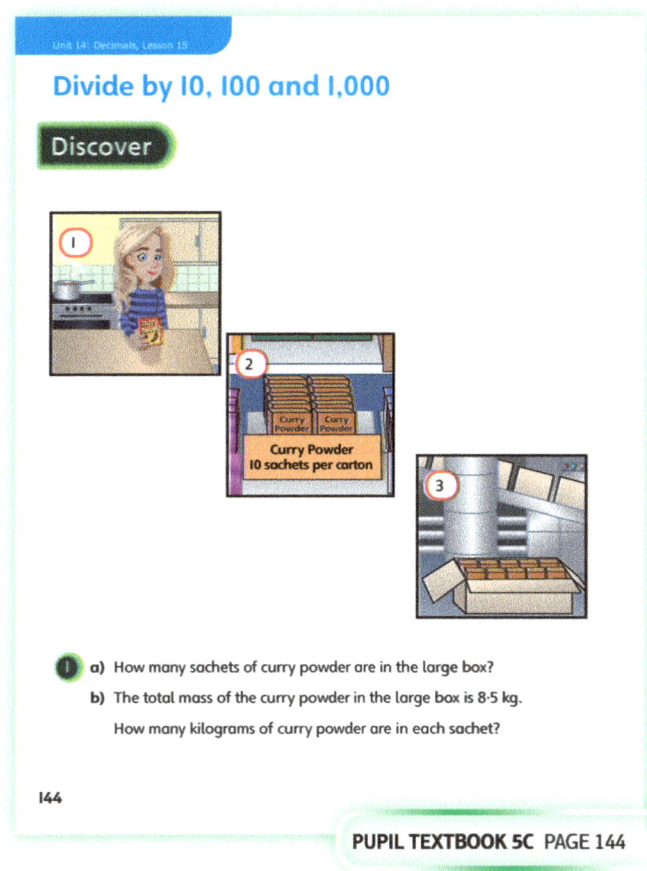

PUPIL TEXTBOOK 5C PAGE 144

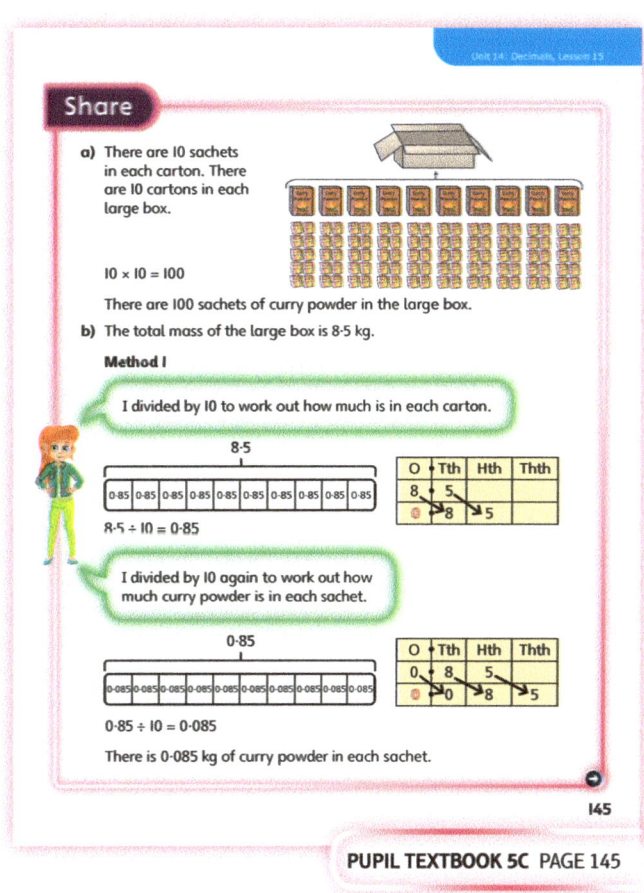

PUPIL TEXTBOOK 5C PAGE 145

Unit 14: Decimals, Lesson 15

Think together

WAYS OF WORKING Whole class teacher led (I do, We do, You do)

ASK

- Question ❶: *How can a place value grid help you find solutions to these calculations?*
- Question ❷: *What do you notice about what happens every time you divide by 100?*
- Question ❸: *How does your knowledge of dividing by 10 help you with this question?*

IN FOCUS Question ❷ offers further opportunity for children to divide by 100. It is important to discuss the method that children could use and clarify any misconceptions. Question ❸ requires children to use place values when dividing by 10, 100 and 1,000, to problem-solve. It will be important to discuss how dividing by 10, then by 10 and 10 again is equivalent to dividing by 1,000.

STRENGTHEN To help children with question ❸, it may be useful for them to use place value counters and a place value grid. Emphasise that when dividing by 100 or 1,000, the numbers will become 100 or 1,000 times smaller.

DEEPEN Deepen children's understanding by asking them to create their own word problems based on the calculations in question ❸. Pay attention to the language they use. Ask: *Can you find the solutions in more than one way?*

ASSESSMENT CHECKPOINT Question ❸ will provide solid evidence of children's understanding of the concepts in this lesson and their ability to link to prior learning from both the previous lesson and their work on place value. Look for children's recognition that, for example, dividing by 10 and then by 10 and 10 again is equivalent to dividing by 1,000.

ANSWERS

Question ❶ a): 12·8 kg ÷ 100 = 0·128 kg   128 ÷ 100 = 1·28
2·52 m ÷ 100 = 0·0252 m   0·9 ÷ 100 = 0·009

Question ❶ b): i) 0·012   iii) 0·718 km
ii) 0·006 m   iv) 7p or £0·07

Question ❷: 4 litres ÷ 100 = 0·04 litres

Question ❸ a): 46 kg ÷ 1,000 = 0·046 kg or 46 g

Question ❸ b): ÷ 10 ÷ 10 ÷ 10 = ÷ 1,000

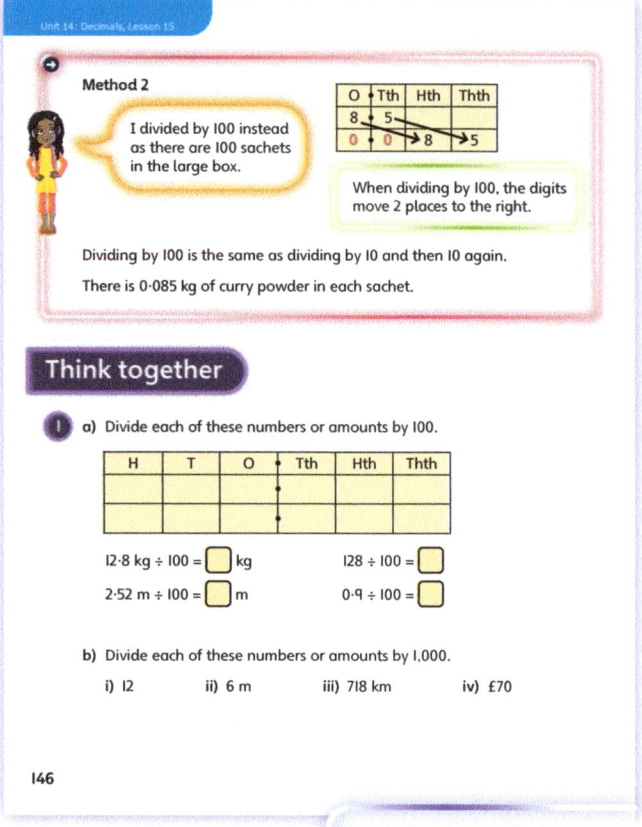

PUPIL TEXTBOOK 5C PAGE 146

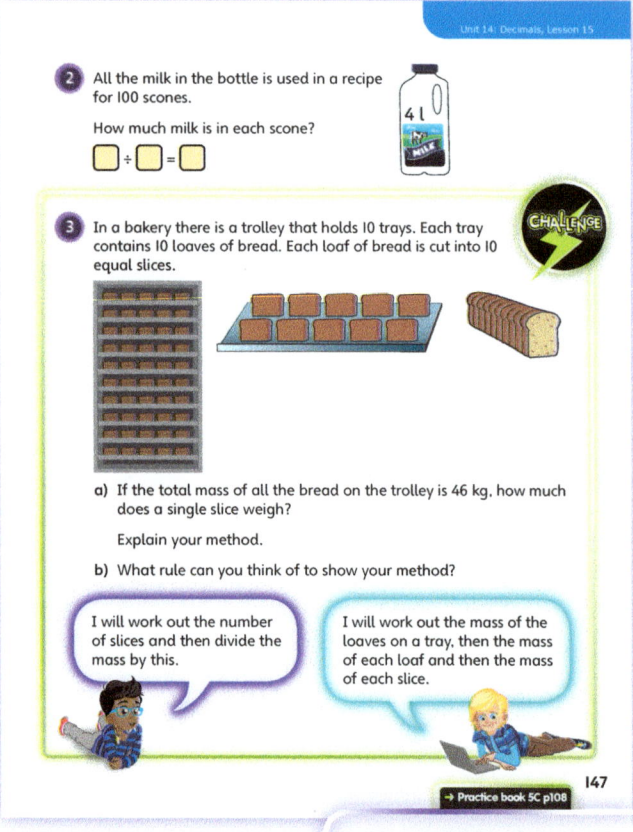

PUPIL TEXTBOOK 5C PAGE 147

Unit 14: Decimals, Lesson 15

Practice

WAYS OF WORKING Independent thinking

IN FOCUS Question ❶ offers children a scaffolded activity where they can begin to use the place value grid to divide by 100 and 1,000, working independently. Question ❷ offers another opportunity for children to generalise the rule that they can use to divide by 100. Discuss with children how they can use the square to prove Bella's theory. Question ❼ is important as it encourages children to think algebraically, considering what they know about each side of the equation and working forwards and backwards to find the missing numbers. This will develop their fluency, reasoning and problem-solving skills.

STRENGTHEN To strengthen understanding when solving question ❻, ask children to use a bar model. Ask: *How can you use the bar model to find the amount of money they save each day? What calculations will you use? Can you find the solution to the question in two ways?*

DEEPEN Explore question ❼ by giving other missing number problems, such as 3·56 ÷ ☐ = 0·0356 × ☐ = 35·6. Encourage children to explain their method clearly using the correct mathematical vocabulary.

ASSESSMENT CHECKPOINT Children are confident in dividing by 10, 100 and 1,000. They use their knowledge of multiplying and dividing decimals to find the missing numbers in problems.

ANSWERS Answers for the **Practice** part of the lesson can be found in the *Power Maths* online subscription.

Reflect

WAYS OF WORKING Independent thinking

IN FOCUS This question will help you assess whether children have fully understood the effects that dividing by 10, 100 and 1,000 have on the place value of digits in a number

ASSESSMENT CHECKPOINT Look for children who use place value accurately and can explain fluently and clearly which calculation is correct and why. The question prompts children to check Reena's answer. Look for children using multiplication and division, and demonstrating their understanding of the inverse relationship between multiplication and division to check their calculations.

ANSWERS Answers for the **Reflect** part of the lesson can be found in the *Power Maths* online subscription.

After the lesson ⏸

- Are children equally confident with dividing by 10, 100 and 1,000 as they were when multiplying?
- Are children confident using the inverse relationship between multiplication and division?
- Can children use place value grids and bar models with confidence to model their thinking?

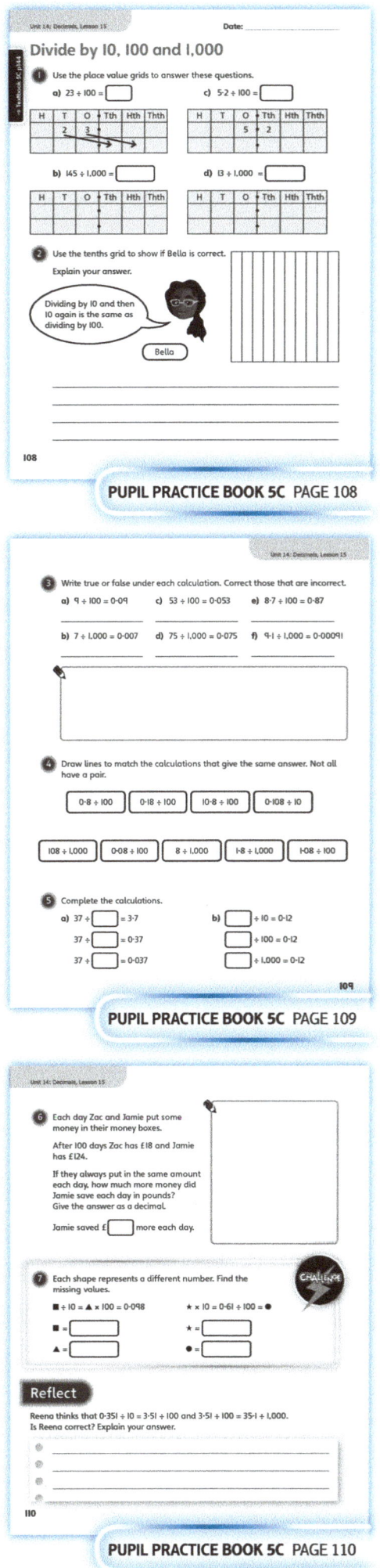

PUPIL PRACTICE BOOK 5C PAGE 108

PUPIL PRACTICE BOOK 5C PAGE 109

PUPIL PRACTICE BOOK 5C PAGE 110

183

End of unit check

> **Don't forget the unit assessment grid in your *Power Maths* online subscription.**

WAYS OF WORKING Group work adult led

IN FOCUS These questions cover the whole unit and are designed to draw out any misconceptions or misunderstandings. Check that children can use the formal method for addition and subtraction and can correctly align the numbers. Children should realise that, to find a missing number, they will need to use an inverse operation. When children are solving word problems, including those with several steps, encourage them to represent each situation using a bar model to help them work out if they need to add or subtract, multiply or divide.

Children should use a place value grid to support their calculations.

ANSWERS AND COMMENTARY Children who have mastered this unit understand the place value of each digit in numbers with three decimal places. They can look for key language to see if they are asked to multiply or divide by 10, 100 or 1,000. They will also have confidence in selecting efficient methods to solve addition and subtraction problems with decimal numbers.

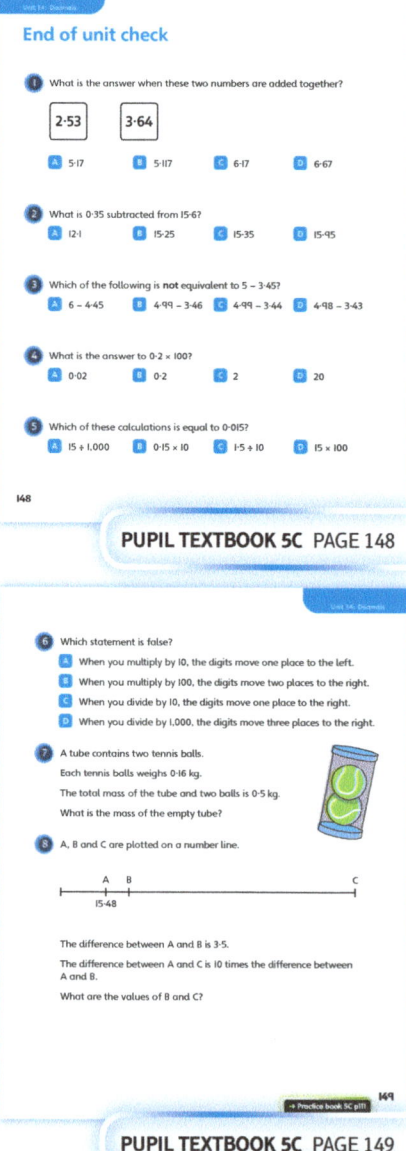

PUPIL TEXTBOOK 5C PAGE 148

PUPIL TEXTBOOK 5C PAGE 149

Q	A	WRONG ANSWERS AND MISCONCEPTIONS	STRENGTHENING UNDERSTANDING
1	C	A suggests the child has forgotten to carry the 1 from adding 0·5 + 0·6. B suggests incorrectly lining up numbers and exchanging.	Use place value grids and counters to support column methods of addition and subtraction. Work through step by step, showing any counter exchanges and linking the place value grid to the column method. Use a bar model to represent each situation and help children decide if they need to add or subtract, multiply or divide.
2	B	A suggests they have lined the numbers up incorrectly.	
3	B	A, C, and D suggest the child has not realised that the difference between the numbers has not changed.	
4	D	A, B or C suggest the child has made a mistake with the place value of the digits in 0·2 or 100.	
5	A	B suggests the child has multiplied 0·15 by 10 instead of dividing.	
6	B	A, C and D suggest insecure knowledge of Dmultiplying and dividing by 10, 100 and 1,000.	
7	0·18 kg	Some children will start with the mass of the tennis balls, some may find the mass of the two balls first.	
8	B is 18·98 C is 50·48	Look for children who are systematic in their approach and find the answers as they move along.	If children are unsure, ask them to re-read the question.

184

Unit 14: Decimals

My journal

WAYS OF WORKING Independent thinking

ANSWERS AND COMMENTARY

Question ❶ a): Ebo could do this as a column subtraction (see example).
Or he could do 12 − 4·35 = 11·99 − 4·34 = 7·65. Encourage children to work out the missing number first using the methods they have encountered in this unit.
Question ❶ b): Ebo needs to remember to line digits up according to their correct place value when using column subtraction. He could also make mistakes when exchanging. Ebo could, for example, ignore the digit '0' and not do any exchange at all or do the wrong exchange.
Question ❷ a): 25 + 2·95 = 27·95; 12·47 + 13·48 = 25·95; 18·3 + 9·65 = 27·95. Each sum has numbers with differing numbers of decimal places. Zero placeholders will be needed for 25 + 2·95 and 18·3 + 9·65.

T	O	•	Tth	Hth
0̸	1̸2̸	•	9̸9̸0̸	10
−	4	•	3	5
	7	•	6	5

Power check

WAYS OF WORKING Independent thinking

ASK

- Are you confident in adding and subtracting using the column method?
- Are you confident in multiplying and dividing decimals by 10, 100 and 1,000?
- Can you solve word problems by working out what information you have and what you need to find?

Power puzzle

WAYS OF WORKING Independent thinking

IN FOCUS Children are presented with two puzzles that require them to start at the top left corner and complete a sequence of calculations to reach their target number at the bottom right corner. They can work independently or in pairs.

ANSWERS AND COMMENTARY Encourage children to double-check the calculations on every row and column and to write all the number sentences.

2	÷ 100	÷ 10	× 100	× 10	÷ 100
÷ 1,000	× 100	× 10	÷ 10	× 100	× 10
× 10	÷ 100	× 10	÷ 10	× 100	÷ 1,000
× 100	÷ 10	× 1,000	× 100	× 10	0·002

2	÷ 100	÷ 10	× 100	× 10	÷ 100
÷ 1,000	× 100	× 10	÷ 10	× 100	× 10
× 10	÷ 100	× 10	÷ 10	× 100	÷ 1,000
× 100	÷ 10	× 1,000	× 100	× 10	2

After the unit

- Can all children confidently add and subtract using formal column methods?
- Can they generalise their method of multiplying and dividing by 10, 100 and 1,000?

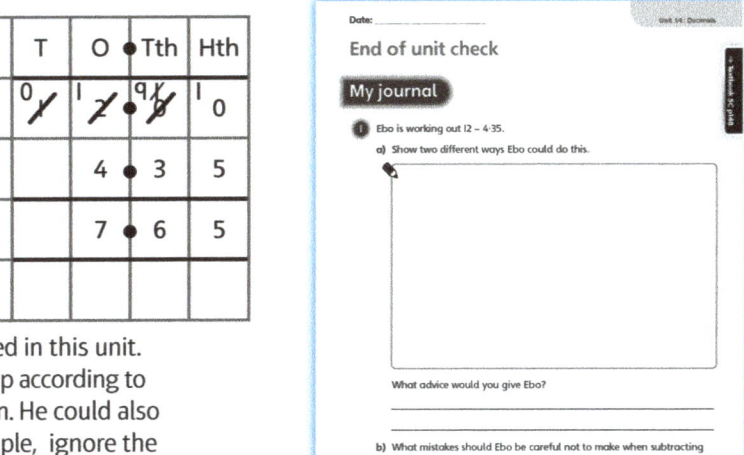

PUPIL PRACTICE BOOK 5C PAGE 111

PUPIL PRACTICE BOOK 5C PAGE 112

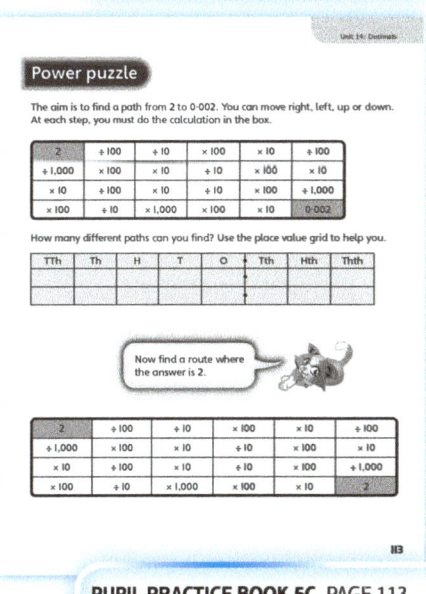

PUPIL PRACTICE BOOK 5C PAGE 113

Strengthen and **Deepen** activities for this unit can be found in the *Power Maths* online subscription.

Unit 15
Negative numbers

Mastery Expert tip! 'I found it important to have a number line that goes from at least ⁻20 to 20 on display in my classroom at all times. Children were then able to use this constantly when counting through 0.'

Don't forget to watch the Unit 15 video!

WHY THIS UNIT IS IMPORTANT

In this unit, children will explore what happens if they go below 0 on the number line. Up to this point children have only met positive values. Some children may be aware that negative numbers exist because they have seen them in real life, however this is the first time they have met them formally. Children will develop their understanding of negative numbers and how they relate and compare to positive numbers. They will use this understanding to calculate with both negative and positive numbers. Finally, they will use their ability to count in regular steps to help them identify rules in number sequences and accurately follow and complete them, including ones that go through 0.

WHERE THIS UNIT FITS

→ Unit 14: Decimals
→ **Unit 15: Negative numbers**
→ Unit 16: Measure – converting units

This unit builds on children's work much earlier in the year in Unit 1, where they explored numbers to 1,000,000 and their representation on place value grids and number lines. It also builds on previous experience of comparing, ordering and rounding numbers from Unit 2.

Before they start this unit, it is expected that children:
- know how to work out what a number line increases by each time
- know how to count in steps of 2, 5 and 10.

ASSESSING MASTERY

Children who have mastered this unit will be able to fluently count back through 0 in 1s, 2s, 5s and 10s. They will also be able to compare any two numbers, including negative numbers and be able to position numbers accurately on a number line. They will understand that negative numbers can be seen in context and will be able to give examples where they might see negative numbers (for example, temperature). They will be able to read negative temperatures from a thermometer.

COMMON MISCONCEPTIONS	STRENGTHENING UNDERSTANDING	GOING DEEPER
Children may read a thermometer or number line incorrectly – for example, between 0 and ⁻10. If children count up from ⁻10, they may say ⁻11, ⁻12 as opposed to ⁻9, ⁻8.	Support children by providing a fully labelled number line. Count aloud with children, paying close attention to the difference between counting on and back from a negative number and counting on and back from a positive number.	Children should count on and back in 2s, 5s and 10s from any number (not just a multiple), especially passing through 0.

Ask them to find the difference between a negative and positive number on a number line. Ask: *How can you do this efficiently?* Encourage children to consider the difference between each number and 0 and to add them to calculate the difference. |
| When counting back in steps of 10 from 33. For example, children may say 23, 13, 3 and ⁻3. They may, incorrectly, think that the pattern is symmetric about 0. | Ask children to slow down when they reach the 0 and think carefully. Ask them to draw the jumps on a number line and they will see that the jump from 3 to ⁻3 is a jump of only 6. | |

Unit 15: Negative numbers

UNIT STARTER PAGES

Use these pages to introduce the unit, with teacher-led discussion. Challenge children to explain as many of the keywords as they can.

STRUCTURES AND REPRESENTATIONS

Number line: Number lines, both horizontal and vertical, are used to help children plot positive and negative numbers and see how they sit on either side of 0.

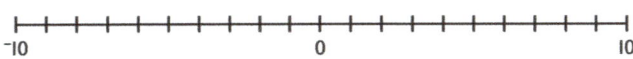

Thermometer: Thermometers are used to help children think about negative numbers in real-life contexts. The scale on a thermometer is much like a real-life number line.

KEY LANGUAGE

Here is some key language that children will need to know as part of the learning in this unit:

- place value
- step, interval
- number line, counting sequence
- negative, positive
- temperature, thermometer
- compare, order
- increase, decrease, ascending, descending
- less than (<), greater than (>), nearest

PUPIL TEXTBOOK 5C PAGE 150

PUPIL TEXTBOOK 5C PAGE 151

Unit 15: Negative numbers, Lesson 1

Understand negative numbers

Learning focus
In this lesson children will be introduced to negative numbers for the first time. They will count back through 0 on number lines using negative numbers.

Before you teach
- Can children count back to 0 from a positive number?
- Can children count in different multiples?
- Can children locate positive numbers on a number line?

NATIONAL CURRICULUM LINKS

Year 5 Number – number and place value

Interpret negative numbers in context, count forwards and backwards with positive and negative whole numbers, including through zero.

ASSESSING MASTERY

Children can count back through zero using negative numbers.

COMMON MISCONCEPTIONS

Children may misunderstand the order of negative numbers. Ask:
- *Can you use a number line to help you count back through 0?*
- *Can you see the symmetry in numbers on either side of 0 to help you 'see' negative numbers?*

When counting on or back through 0, children may count the numbers rather than the jumps. For example, if a lift is on floor ⁻3 and travels up 4 floors, children may incorrectly count ⁻3, ⁻2, ⁻1, 0 and conclude that the lift is at floor 0. Ask:
- *Do you count the numbers or the jumps between the numbers?*

STRENGTHENING UNDERSTANDING

To strengthen understanding, encourage children to use a number line to support their counting. Display the number line both horizontally and vertically to physically represent negative numbers, and encourage children to count aloud through 0, on and back, to reinforce the counting pattern.

GOING DEEPER

Prompt children to count through 0 in different multiples and to count through 0 both on and back.

KEY LANGUAGE

In lesson: down, jumps, **negative**, horizontal, sequences, count on, count back, thermometer

Other language to be used by the teacher: multiples, vertical, less than zero (< 0), greater than zero (> 0), positive

STRUCTURES AND REPRESENTATIONS

Vertical and horizontal number lines that pass through 0

RESOURCES

Optional: thermometers

 In the eTextbook of this lesson, you will find interactive links to a selection of teaching tools.

Quick recap

Ask children to find missing numbers on number lines that go from 0 to 10, then 0 to 20, 0 to 50 and 0 to 100. On each number line there should be 10 intervals marked, but only the start and end numbers labelled. So, for example, ask children to mark 55 on the number line from 0 to 100. Increase the challenge as appropriate for the group.

Unit 15: Negative numbers, Lesson 1

Discover

WAYS OF WORKING Pair work

ASK

- Question 1 a): *What numbers can you see? Do you know how to read these numbers? Why might a number have a negative symbol in front of it? Do you know what to call numbers with a negative sign in front of them? Have you seen negative numbers anywhere before?*
- Question 1 b): *Which floor is the restaurant on? What do you think a negative floor number means?* [below ground]

IN FOCUS You might want to discuss question 1 a) as a class beforehand. Ask children to look at the diagram. Discuss with them why negative numbers might be below ground. Check if they have seen negative numbers anywhere before. Question 1 b) focuses children on the end point after counting back through 0. Look for children counting aloud with their partner and counting each step.

PRACTICAL TIPS Children could make a human number track where each child represents a number. They could also contribute to making, and then counting back on, a large number line in the classroom or playground. This will reinforce their understanding of the number lines they use in class. Number lines in class should be horizontal and vertical, as vertical number lines can be especially helpful when representing negative numbers and counting back. Children could move counters on and back on the number lines to represent each question.

ANSWERS

Question 1 a): Reception is on the ground level, so floor 0.
The Restaurant is up one level from Reception.
Car Park A is one level down from Reception.
The Restaurant is on the first floor, so floor 1.
Car Park A is one level lower than ground level, so floor ⁻1.

Question 1 b): The waiter is in the Kitchen on floor ⁻3.

Share

WAYS OF WORKING Whole class teacher led

ASK

- Question 1 a): *What do you notice about the numbers? What do you think the negative means in this case?*
- Question 1 b): *Where do we start? Why? How many do we count back? Why? What floor are we on now? Do we count the floor we are currently on?*

IN FOCUS In question 1 a), explain to children what the negative sign means in front of the 1. Discuss what it means in the context of the situation. Children have already discussed in **Discover** other places they may have seen ⁻1. Question 1 b) focuses on how to use a number line effectively to count back through 0. It introduces the idea of negative and positive numbers. Show children a variety of different number lines and explain how they can be vertical and horizontal. Compare them. Ask children if their answers are dependent on which number lines they use. Discuss how, when counting back, you don't count the number you are already on.

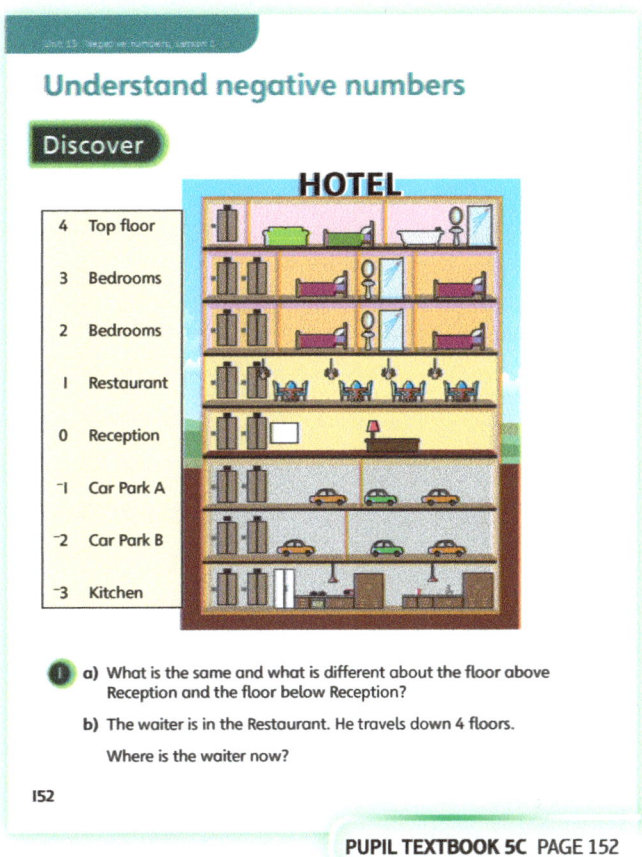

PUPIL TEXTBOOK 5C PAGE 152

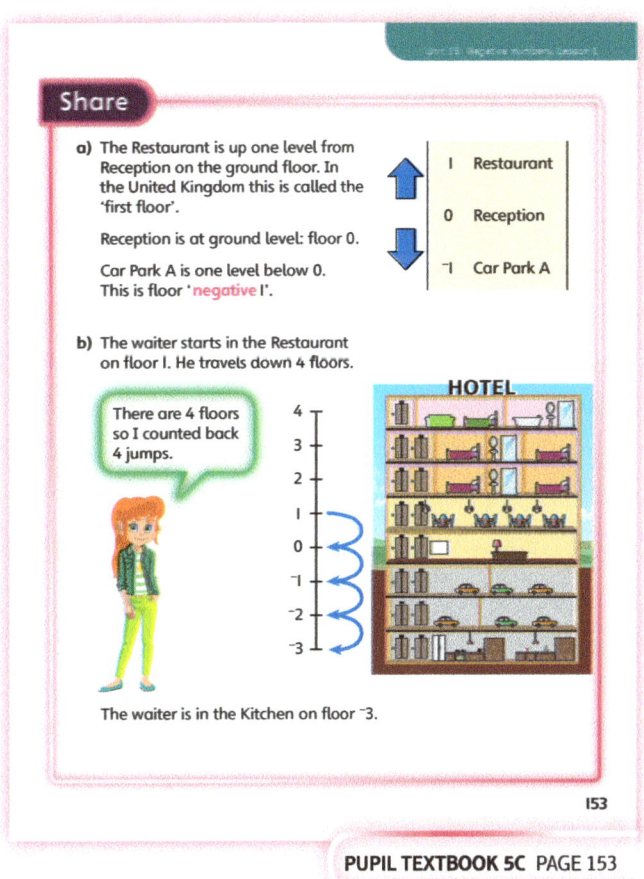

PUPIL TEXTBOOK 5C PAGE 153

Unit 15: Negative numbers, Lesson 1

Think together

WAYS OF WORKING Whole class teacher led (I do, We do, You do)

ASK

- Question ❶: *Which thermometers show temperatures that are below freezing? How do you know?*
- Question ❷: *What comes after 10 if we count back? What mistake has Ash made?*
- Question ❸: *Can you use a number line to help you find the next two numbers in the sequences? What do the numbers increase or decrease by each time?*

IN FOCUS In question ❶, children are reading numbers from thermometers, which they have already done before – however this time some of the values are negative. Discuss with children values that are below freezing. For question ❷, practise counting out loud with children. Listen carefully to what they say when they reach 0 and continue to count back.

In question ❸, some of the sequences increase and some decrease. All the sequences change by 1 each time so children get used to counting on or back using both positive and negative numbers. In question ❹, children work out the position of numbers on the number line. Again, encourage counting from a known number, making sure children count in the right direction.

STRENGTHEN Continue to use a number line to support counting into negative numbers. For children who are struggling to count back in certain multiples, ask them to count back in 1s and pause after each count. Stress to children that they should not count the number they start on, which some children may want to do.

DEEPEN Ask children to draw a number line starting at 10 at the right-hand side and counting down in 2s to ⁻10. Children should then try drawing other lines that go down in other multiples, not just in 1s.

ASSESSMENT CHECKPOINT Use questions ❶ to ❸ to check that children are confident counting back through 0 and can say the count correctly. Use question ❹ to assess whether children can identify negative numbers on a number line where not all the numbers are labelled.

ANSWERS

Question ❶ a): 2 °C   Question ❶ c): ⁻1 °C
Question ❶ b): 0 °C   Question ❶ d): ⁻4 °C
Question ❷ a): Children count: 10, 9, 8, 7, 6, 5, 4, 3, 2, 1, 0, ⁻1, ⁻2, ⁻3, ⁻4, ⁻5, ⁻6, ⁻7, ⁻8, ⁻9, ⁻10
Ash has missed out zero.
Question ❷ b): Children count: ⁻10, ⁻9, ⁻8, ⁻7, ⁻6, ⁻5, ⁻4, ⁻3, ⁻2, ⁻1, 0, 1, 2, 3, 4, 5, 6, 7, 8, 9, 10
Question ❸ a): ⁻2, ⁻3   Question ❸ c): ⁻1, 0
Question ❸ b): ⁻5, ⁻6
Question ❹ a): ⁻2   Question ❹ c): ⁻16
Question ❹ b): ⁻7

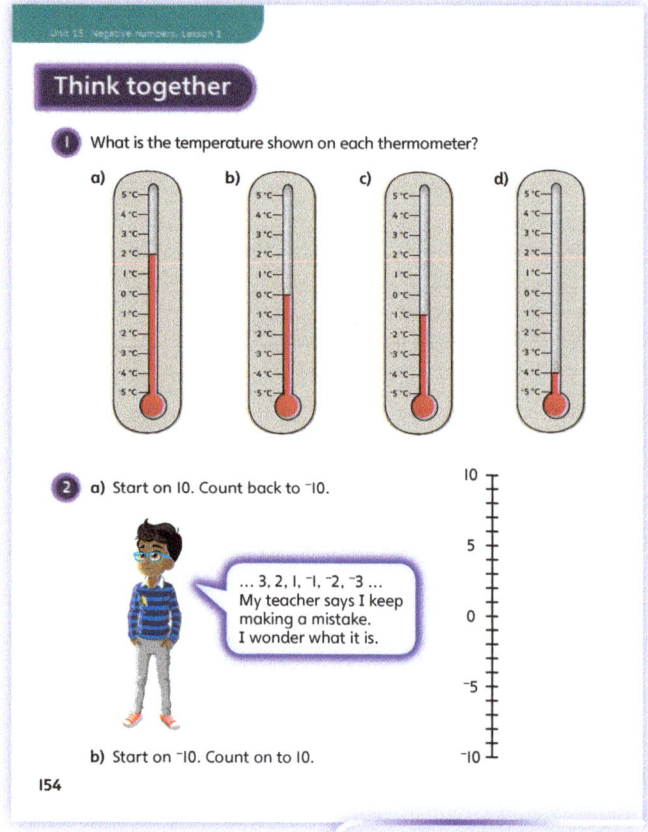

PUPIL TEXTBOOK 5C PAGE 154

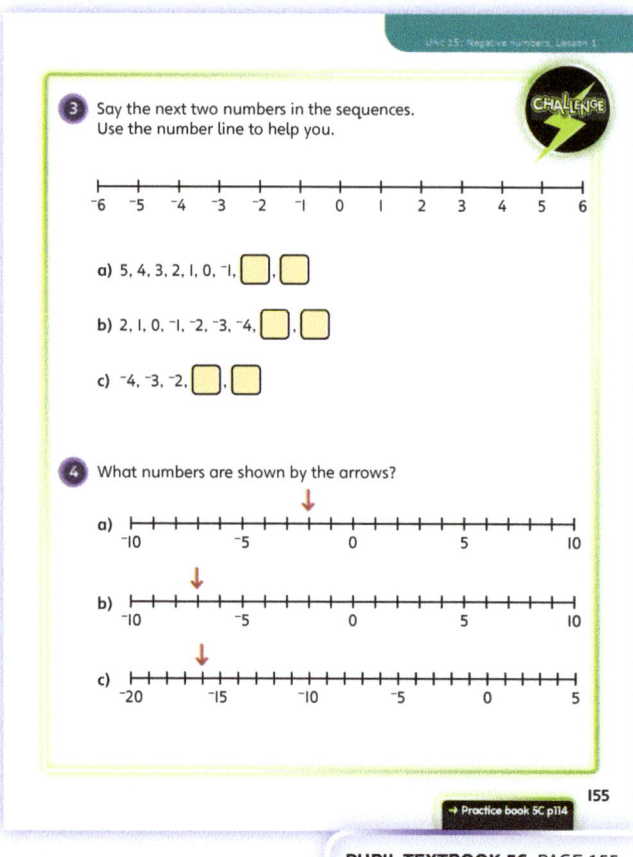

PUPIL TEXTBOOK 5C PAGE 155

Practice

WAYS OF WORKING Independent thinking

IN FOCUS Questions in this **Practice** section use a mixture of horizontal and vertical number lines to support children and to help them practise counting on and back through 0.

Questions 1 and 2 practise using negative numbers in context: it is important to highlight to children how counting in negative numbers is a useful and practical skill.

In questions 3 and 4, children practise counting on and back. Question 5 is an extension of the skills practised in the **Challenge** question of the Textbook.

STRENGTHEN Continue to encourage children to use a number line to support counting on and back through 0. Model and prompt counting aloud, both on and back, in singles and multiples, until children can securely demonstrate this important skill.

DEEPEN Can children count back in steps of 2, 4, 5, 10 and so on from any given number? Challenge children to practise this skill by asking them to count back in steps of 10 from 42. What is the first negative number that they meet?

ASSESSMENT CHECKPOINT Use questions 1 to 4 to assess whether children can count on and back through 0 in 1s.

ANSWERS Answers for the **Practice** part of the lesson can be found in the *Power Maths* online subscription.

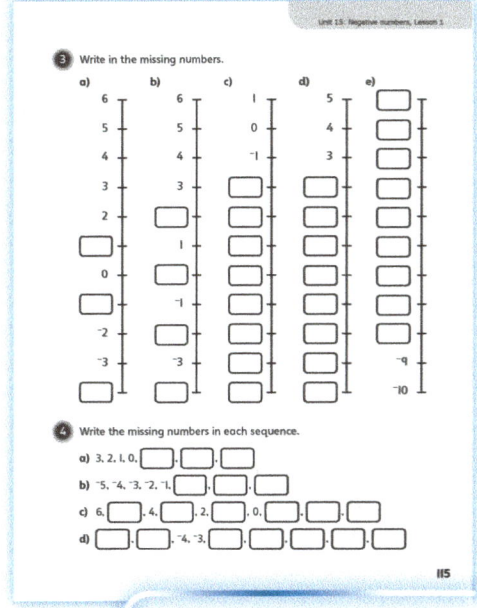

PUPIL PRACTICE BOOK 5C PAGE 114

PUPIL PRACTICE BOOK 5C PAGE 115

Reflect

WAYS OF WORKING Independent thinking and pair work

IN FOCUS Children play a game using their knowledge of counting on and back through negative numbers.

ASSESSMENT CHECKPOINT Assess whether children can confidently count on and back on the number track. Can children predict where they will land without counting?

ANSWERS Answers for the **Reflect** part of the lesson can be found in the *Power Maths* online subscription.

After the lesson

- Do children rely on the number line to count back through 0?
- Can children spot patterns when counting on and counting back in multiples?
- Can children explain what happens when you count back into negative numbers?

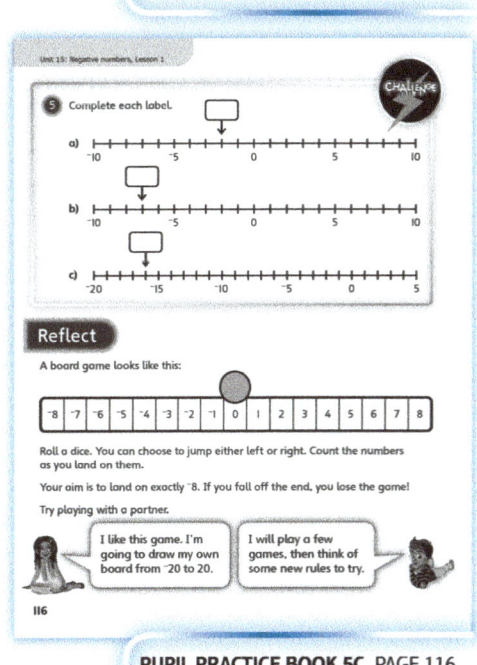

PUPIL PRACTICE BOOK 5C PAGE 116

Unit 15: Negative numbers, Lesson 2

Count through zero

Learning focus
In this lesson, children will look at negative numbers in context. They will count back through 0 on number lines using negative numbers.

Before you teach
- Can children count back through 0?
- Can children count in different multiples?
- Can children estimate numbers on a positive number line?

NATIONAL CURRICULUM LINKS

Year 5 Number – number and place value

Interpret negative numbers in context, count forwards and backwards with positive and negative whole numbers, including through zero.

ASSESSING MASTERY

Children can count back through zero in context, using negative numbers. They can read negative numbers on a number line and can find the difference between a positive and a negative number.

COMMON MISCONCEPTIONS

If children misunderstand the order of negative numbers, ask:
- *Can you use a number line to help you count back through 0?*
- *Can you see the symmetry in numbers on either side of 0 to check your counting?*

Children may continue to miscount by counting the numbers rather than the number of jumps back. Ask:
- *What strategies did you use for counting back in the last lesson?*

When using number lines, children may not realise that some number lines do not go up and down in intervals of 1. Remind children to check the intervals on number lines before using them by asking questions that prompt this thinking. For example, if there are five increments between 0 and 10, ask:
- *Does this number line count up in 1s? Does it count up in 2s? How can you tell?*

STRENGTHENING UNDERSTANDING

To strengthen understanding, encourage children to use a number line to support their counting. Show the number line both horizontally and vertically and ask children to slowly count aloud with you, to help embed the rhythm and pattern of the numbers.

GOING DEEPER

To deepen thinking and give further opportunities to practise working in negative numbers, ask children to complete the start and end numbers on a variety of number lines using the position of the other numbers on the line to help them.

KEY LANGUAGE

In lesson: difference, intervals, degrees Celsius, estimate

Other language to be used by the teacher: negative, positive, forwards, backwards, position

STRUCTURES AND REPRESENTATIONS

Vertical and horizontal number lines that pass through 0

RESOURCES

Optional: thermometers

 In the eTextbook of this lesson, you will find interactive links to a selection of teaching tools.

Quick recap

On the board display a number line from ⁻10 to 10. Pick a start number and end number and ask children to count out loud as a class from the start number to the end number. Check that children are confident saying the numbers as they go through 0.

Unit 15: Negative numbers, Lesson 2

Discover

WAYS OF WORKING Pair work

ASK

- Question 1 a): *How do you read the temperature on a thermometer? How is it similar to a number line? What unit do you measure temperature in?*
- Question 1 b): *What is the temperature in each 'world'?*

IN FOCUS In questions 1 a) and b), children begin to see where number lines are used in context. Explore the connection between number lines and thermometers. Children may sometimes read the negative temperatures as positive. Prompt children to check the intervals on the thermometer before they try to answer the question – this is good practice.

PRACTICAL TIPS Children could use real thermometers to look at how to read temperature.

ANSWERS

Question 1 a): The thermometers go up in jumps of 2 °C.

Question 1 b): Arctic World is at exactly ⁻8 °C.
Nocturnal World is ⁻3 °C.
Oceanic World is 13 °C.

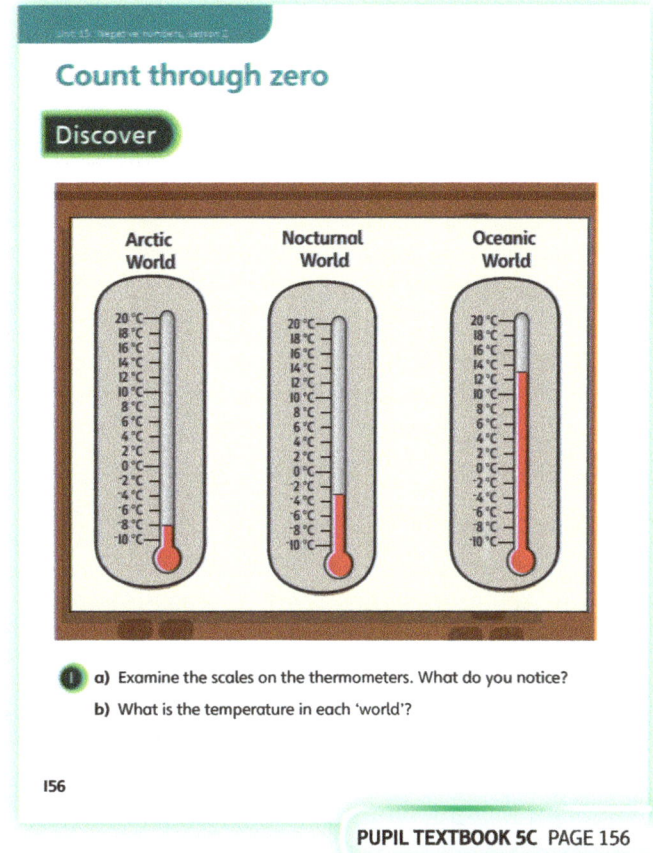

PUPIL TEXTBOOK 5C PAGE 156

Share

WAYS OF WORKING Whole class teacher led

ASK

- Question 1 a): *What do thermometers help us to do? How do you read a thermometer? What intervals does each thermometer go up in? How can you check?*
- Question 1 b): *How can we work out each temperature? What do we need to count in?*

IN FOCUS In question 1 a), discuss with children the difference between these thermometers and the ones that they saw in the last lesson. Question 1 b) then focuses on reading a scale and finding missing numbers on a scale, including negative numbers. Children may think that the thermometer goes up in 1s: encourage them to check this by counting aloud and realising that this does not work. Ask if the intervals go up in 2s and encourage children to check this for themselves.

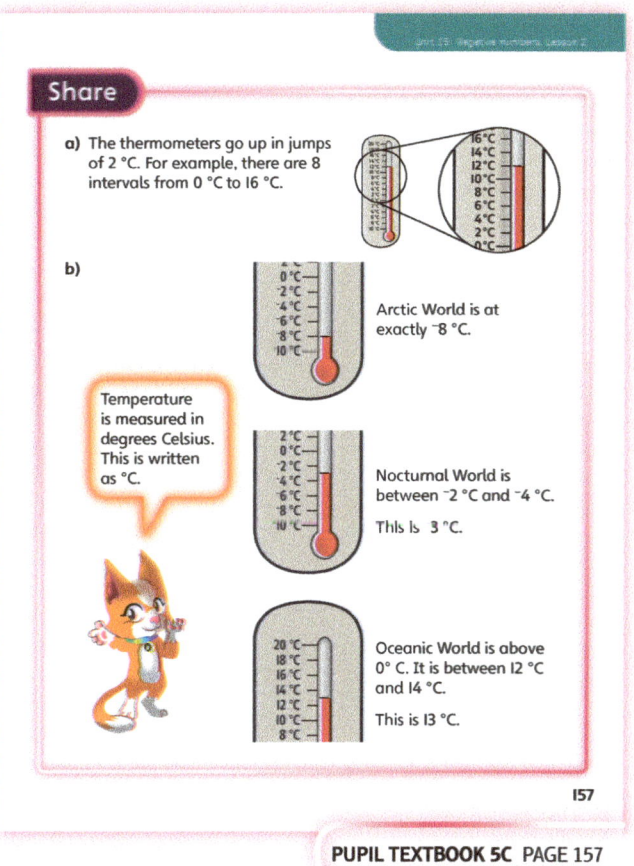

PUPIL TEXTBOOK 5C PAGE 157

193

Unit 15: Negative numbers, Lesson 2

Think together

WAYS OF WORKING Whole class teacher led (I do, We do, You do)

ASK

- Question ❶: *What do these thermometers increase by each time? How do you know? How can you check? What are the temperatures?*
- Question ❷: *What are we counting back in? What happens when we get to 0?*
- Question ❸: *Can you see a pattern when counting back in 5s into negative numbers? Which is the most difficult?*

IN FOCUS In question ❶, children practise reading different scales. Discuss with children how they can check what the scale increases by each time. Discuss negative temperatures, to make sure they don't read this incorrectly.

In question ❷, ask children to count aloud in pairs together on the number line and then count back as a whole class. Slow down when you get to 0.

Question ❸ requires children to count in different multiples. What different methods do children use here? Some children may remember number patterns or be able to jump straight back to the numbers, and immediately be able to confirm if the sentence is true or false. Others may need to count back each number to check.

STRENGTHEN Continue to use a number line to support counting back from 0 into negative numbers. For children who are struggling to count back in certain multiples, ask them to count back in 1s and pause after each count of 2, 5 or 10, depending on which multiple they are counting back in.

DEEPEN Extend thinking further by expanding question ❸ and asking children to count back in more challenging multiples, such as 3 or 6. Do children use their knowledge of number patterns, timestables and counting back through zero into negative numbers to make informed decisions about whether they will say ⁻20?

ASSESSMENT CHECKPOINT Use the questions to check that children can read positive and negative values on a number line that goes up in 2s, 5s and 10s.

ANSWERS

Question ❶ a): ⁻5 °C

Question ❶ b): ⁻10 °C

Question ❶ c): ⁻15 °C

Question ❷ a): Children count: ⁻50, ⁻40, ⁻30, ⁻20, ⁻10, 0, 10, 20, 30, 40, 50;
50, 40, 30, 20, 10, 0, ⁻10, ⁻20, ⁻30, ⁻40, ⁻50

Question ❷ b): Children count: ⁻50, ⁻45, ⁻40, ⁻35, ⁻30, ⁻25, ⁻20, ⁻15, ⁻10, ⁻5, 0, 5, 10, 15, 20, 25, 30, 35, 40, 45, 50;
50, 45, 40, 35, 30, 25, 20, 15, 10, 5, 0, ⁻5, ⁻10, ⁻15, ⁻20, ⁻25, ⁻30, ⁻35, ⁻40, ⁻45, ⁻50

Question ❸: Lee will say ⁻20: 10, 5, 0, ⁻5, ⁻10, ⁻15, ⁻20.
Emma will say ⁻20: 8, 4, 0, ⁻4, ⁻8, ⁻12, ⁻16, ⁻20.
Zac will not say ⁻20. He will say: 5, 3, 1, ⁻1, ⁻3, …, ⁻17, ⁻19, ⁻21. All of Zac's numbers are odd.

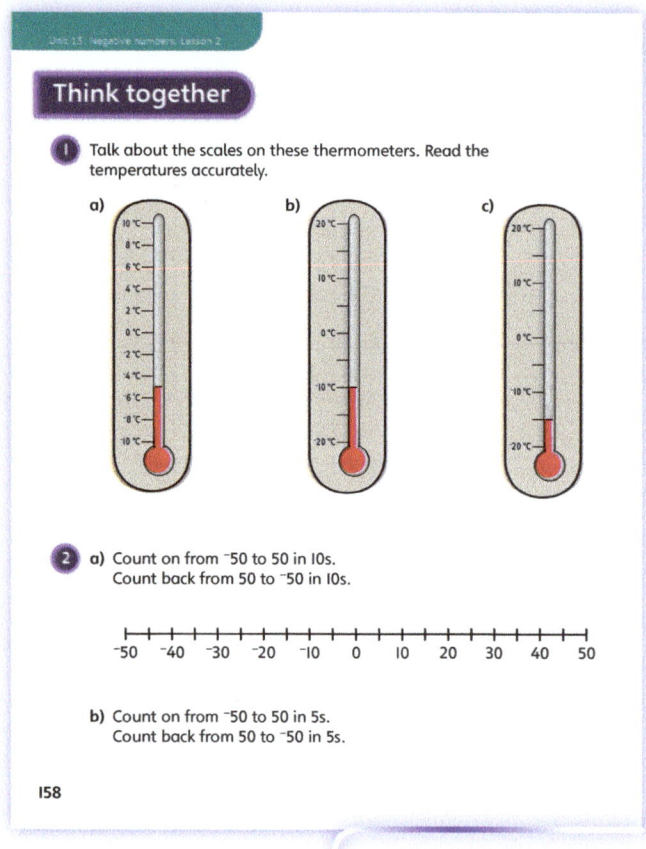

PUPIL TEXTBOOK 5C PAGE 158

PUPIL TEXTBOOK 5C PAGE 159

Unit 15: Negative numbers, Lesson 2

Practice

WAYS OF WORKING Independent thinking

IN FOCUS Questions in this **Practice** section use a mixture of horizontal and vertical number lines to support children and to help them practise counting on and back through 0. In question **1**, children work out the missing numbers on thermometers that don't all go up in 1s. First, children need to work out what the thermometers increase by each time, looking at the values given. In questions **2** and **3**, children mark missing numbers on number lines. They need to be careful when they go below zero as it is easy to get the numbers the wrong way around. In question **5**, children count on and back in different multiples but the starting number doesn't necessarily fall on a multiple of that count interval. Children should use a number line for support. This can pose difficulties when they get to 0 as it is easy to keep the same pattern going (for example, in the sequence 33, 23, 13, 3, children may say ⁻3 next, but the correct count is ⁻7).

STRENGTHEN Continue to encourage children to use a number line to support counting on and back through 0. Refer children to some of the number lines you have made in class to help support their answers in this section.

DEEPEN Extend children's thinking further by asking them to complete the start and end numbers on a variety of 10-interval number lines. Explain that each number line must cross 0 and have a different number as a starting point. Do children consistently complete the number line correctly? Can they add numbers in the correct sequence using a range of intervals?

ASSESSMENT CHECKPOINT Assess whether children count on and back through 0 to complete a variety of number lines which use different intervals. Can they reason what numbers are needed to complete number lines, with an awareness of intervals, rather than using trial and error?

ANSWERS Answers for the **Practice** part of the lesson can be found in the *Power Maths* online subscription.

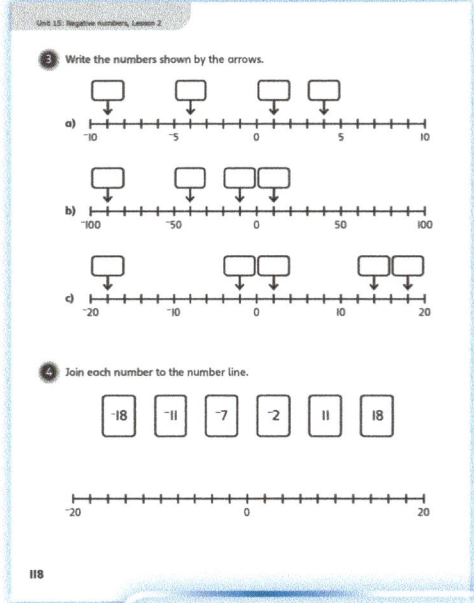

PUPIL PRACTICE BOOK 5C PAGE 117

PUPIL PRACTICE BOOK 5C PAGE 118

Reflect

WAYS OF WORKING Independent thinking

IN FOCUS If counting on in 2s, you may wish to start from a number closer to 0 than ⁻100 (for example, ⁻20). Ask children to start counting on and take it in turns to say each number – with other children actively listening and checking they are correct.

ASSESSMENT CHECKPOINT Check that children can count back through zero in steps of 2, 5 and 10.

ANSWERS Answers for the **Reflect** part of the lesson can be found in the *Power Maths* online subscription.

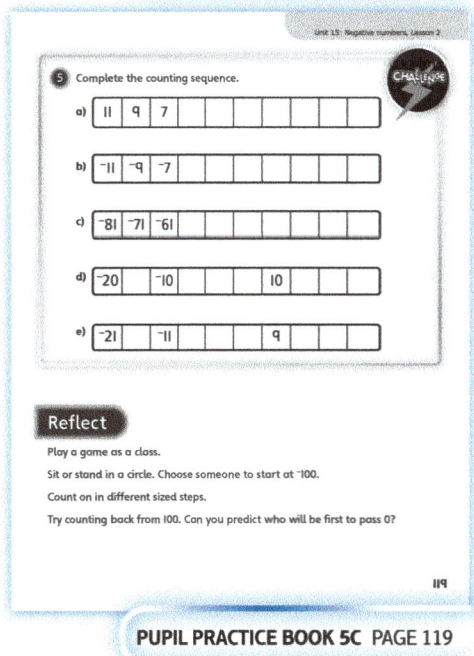

PUPIL PRACTICE BOOK 5C PAGE 119

After the lesson

- Can children identify the intervals that a number line increases by each time?
- Can children count back through 0 in multiples of 2, 5 and 10?
- Can children work out missing numbers on number lines that increase in multiples of 2, 5 and 10?

195

Unit 15: Negative numbers, Lesson 3

Compare and order negative numbers

Learning focus
In this lesson, children compare any two numbers, including negative numbers.

Before you teach
- Check that children can compare any two numbers by comparing the place value.
- Children understand how to correctly use the < and > symbols.

NATIONAL CURRICULUM LINKS

Year 5 Number – number and place value

Interpret negative numbers in context, count forwards and backwards with positive and negative whole numbers, including through zero.

ASSESSING MASTERY

Children can compare any two numbers, including negatives, using a number line or generalisations for support. They understand that a negative number is always smaller than a positive number and how to work out which number is smaller if both numbers are negative.

COMMON MISCONCEPTIONS

When comparing two negative numbers, such as ⁻10 and ⁻18, children often think that ⁻10 is smaller than ⁻18 as they compare the absolute values (that is, 10 and 18). Because 10 is smaller than 18, they think that ⁻10 is smaller than ⁻18. Provide children with a number line and ask:

- *Where does ⁻10 lie on a number line? What about ⁻18? What do we know about the position of smaller numbers?*

Explain that the further the number is from 0 in the negative direction, the smaller the number is. You could also support this by giving them real-life examples, such as temperature. Explain that ⁻18 is colder (and therefore smaller) than ⁻10.

STRENGTHENING UNDERSTANDING

Provide children with a printed number line to support their understanding. Ask children to mark each of the values on the number line. Explain that the smaller the number, the further left (or down on a vertical number line) it is.

GOING DEEPER

Give children a positive and negative number and ask them to find the difference. Ask them if they can think of a general way of working it out, where they don't have to use a number line. For example, can they work out the difference between ⁻1,500 and 790? Ask them to explain how they did it and to justify their reasoning. You might prompt them to first consider the difference between each of the two numbers and 0.

KEY LANGUAGE

In lesson: positive, number line, negative, difference, greater, smaller, less, more, greatest, smallest

Other language to be used by the teacher: label, coldest, hottest, colder, warmer/hotter

STRUCTURES AND REPRESENTATIONS

Number lines that pass through 0

RESOURCES

Optional: thermometers

 In the eTextbook of this lesson, you will find interactive links to a selection of teaching tools.

Quick recap

Show children a ⁻10 to 10 number line on the board with 0 marked, but no other points labelled or marked. Ask children if they can place the number 5. Ask them to discuss and agree with a partner, giving their reasons. Repeat with other numbers between ⁻10 and 10.

Discover

WAYS OF WORKING Pair work

ASK

- Question 1 a): *Why might Alex think that ⁻18 is greater/warmer than 4? Why is it unlikely the freezer is warmer than the fridge?*
- Question 1 b): *What can you draw to help you? Where would you place these numbers on a number line? How far from 0 is each number?*

IN FOCUS In question 1 a), children discuss amongst themselves why Alex might think that it is colder in the fridge. If children are struggling, you could ask them to consider the number without the sign in front of it. In question 1 b), children could use a number line to support their understanding. They start to see that the number (positive or negative) tells them how far away from 0 the number is. If it is negative then it is less than 0 and if positive it is greater than 0.

PRACTICAL TIPS Have a large number line available for children to see.

ANSWERS

Question 1 a): 18 is greater than 4, but a negative number is always less than a positive number. ⁻18 < 4, so ⁻18 °C < 4 °C.

Question 1 b): 4 °C is 4 degrees above 0 °C.
⁻18 °C is 18 degrees below 0 °C.

PUPIL TEXTBOOK 5C PAGE 160

Share

WAYS OF WORKING Whole class teacher led

ASK

- Question 1 a): *Why do you think Alex thought what she did? Why is this not the case? What do we know about negative numbers?*
- Question 1 b): *Where are ⁻18 and 4 on the number line? How far away from 0 is ⁻18? In what direction? What about 4?*

IN FOCUS In question 1 a), use the **Share** section and comments to help children understand why a negative number is always less than a positive number. Explain that Alex was thinking that because 18 is greater than 4, then ⁻18 must be greater than 4. Explain that with negatives this is not the case and the opposite happens. In question 1 b), encourage children to draw/use a number line to see the position of numbers. Help them to see that the value of the number tells them how far away from 0 they are. Agree that positioning numbers on a number line helps you compare numbers.

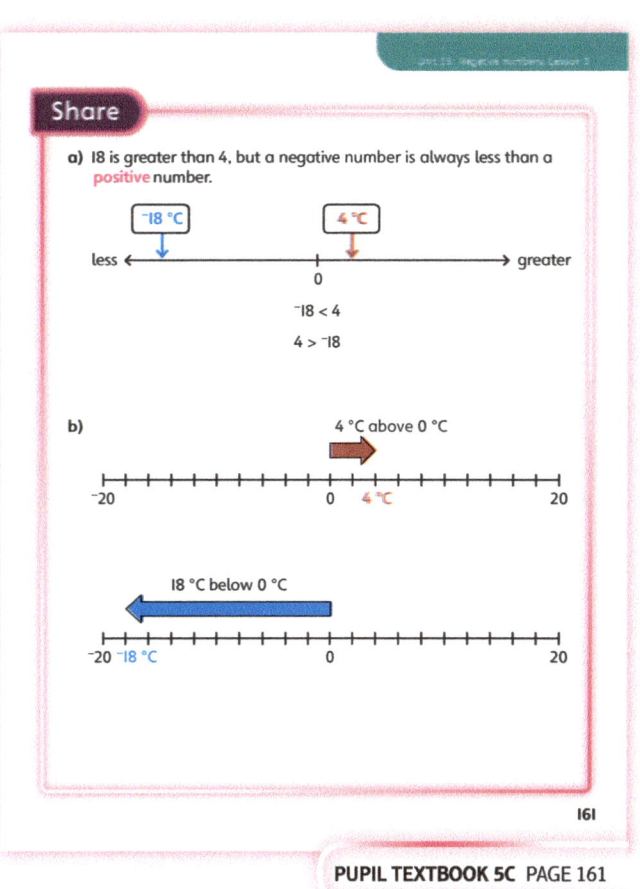

PUPIL TEXTBOOK 5C PAGE 161

197

Think together

WAYS OF WORKING Whole class teacher led (I do, We do, You do)

ASK

- Question ❶: *Where do the numbers lie on the number line? Which is the greater? Which is smaller? How do you know?*
- Question ❷: *What could you draw to help you order these numbers? Where would they lie on a number line? Can you order the numbers without drawing a line?*
- Question ❸: *Do you feel more confident estimating the numbers in part a), part b) or part c)? How do you know whether a number is positive or negative? Can you mark other points on each number line to help you make more accurate estimates (for example, in part a), can you mark ⁻5 and 5)?*

IN FOCUS In question ❶, children use the < and > symbols to compare numbers. The number line has been provided to support their understanding. Work through an example and then ask the children to complete the others.

In question ❷, children are asked to order a set of numbers. Some children may want or need to draw a number line to support them, whereas other children will, by this stage, be able to do it without. Watch out for the common misconception where children order the negative numbers in the wrong order because they are looking at the absolute values of the numbers.

Finally, in question ❸, children use their knowledge of the order of numbers to help them estimate which numbers the arrows are pointing to.

STRENGTHEN For those who need it, provide children with a printed number line to support their understanding. Ask children to mark each of the values on the number line. Explain that the smaller the number, the further left (or down on vertical number line) it is.

DEEPEN Give children a positive and negative number and ask them to find the difference. Ask children to see if they can come up with a general way of working it out, where they don't have to use a number line.

ASSESSMENT CHECKPOINT Use questions ❶ and ❷ to check that children can confidently order numbers, including negatives.

ANSWERS

Question ❶ a): 0 > ⁻10

Question ❶ b): ⁻5 > ⁻10

Question ❶ c): 10 > ⁻10

Question ❶ d): ⁻10 < ⁻1

Question ❶ e): ⁻10 < 1

Question ❶ f): ⁻9 > ⁻10

Question ❷: ⁻40 < ⁻30 < ⁻25 < 45 < 70

Question ❸: Children's answers to a), b) and c) should all be related.
    a) A ⁻9   B ⁻6   C ⁻2   D 2   E 8
    b) A ⁻90   B ⁻60   C ⁻20   D 20   E 80
    c) A ⁻0·9   B ⁻0·6   C ⁻0·2   D 0·2   E 0·8

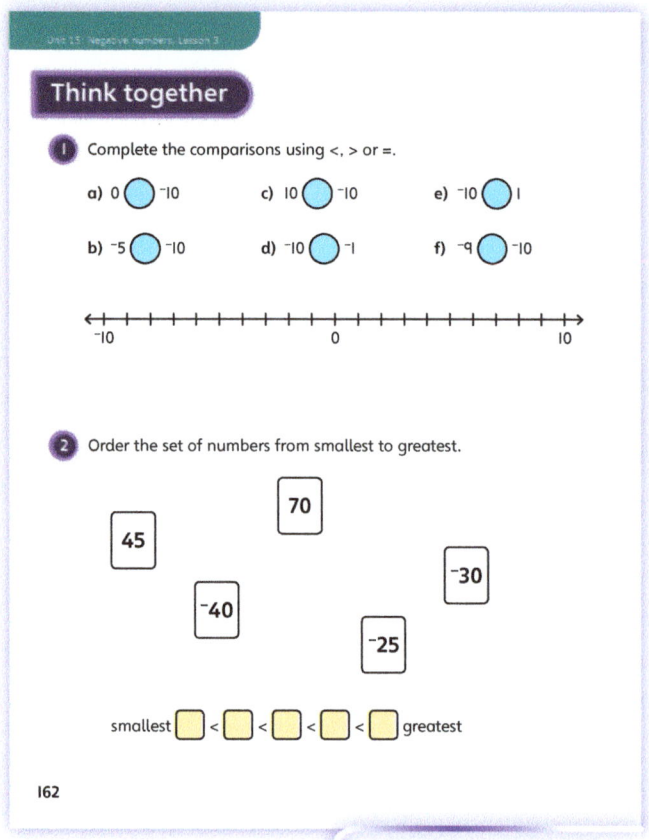

PUPIL TEXTBOOK 5C PAGE 162

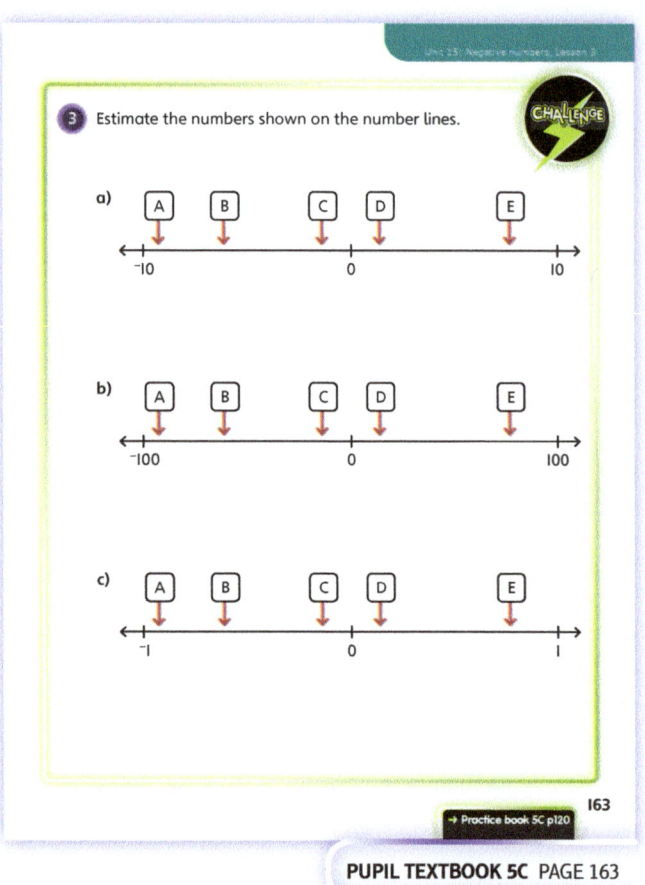

PUPIL TEXTBOOK 5C PAGE 163

Unit 15: Negative numbers, Lesson 3

Practice

WAYS OF WORKING Independent thinking

IN FOCUS In question ❶, children compare numbers supported by a number line. Then, in question ❷, this support is removed. In questions ❸ and ❹, children use their knowledge to order numbers or find possible numbers. In the **Challenge** question, two number lines are drawn with common misconceptions and children have to describe the mistakes that have been made.

STRENGTHEN For those who need it, use a number line to support children. Discuss with them that the smaller the number, the further left on the number line that number is. The greater the number, the further right. Build on children's knowledge of ordering positive numbers.

DEEPEN Ask children to find solutions to inequalities such as $^-4 < \square < 2$. What are all the possible whole number solutions?

ASSESSMENT CHECKPOINT Questions ❶ to ❸ will allow you to assess whether or not children can compare and order numbers including negatives.

ANSWERS Answers for the **Practice** part of the lesson can be found in the *Power Maths* online subscription.

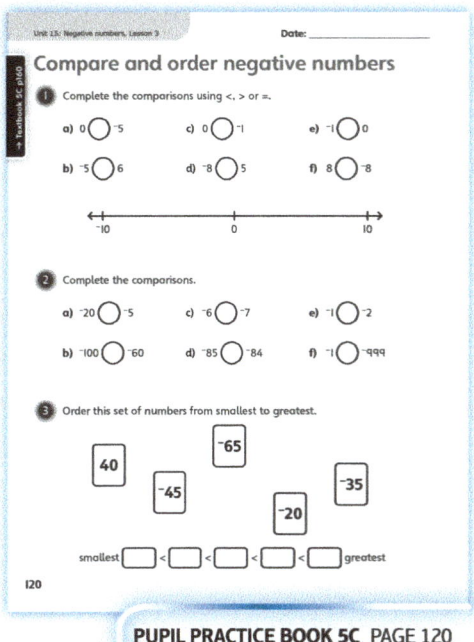

PUPIL PRACTICE BOOK 5C PAGE 120

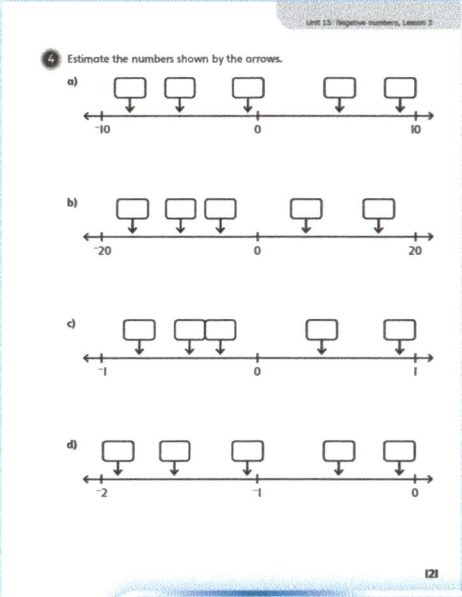

PUPIL PRACTICE BOOK 5C PAGE 121

Reflect

WAYS OF WORKING Independent thinking

IN FOCUS Children focus on their understanding of the position of negative numbers in order to explain why Max is wrong. They may explain in words, or they may use a number line to illustrate their thinking.

ASSESSMENT CHECKPOINT Assess whether children can clearly explain why Max is wrong, using accurate vocabulary or models to support their answer.

ANSWERS Answers for the **Reflect** part of the lesson can be found in the *Power Maths* online subscription.

After the lesson

- Can children compare numbers, including negatives?
- Can children order a list of numbers that includes negatives?

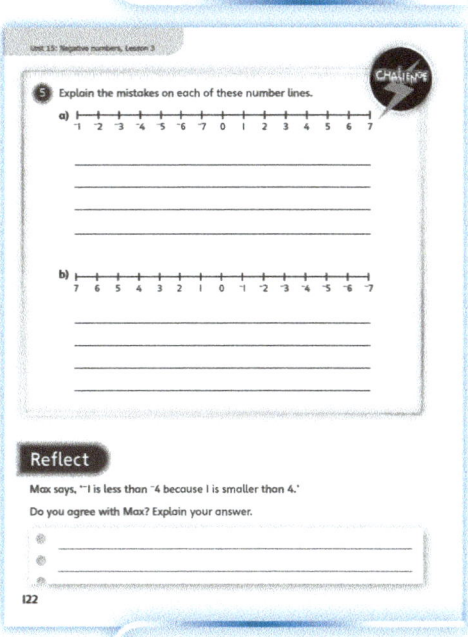

PUPIL PRACTICE BOOK 5C PAGE 122

199

Unit 15: Negative numbers, Lesson 4

Find the difference

Learning focus
In this lesson children will continue to learn about the position of positive and negative numbers on a number line and begin to find the difference between two numbers where one number is positive and the other is negative, or both are negative.

Before you teach
- In what contexts might children have met negative numbers before?

NATIONAL CURRICULUM LINKS

Year 5 Number – number and place value

Interpret negative numbers in context, count forwards and backwards with positive and negative whole numbers, including through zero.

ASSESSING MASTERY

Children can find the difference between a positive and negative number or two negative numbers using a number line. They can apply their understanding to a context such as temperature to work out answer to questions, such as 'How much warmer/colder?'.

COMMON MISCONCEPTIONS

Children may assume that negative numbers work in a similar way to positive numbers, for example, that ⁻5 must be greater than ⁻1 because 5 is greater than 1. Show children a number line and ask:
- Plot the numbers from ⁻10 to 10 on this number line. Which end shows the greatest number? Which end shows the smallest? What can you say about the difference between ⁻5 and ⁻1? Explain how you know.

Children may miscalculate the difference between positive and negative numbers (for example, they may calculate the difference between 7 and ⁻6 as 1). Ask:
- Show me these two numbers on a number line. Count the difference between them.

STRENGTHENING UNDERSTANDING

To support children to find the difference between two numbers, ask children to first draw a number line and, using arrows, to mark the two numbers they are finding the difference between. Then, ask them to count up in ones from the smaller number to the greater number, or count down from the greater number to the smaller number.

GOING DEEPER

Encourage children to create their own word problems requiring the use of negative numbers. Discuss with children the ways in which negative numbers are used in real-life contexts.

KEY LANGUAGE

In lesson: °C, minus (⁻), negative

Other language to be used by the teacher: positive (⁺)

STRUCTURES AND REPRESENTATIONS

Number lines that pass through 0

RESOURCES

Optional: laminated paper thermometers, number cards

 In the eTextbook of this lesson, you will find interactive links to a selection of teaching tools.

Quick recap

Before introducing differences with negative numbers, ask children to find the difference between 12 and 25. Discuss the different methods that they use. Encourage children to use a number line. Repeat for other pairs of positive numbers (for example, 98 and 133).

Unit 15: Negative numbers, Lesson 4

Discover

WAYS OF WORKING Pair work

ASK
- Question 1 a): *How is a thermometer similar to a number line? How is it different? How can you use the thermometer to find the difference between the two temperatures?*
- Question 1 b): *How do you know which two temperatures have the greatest difference?*

IN FOCUS This section offers children an opportunity to begin calculating with negative numbers in context, in this instance finding the difference between two temperatures. The thermometers can be used like a number line and provide scaffolding.

PRACTICAL TIPS Provide children with laminated thermometers to colour in with dry-wipe pens. Give them a temperature to mark on their thermometer and ask them to show what the thermometer would look like if it got *x* degrees colder. Discuss what happens when the temperature drops below 0 °C.

ANSWERS

Question 1 a): In Tomsk, May is 13 °C warmer than March.

Question 1 b): The two months that have the greatest temperature difference are January and July. The temperature difference is 27 °C.

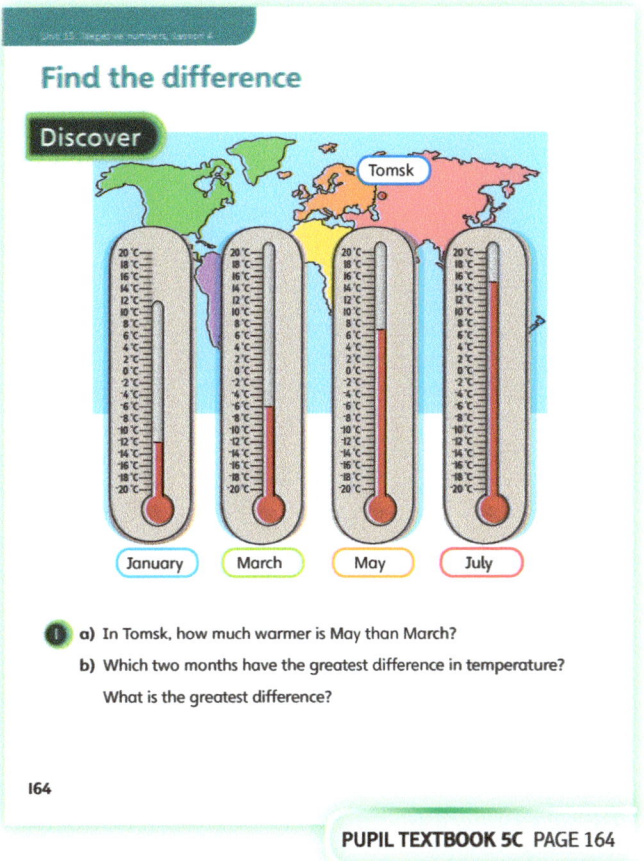

PUPIL TEXTBOOK 5C PAGE 164

Share

WAYS OF WORKING Whole class teacher led

ASK
- Question 1 a): *What difference did you find? Why was the difference not 1 °C?*
- Question 1 b): *How did you know which temperature was greatest and which was the lowest? How did you find the difference between the two temperatures? Why does Dexter's method work?*

IN FOCUS Questions 1 a) and b) link the thermometers seen in **Discover** to number lines, which children are very familiar with. As a class, establish an understanding of the method to find the difference between a positive and a negative number: count on from the negative number to zero, then count on from 0 to the positive number, and add the answers together.

Children may want to write down the size of the two jumps, and then choose a suitable method for addition (mentally, or column method, etc).

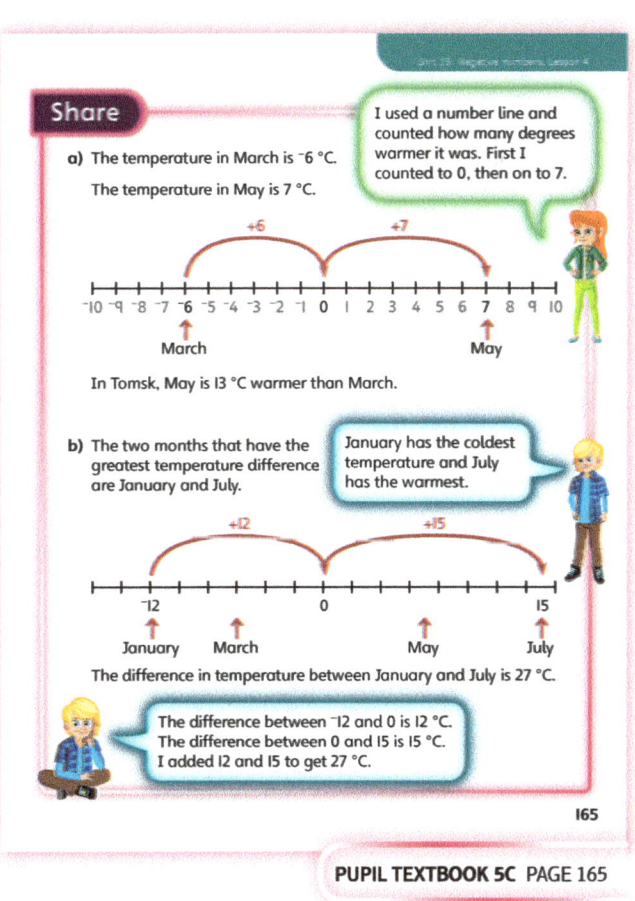

PUPIL TEXTBOOK 5C PAGE 165

Unit 15: Negative numbers, Lesson 4

Think together

WAYS OF WORKING Whole class teacher led (I do, We do, You do)

ASK

- Question ❶: *How could you find the difference between these temperatures using a number line? Which temperature is warmer? How do you know?*
- Question ❷: *How many floors do you predict Mrs Dean will have travelled down? Explain your prediction.*
- Question ❸: *How could you represent the temperatures in a different way? Can you find the difference between two of the temperatures?*

IN FOCUS Question ❸ deepens children's understanding as it gives them an opportunity to draw their own conclusions from a set of data that includes negative numbers. Children should be encouraged to justify their ideas with evidence or proof.

STRENGTHEN Encourage children to recognise that there is a number line hidden within the floor list in question ❷. Ask: *Could you show (or find) these numbers on a number line? What is the same about the number line and the lift buttons? What is different? How could you use a number line to find the difference between the two floors?*

DEEPEN Deepen learning in question ❶ by asking children to research some temperatures from cities around the world. They could create a table showing the temperature in each of ten cities. These should include both negative and positive temperatures. They can ask a partner to find the difference between the temperatures in two given cities.

ASSESSMENT CHECKPOINT Check that children can find the difference between a negative number and a positive number. Use children's conclusions in question ❸ to assess whether they can find differences between numbers, including two negative numbers.

ANSWERS

Question ❶: It is 21 °C warmer in Cairo than in New York.

Question ❷: Mrs Dean travels 19 floors down.

Question ❸: Children may give various answers, such as:
The temperature rises between 1 am and 3 am, it increases by 11 degrees.
The temperature rises between 3 am and 1 pm, it increases by 21 degrees.
The temperature falls between 1 pm and 6 pm, it decreases by 15 degrees.
It was 32 degrees warmer at 1 pm than at 1 am.

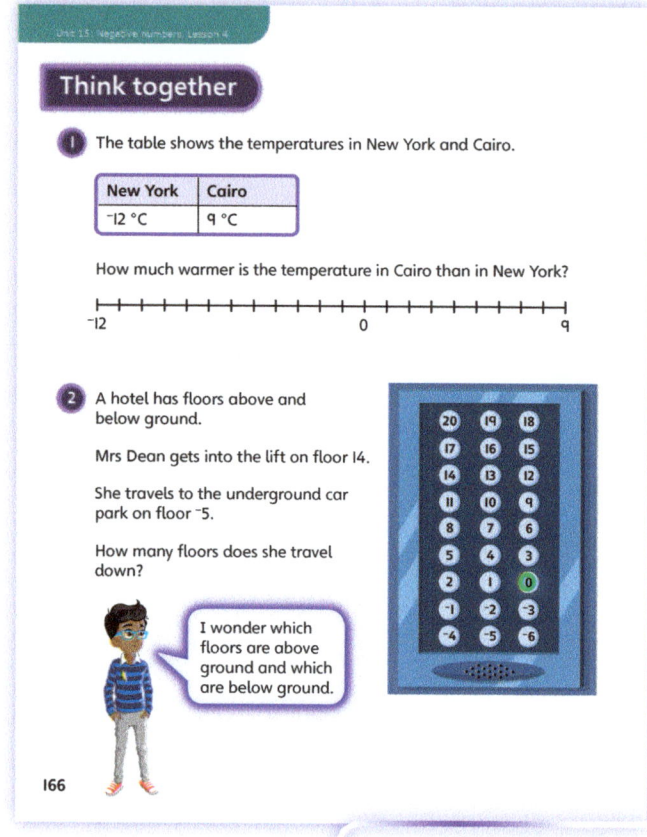

PUPIL TEXTBOOK 5C PAGE 166

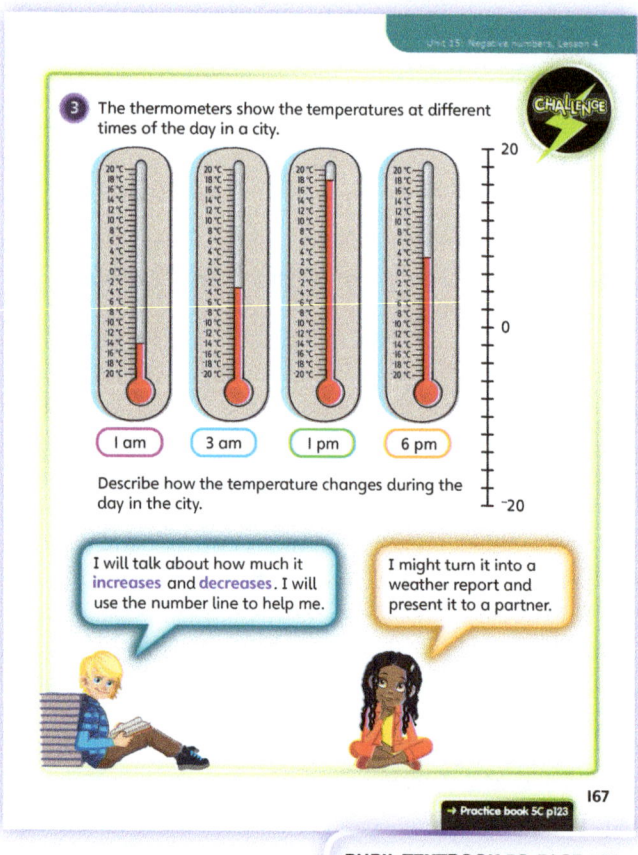

PUPIL TEXTBOOK 5C PAGE 167

Practice

WAYS OF WORKING Independent thinking

IN FOCUS Questions ❶ and ❷ will develop children's ability to recognise negative numbers independently and place them along a number line. Children will also have the opportunity to practise finding the difference between a positive and negative number, using a number line to support their thinking. Question ❸ allows children to practise finding negative numbers along a number line and to use this skill to find differences between and to order negative numbers.

STRENGTHEN For question ❸, provide laminated paper thermometers (from the **Discover** activity) and whiteboard pens, and ask children to shade a thermometer to show each temperature from the question. Then ask children to place the thermometers in order, from coldest to warmest. Encourage children to discuss how they decided which thermometer was showing the coldest temperature. This will cement their understanding of how to order negative numbers.

DEEPEN Question ❻ requires children to solve a problem with multiple steps within a different real-life context. Deepen learning by changing the digits. For example, change the iceberg to be 14 metres above sea level and ask children to work out the depth of the iceberg if it is 9 times (or 10 times) the height of the iceberg above sea level.

THINK DIFFERENTLY Question ❺ makes children aware of another real-life context for negative numbers. This question will challenge children's thinking as the 0 point (sea level) is an imaginary line, not actually below the mountain itself but some way up it.

ASSESSMENT CHECKPOINT Check that children are able to identify negative numbers on a number line accurately and find the difference between positive and negative numbers. Can they correctly order a list of numbers that includes negative numbers? Assess whether children can add a positive number to a negative number.

ANSWERS Answers for the **Practice** part of the lesson can be found in the *Power Maths* online subscription.

Reflect

WAYS OF WORKING Independent thinking or pair work

IN FOCUS This **Reflect** question gives a final opportunity to assess children's fluency with negative numbers. Encourage children to work with a partner, comparing ideas and discussing their conclusions.

ASSESSMENT CHECKPOINT Children should be able to use the concepts they have learnt, including their ability to calculate differences and order negative numbers, to write two sentences describing how the temperature changed during the day

ANSWERS Answers for the **Reflect** part of the lesson can be found in the *Power Maths* online subscription.

After the lesson

- Were children able to use accurate mathematical vocabulary to explain the key differences between positive and negative numbers?
- How confident were children in calculating with negative numbers? Was there an operation they were less confident with? How will you support this in future lessons?

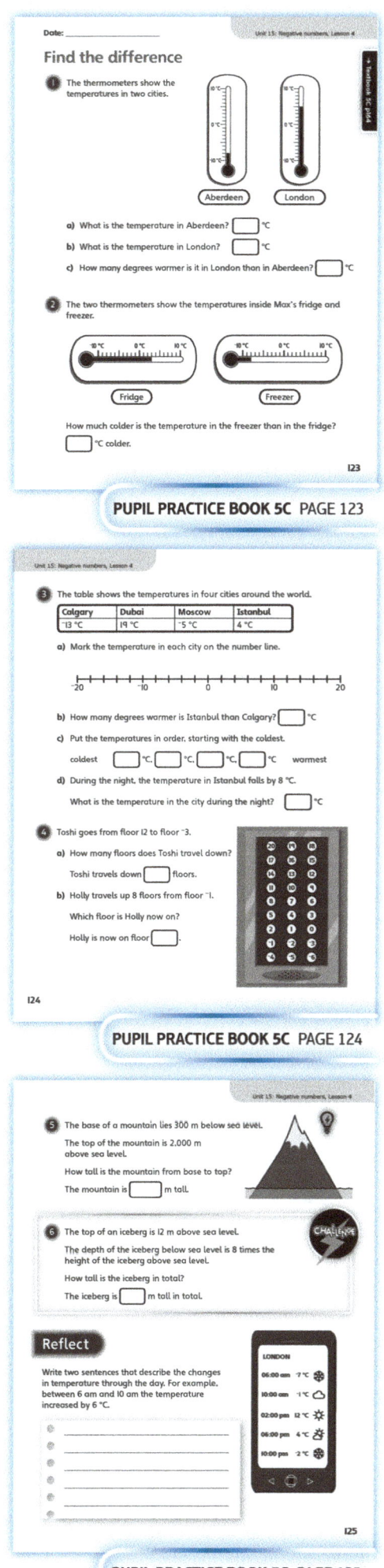

Unit 15: Negative numbers

End of unit check

> Don't forget the unit assessment grid in your *Power Maths* online subscription.

WAYS OF WORKING Group work adult led

IN FOCUS These questions are designed to draw out misconceptions or misunderstandings.
- Question ❶ assesses whether children can read a negative value from a number line scale that goes up in 2s.
- Question ❷ assesses children's ability to work out the next number in a counting sequence that goes down in 5s through zero.
- Question ❸ asks children to read off a negative value from a number line that goes up in 10s.
- In question ❹, children work out which comparison statement is incorrect.
- In question ❺, children apply their knowledge of the position of negative numbers to find the difference between a positive and a negative number.
- Question ❻ is a more challenging question, where children count back in 5s, through zero, from a non-multiple of 5. A common misconception in this question is that children get to 1 and then say ⁻1, as they believe the count should be symmetric on either side of zero.

ANSWERS AND COMMENTARY Children who have mastered the concepts in this unit will be able to demonstrate fluency in reading numbers off number scales, including negative numbers. They will be able to compare and order numbers and solve simple problems involving finding the difference between negative numbers. They will also be able to talk about instances where they have seen and used negative numbers in real life.

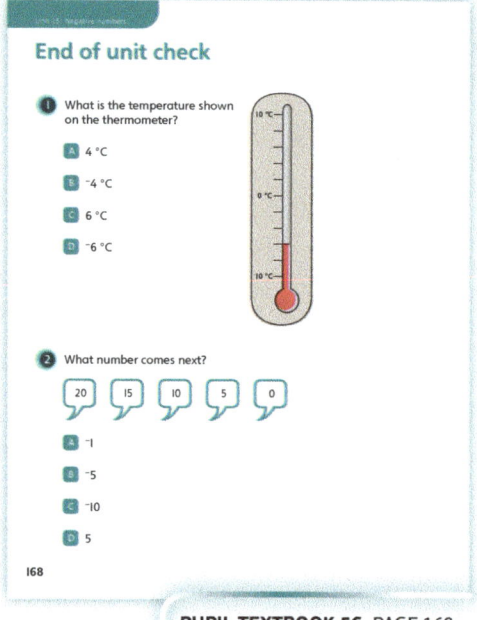

PUPIL TEXTBOOK 5C PAGE 168

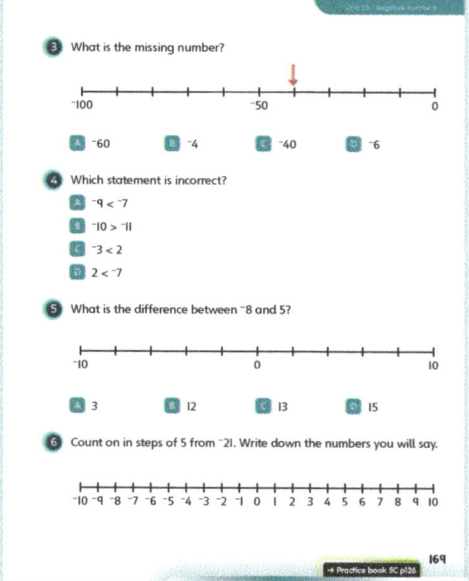

PUPIL TEXTBOOK 5C PAGE 169

Q	A	WRONG ANSWERS AND MISCONCEPTIONS	STRENGTHENING UNDERSTANDING
1	D	A or B may suggest that children understand that the mercury is 4 units above ⁻10, but have not counted on from ⁻10 to get to ⁻6.	To support children throughout, encourage them to use a number line to help them work out missing values or to help them compare numbers. When finding the difference between a negative number and any other number, offer children who need support the opportunity to use a number line to help them. Ask: • Can you find the two numbers on the number line? • Can you count the difference between them?
2	B	A suggests children think that the sequence decreases in 1s. D suggests children may think the sequence is now going back up.	
3	C	A implies that children are reading the number line as they would read a positive number line. B suggests children think the number line goes up in 1s.	
4	D	A, B or C suggest children have misunderstood how to order negative numbers.	
5	C	A suggests that children have calculated 8 − 3. B or D suggests a miscalculation.	
6	⁻21, ⁻16, ⁻11, ⁻6, ⁻1, 4, 9, 14, 19 Look out for children getting to 0 and then saying 1, 6, 11, etc. The count is not symmetric about 0 as the difference between 4 and ⁻1 is 5.		

204

Unit 15: Negative numbers

My journal

WAYS OF WORKING Independent thinking

IN FOCUS Children use their knowledge of the number line to help them estimate the values of A, B, C and D. Encourage children to think about where the centre of the line will be and what value will be at the centre (0). They then start to consider whether the arrows lie below 0 or above. Children may want to try to divide each side of 0 into 2 parts or 10 parts, for example, to help them estimate the other values.

To strengthen children's understanding, ask:
- If the start of the number line is ⁻10 and the end is 10, what number is the half-way point?
- Are there any clues you can use to estimate the numbers?

ANSWERS AND COMMENTARY

Various explanations may be given for children's estimates, for example:

I estimate **⁻7·5** for A because it is about half-way between ⁻10 and B.

I estimate **⁻5** for B because it is about a quarter of the way along the line.

I estimate **⁻2·5** for C because it is about half-way between ⁻5 (B) and 0 (D).

I estimate **0** for D because it is about half-way between ⁻10 and 10.

Power check

WAYS OF WORKING Independent thinking

ASK
- What are negative numbers?
- How confident are you with negative numbers?
- Could you explain to someone how to place negative numbers on a number line?
- Are you confident in your ability to compare and order negative numbers?

Power puzzle

WAYS OF WORKING Independent thinking or pair work

IN FOCUS Use this **Power puzzle** to assess whether children can recognise and follow number sequences. Understanding that number sequences follow a rule will help children to see that the numbers in a sequence will change in a regular way.

ANSWERS AND COMMENTARY Can children recognise consistent and regular number sequences in the given cards? The sequences are:
⁻22, ⁻16, ⁻10, ⁻4, 2 and 7, 4, 1, ⁻2, ⁻5, ⁻8 or
2, ⁻4, ⁻10, ⁻16, ⁻22 and ⁻8, ⁻5, ⁻2, 1, 4, 7

If children need support, ask:
- Which number is likely to be the first number in the sequence that gets bigger? Why?

PUPIL PRACTICE BOOK 5C PAGE 126

PUPIL PRACTICE BOOK 5C PAGE 127

After the unit

- How confident are you that children have a deep conceptual understanding of negative numbers?
- What one strategy or teaching approach worked particularly well in this unit? How can you apply the same strategy in other areas of your mathematics teaching?

Strengthen and **Deepen** activities for this unit can be found in the *Power Maths* online subscription.

Unit 16
Measure – converting units

Mastery Expert tip! 'This unit lends itself to so many opportunities for meaningful, real-life applications of maths concepts. Before starting the unit I identified as many instances as I could to make the concepts significant to my class, for example using local bus timetables and even a cookery lesson when I gave children a recipe entirely in imperial units.'

Don't forget to watch the Unit 16 video!

WHY THIS UNIT IS IMPORTANT

This unit consolidates children's existing knowledge of units of measurement and develops it further. It is a very practical unit in the sense that the skills children learn will be clearly applicable to real-life measurement situations. Many of the problems will be ones children will face at some point in life. For example, conversion between units (both metric and imperial), including conversion and scaling of amounts; and, when using timetables, including converting between units of time when the conversion does not result in a whole number answer.

WHERE THIS UNIT FITS

→ Unit 15: Negative numbers
→ **Unit 16: Measure – converting units**
→ Unit 17: Measure – volume

This unit builds on the concepts of measurement learnt in Year 4, particularly conversion between metric units. Children link their prior knowledge to bar models that will help them use these facts to convert units.

Before they start this unit, it is expected that children:
- can convert between different metric units of measure (whole number amounts) and between units of time
- can read and understand 24-hour clock times and convert between 24-hour and 12-hour clocks
- can calculate the duration between two times.

ASSESSING MASTERY

Children can convert between units of mass, length and capacity (metric → metric, imperial → imperial and metric ↔ imperial), and units of time, including where there is a remainder, and confidently apply this knowledge to solve problems. For each type of conversion, children can use reasoning to explain their methodology. Children can read information from timetables and use it to solve time-based problems.

COMMON MISCONCEPTIONS	STRENGTHENING UNDERSTANDING	GOING DEEPER
Children may confuse the operation needed to convert from one unit to another. Often this can be because they think that larger unit → smaller unit means larger amount → smaller amount (and so divide instead of multiplying).	Give children opportunities to measure simple lengths, masses and capacities using different units. Where possible, provide equipment that shows different units on the scales – for example, different metric units or metric and imperial units. Emphasise the equivalence between units using bar models.	Encourage children to practise converting between unusual combinations of units. For example, ask: *How tall is a 6-foot person in millimetres?* When working through several conversions in this way, ask children to draw a function machine or flow chart to show the steps they need to take. In the above example, convert feet into inches (× 12), inches into centimetres (× 2·5) and then centimetres into millimetres (× 10). Finally, challenge children to find one conversion that could replace the separate steps: in this case × 300 will convert directly from feet to millimetres.
Children may think that they can solve a problem without converting any units, just using the given numbers even though they may be in different units.	Ask children to circle or point to all mentions of units of measurement in the question. Where possible, provide examples of the measurements so they can see that they need to convert to use the numbers together. Discuss which units it makes sense to convert to in each problem.	

Unit 16: Measure – converting units

UNIT STARTER PAGES

Use these pages to introduce the unit focus to children. Can children see how the number line helps them convert easily between two units of measurement? Find out how many words they recognise from the list of vocabulary, especially the imperial units.

STRUCTURES AND REPRESENTATIONS

Bar model: This model helps children to represent the equivalence between different units of measure. Children can then see the calculation that they need to do to convert one unit into another.

3 feet		
1 foot	1 foot	1 foot
12 inches	12 inches	12 inches

Number line: This model also helps children in considering the equivalence of units. It can help them to convert between two units quickly or recognise where a measurement comes in terms of whole measures and parts (for example, 192 seconds is between 180 and 240 seconds, and so comes between 3 and 4 minutes). Number lines are also useful for working out durations between two times.

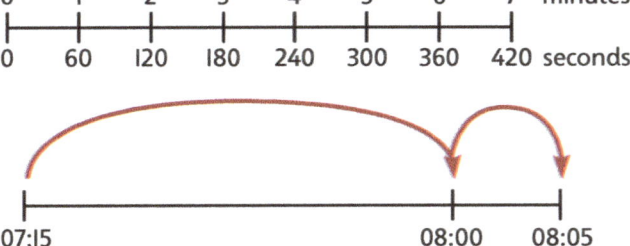

KEY LANGUAGE

There is some key language that children will need to know as part of the learning in this unit:

→ mass, capacity, length, time, quantity

→ metric units, gram (g), kilo, kilogram (kg), milli, millilitre (ml), litre (l), millimetre (mm), centimetre (cm), metre (m), kilometre (km)

→ imperial units, ounce (oz), pound (lb), stone (st), pint (pt), gallon, inch (in), foot (ft), yard (yd)

→ second, minute, hour, day, week, month, year

→ convert, equal to, equivalent, approximately, per, measure, remainder, multiple

→ timetable, 24-hour, digital, duration

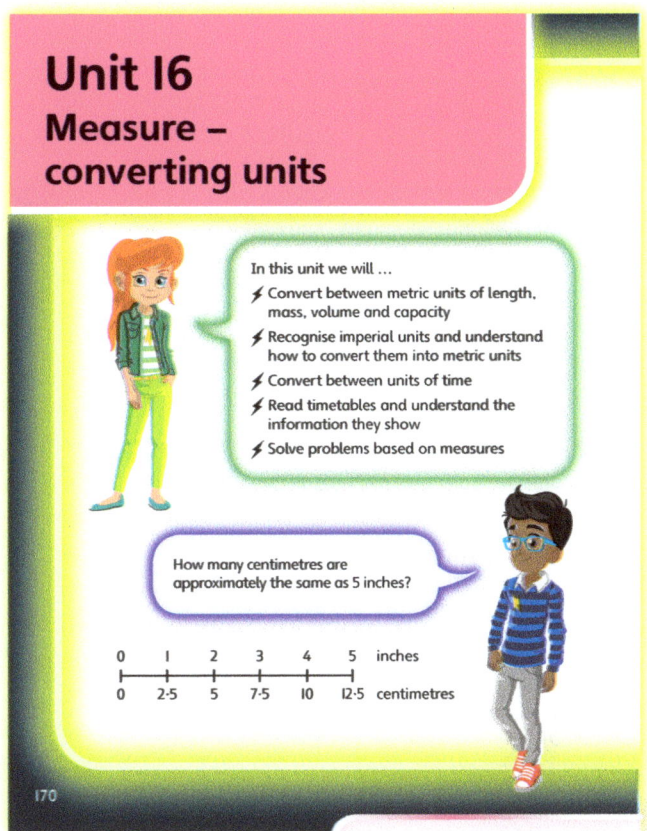

PUPIL TEXTBOOK 5C PAGE 170

PUPIL TEXTBOOK 5C PAGE 171

Unit 16: Measure – converting units, Lesson 1

Kilograms and kilometres

Learning focus

In this lesson, children will focus on metric units that begin with the prefix 'kilo'. They will apply their knowledge of place value to convert between kilograms and grams and vice versa.

Before you teach

- Are there ways you can adapt this lesson to link it to other lessons or curriculum work?
- How will you improve children's fluency when multiplying and dividing by 1,000?

NATIONAL CURRICULUM LINKS

Year 5 Measurement

Convert between different units of metric measure (for example, kilometre and metre; centimetre and metre; centimetre and millimetre; gram and kilogram; litre and millilitre).

ASSESSING MASTERY

Children recognise the relationship between kilometres and metres and between kilograms and grams. They can then apply this information to convert from one unit of measure to another, choosing to multiply or divide according to the units involved.

COMMON MISCONCEPTIONS

Children may confuse the operation needed to convert from one unit to another. Often this can be because they think a larger unit must become smaller when converting to a smaller unit (and so they divide instead of multiplying) and vice versa. Ask:

- *Would you expect the number itself to increase or decrease when you convert from cm to mm or litres to millilitres? Do you think you should multiply or divide?*

Children may think that they can multiply by 1,000 by writing three zeros on the end of a number, which does not work with decimals and can lead to confusion if encouraged as a strategy. Ask:

- *How does the value of a digit change when you multiply it by 1,000? … divide by 1,000?*

STRENGTHENING UNDERSTANDING

To strengthen understanding, provide practical opportunities for children to measure and use kilometres and metres and kilograms and grams in context. For example, provide metre measuring wheels and ask children to measure various distances around the school grounds in kilometres. Give them measuring scales set to kilograms and ask them to measure various objects in grams. Encourage children to work with measurements that involve decimals as well as whole numbers.

GOING DEEPER

Provide children with real-life data that shows distances of flights between cities in kilometres. Challenge children to design their own departure board for an airport where the distances are given in metres. Extend the task by providing data that has a mixture of units of measurement.

KEY LANGUAGE

In lesson: **kilo**, kilometre, metre, kilogram, gram, unit, convert, thousands, hundreds, tens, ones, tenths, mass, length

Other language to be used by the teacher: place value, hundredths

STRUCTURES AND REPRESENTATIONS

Bar model, place value grid

RESOURCES

Optional: measuring wheels, sets of scales, digit cards, suitcase (with a selection of items to fill it)

 In the eTextbook of this lesson, you will find interactive links to a selection of teaching tools.

Quick recap

Ask children to multiply whole numbers by 1,000. They should multiply 1-, 2- and 3-digit numbers. Then ask children to multiply numbers with 1 decimal place. Discuss the different methods that children use. Show on a place value grid how the digits of the number move.

Unit 16: Measure – converting units, Lesson 1

Discover

WAYS OF WORKING Pair work

ASK

- Question 1 a): *What are the different units of measurement you can see in the image? Why do you think long distances are usually measured in kilometres and not metres?*
- Question 1 a): *Would you expect the number of metres from London to Berlin to be a larger or smaller number than the number of kilometres? Why?*
- Question 1 b): *'6,000 is a lot more than 7, so Jen's bag is far too heavy.' What is wrong with this sentence?*
- Question 1: *What do you need to do to convert from one unit into another? Why?*

IN FOCUS Use the picture to discuss whether any children have experience of travelling on an aeroplane and, in particular, of weighing luggage prior to boarding a plane. Discuss what the luggage limit means and why aeroplanes have luggage limits for passengers.

PRACTICAL TIPS Show children a bag and talk about the sorts of things they might like to take with them onto an aeroplane. Have about 20 different items at the front of the class (for example, a book, a packet of sweets, a bottle of water, a pillow). Encourage children to choose a selection of these items and then weigh the bag to see whether they are over the luggage limit or not. This activity will not only help children to understand the scenario, but will also provide them with estimation practice and, if the scales show grams, a chance to discuss the need to convert between units.

ANSWERS

Question 1 a): It is 930,000 metres from London to Berlin.

Question 1 b): 6 kg < 7 kg, so Jen can take her bag onto the plane.

PUPIL TEXTBOOK 5C PAGE 172

Share

WAYS OF WORKING Whole class teacher led

ASK

- Question 1: *What is similar about the two units of measurement being used at the airport? Do their names give you a clue about how to convert them?*
- Question 1: *How does a place value grid help when multiplying or dividing by 1,000? Are there any other numbers you could use a place value grid to help to multiply or divide by?*
- Question 1: *Think about the two conversions that you needed to do to work out the answers to both questions. What is the same? What is different? Why?*

IN FOCUS Both questions 1 a) and b) use place value grids in order to model how to multiply and divide by 1,000. Check that children understand why this is such a useful way of representing the change in a number. They should be able to describe the effect on the digits when a number is multiplied or divided by 1,000.

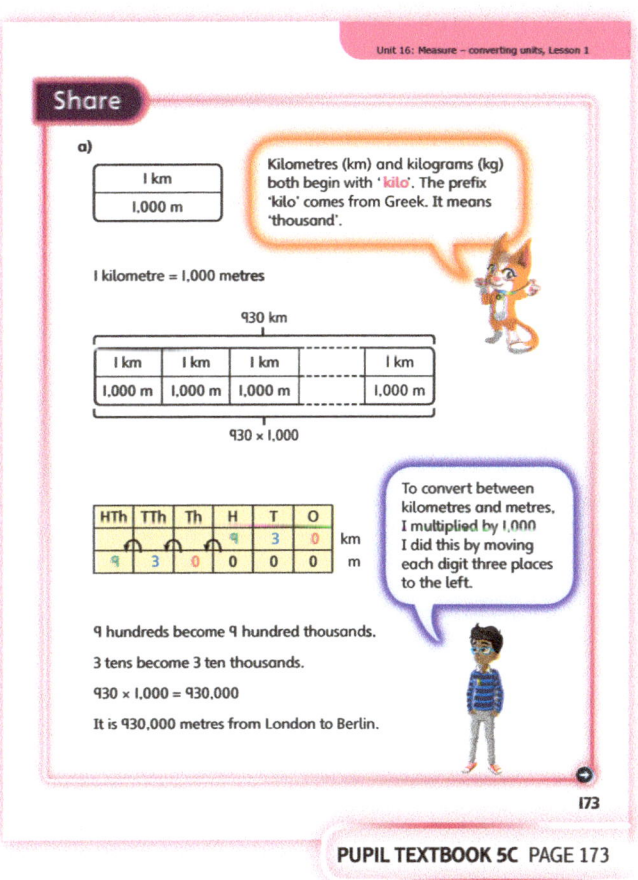

PUPIL TEXTBOOK 5C PAGE 173

209

Unit 16: Measure – converting units, Lesson 1

Think together

WAYS OF WORKING Whole class teacher led (I do, We do, You do)

ASK

- Question 1 a): *What unit of measure do the scales measure in? Why do you multiply by 1,000 when converting from kilograms to grams?*
- Question 1 b): *Are these masses in grams or kilograms? How can you work out the mass in kg?*
- Question 2: *What facts do you need to know to help you complete these conversions? Which do you multiply for and which do you divide for?*

IN FOCUS In question 1 a), children convert from kg to g. Help children to see that they need to use the same method of multiplying by 1,000 that they used in **Discover** and **Share** question 1 a). Remind children that 'kilo' means 1,000, so if they want to change from kg to g, or from km to m, they need to multiply by 1,000. Question 2 provides some fluency practice to help give children confidence when converting units from km to m, or vice versa.

STRENGTHEN To strengthen understanding of the effect of multiplying or dividing by 1,000, encourage children to draw place value grids and provide them with digit cards. Begin by modelling the multiplying of simple whole numbers by 1,000. Emphasise the three-place shift to the left in each digit. Ask children what they notice about how the value of each digit changes. Challenge them to predict the new value of a digit when a number is multiplied or divided by 1,000.

DEEPEN Provide children with measurements given in different units. For example, the mass of a rucksack is 5·9 kg and the mass of a suitcase is 5,200 g. Challenge them to compare the measurements and write them in order. Ask children whether it matters which unit they decide to convert, and whether it affects the answer.

ASSESSMENT CHECKPOINT Use questions 1 and 2 to assess whether children can convert between grams and kilogrammes, and betwen metres and kilometres. They should display a growing confidence when converting. Ensure children can explain clearly why and how place value grids are helpful.

ANSWERS

Question 1 a): The scales will show 5,000 g when the rucksack is placed on them.

Question 1 b): Green rucksack: 12 kg
Pink suitcase: 42 kg

Question 2 a): 3 km = 3,000 m   4·8 km = 4,800 m
5 km = 5,000 m   11·3 km = 11,300 m
17 km = 17,000 m   0·6 km = 600 m

Question 2 b): 6,000 m = 6 km   7,600 m = 7·6 km
19,000 m = 19 km   750 m = 0·75 km
260,000 m = 260 km   26,500 m = 26·5 km

Question 3: Multiplying by 1,000 involves shifting digits 3 places to the left.
Lee has only added three zeros which is incorrect. 8·3 kg = 8,300 g

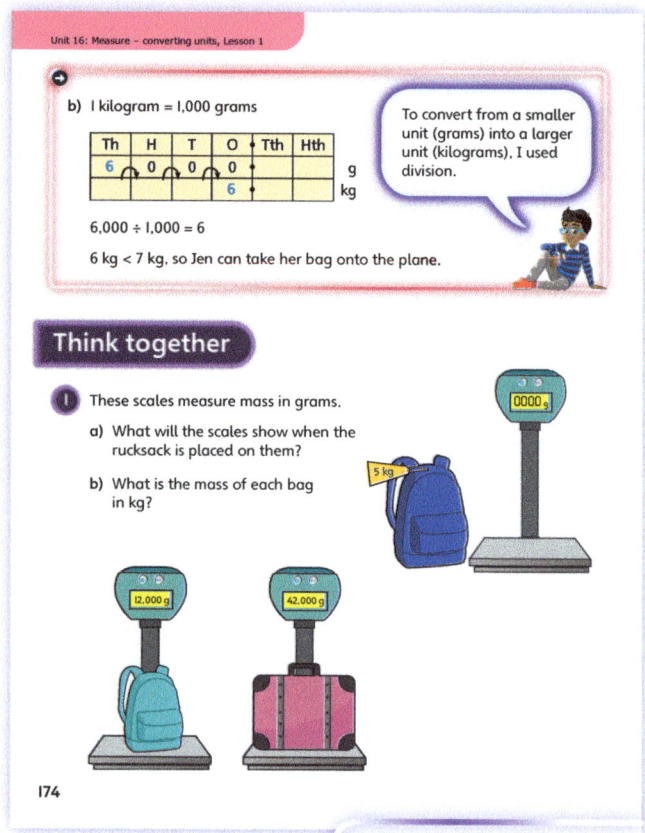

PUPIL TEXTBOOK 5C PAGE 174

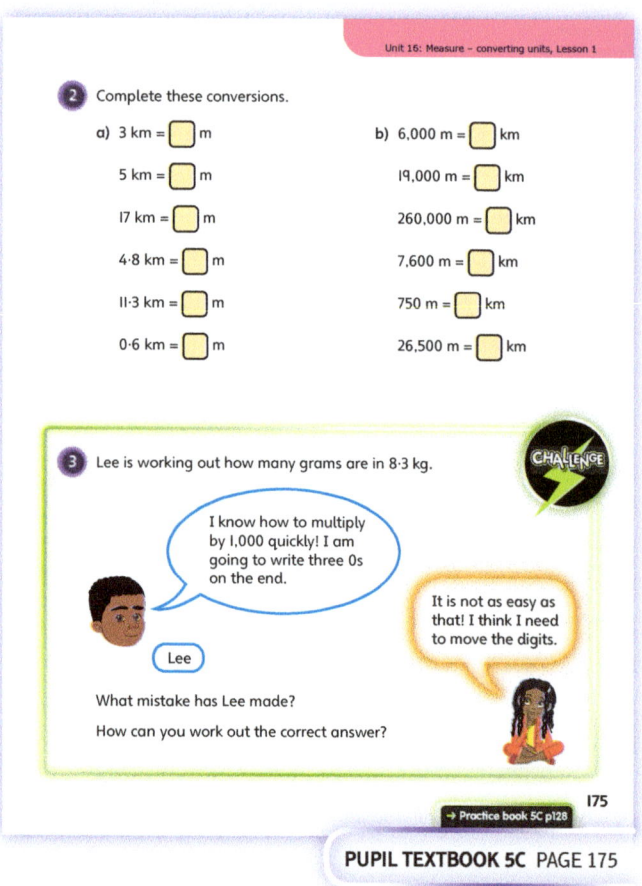

PUPIL TEXTBOOK 5C PAGE 175

Unit 16: Measure – converting units, Lesson 1

Practice

WAYS OF WORKING Independent thinking

IN FOCUS Questions ① and ② provide children with fluency practice, completing boxes to show what various measurements in kilograms and grams are equivalent to when converted. Question ③ assesses whether children understand when they need to multiply and divide by 1,000. They should see that when they move from a larger unit of measure to a smaller one they multiply, but divide when doing the opposite. Question ④ parts e) and f) are designed to show that there can be two ways to represent the same measurements. Children should therefore convert into grams and into kilograms separately.

STRENGTHEN For question ③, encourage children to express each conversion as larger units → smaller units, or vice versa. Ask them to say whether they would expect the number to get larger or smaller. Check that children do not confuse the move from larger to smaller units with a change from larger to smaller numbers. Provide large sorting circles and the statements on pieces of card for children to discuss each one as a group.

DEEPEN Use question ⑦ to deepen children's reasoning skills. Observe whether any children can answer the question using mental methods. Then challenge them to devise a rule that explains how they know the answer. For example, 'If an amount of grams is a multiple of 1,000 it will make a whole number of kilograms when it is converted.'

THINK DIFFERENTLY In question ⑥, children need to express in kilometres a distance that is not a whole number of kilometres using only the digits 4, 5 and 0. They should then give the possible distances in metres. Children should recognise that the 0 does not belong in the tenths position. Suggest that they find all possible distances before converting. Challenge children to do the conversions mentally.

ASSESSMENT CHECKPOINT Use questions ①, ④ and ⑦ to check that children are confident when applying their knowledge of converting between kilometres and metres, and kilograms and grams, in problem-solving contexts and in abstract conversions. Children should display reasoning skills, explaining appropriate methods.

ANSWERS Answers for the **Practice** part of the lesson can be found in the *Power Maths* online subscription.

Reflect

WAYS OF WORKING Independent thinking

IN FOCUS This question provides an opportunity to check children's methodology. Children should recognise that they need to divide by 1,000 in order to convert from grams into kilograms and should be able to explain how this is done by shifting the digits 3 places to the right.

ASSESSMENT CHECKPOINT Look for children who can describe clearly and accurately how to convert 12,500 g into kilograms.

ANSWERS Answers for the **Reflect** part of the lesson can be found in the *Power Maths* online subscription.

After the lesson

- How well did the prompts and questions promote learning and how will you learn from this in the next lesson on metric units?
- How did children respond to the materials and approaches used? Were they adequately (or excessively) challenged by them?

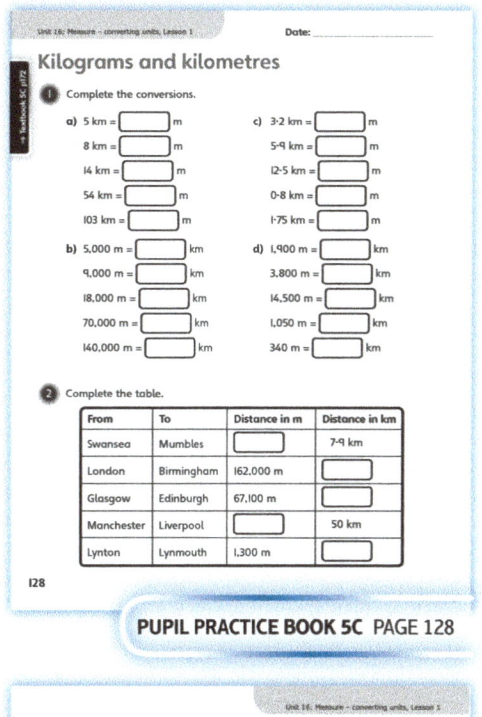

PUPIL PRACTICE BOOK 5C PAGE 128

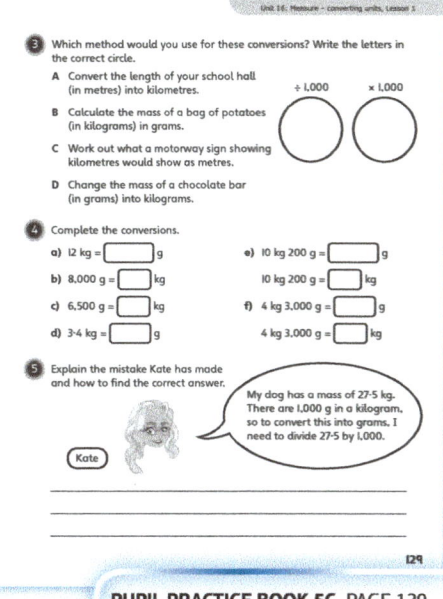

PUPIL PRACTICE BOOK 5C PAGE 129

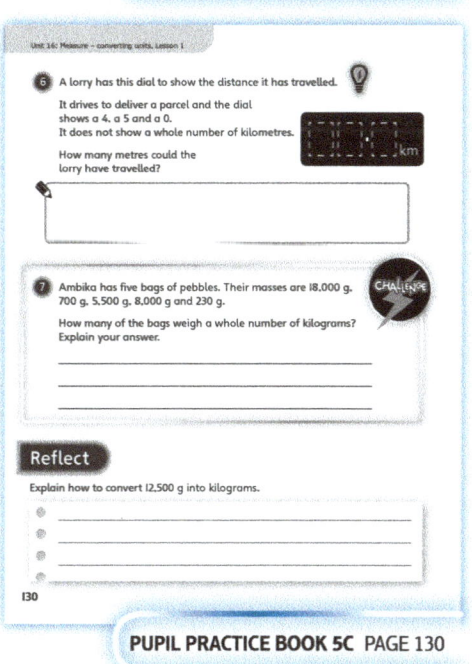

PUPIL PRACTICE BOOK 5C PAGE 130

Unit 16: Measure – converting units, Lesson 2

Millimetres and millilitres

Learning focus
In this lesson, children will focus on metric units that begin with the prefix 'milli'. They will apply their knowledge of place values to convert between millimetres and metres or centimetres, and between millilitres and litres.

Before you teach
- How might you scaffold questioning to help children to reflect on their assumptions?
- How will you provide practical opportunities to convert between metric units in this lesson?
- Are children confident multiplying and dividing by 1,000?

NATIONAL CURRICULUM LINKS

Year 5 Measurement

Convert between different units of metric measure (for example, kilometre and metre; centimetre and metre; centimetre and millimetre; gram and kilogram; litre and millilitre).

ASSESSING MASTERY

Children understand the relationships between millimetres and metres, millimetres and centimetres, and millilitres and litres. They are able to apply their knowledge when converting between these units of measure, choosing to multiply or divide according to the units involved.

COMMON MISCONCEPTIONS

Children may think that only 'kilo' units have a link with 1,000, because millimetres (and millilitres) are so small that it is hard to visualise how many fit into 1 metre (or litre), or possibly because they link 'milli' with one million. Ask:
- *What does 1 millimetre look like? How many millimetres do you think there are in 1 metre? Why?*

Children may confuse the operation needed to convert from one unit to another. They may link moving from a larger unit to a smaller unit to moving from a larger number to a smaller one (and so divide when they need to multiply). Ask:
- *Would you expect the number to increase or decrease when you convert from millilitres to litres? Do you think you should multiply or divide?*

STRENGTHENING UNDERSTANDING

To strengthen understanding, provide opportunities for children to measure and use millimetres and millilitres in context. As an example, you could give children water bottles (for example, 500 ml in capacity) and ask them to fill the bottles with water. Children should then use a 1-litre measuring jug (marked in parts of a litre: 0·1, 0·2 and so on) to find out their equivalence in litres.

GOING DEEPER

Consider fractions of litres, centimetres and metres in terms of millilitres and millimetres. Ask children to express $\frac{1}{4}$ of a litre or $\frac{3}{4}$ of a metre in smaller units. Challenge children to identify more complex fractions (for example, $\frac{3}{10}$ of a metre and $\frac{4}{1,000}$ of a litre).

KEY LANGUAGE

In lesson: milli, millimetre, centimetre, metre, millilitre, litre, unit, convert, place value, thousands, tens, tenths, ones, thousandths, length, capacity

Other language to be used by the teacher: hundreds, hundredths

STRUCTURES AND REPRESENTATIONS

Bar model, place value grid

RESOURCES

Optional: bottles with different capacities in ml, rulers marked in cm and rulers marked in mm, metre rulers marked in mm, large sheets of paper, measuring jug

 In the eTextbook of this lesson, you will find interactive links to a selection of teaching tools.

Quick recap
Check that children can divide 5- and 4-digit numbers by 1,000. Start with numbers that are multiples of 1,000 and then numbers that will lead to a decimal answer with up to 2 decimal places.

Unit 16: Measure – converting units, Lesson 2

Discover

WAYS OF WORKING Pair work

ASK

- Question ❶ a): *Which part of the flower bed is 2 metres long? The flower bed is measured in metres and the border fencing is measured in millimetres. Without unravelling the fencing, how can Ebo work out whether he has enough fence to go across the length of the flower bed?*
- Question ❶: *Can you think of any real-life objects that have about the same length or capacity of some of the things in the picture?*

IN FOCUS Discuss the situation that Ebo is in. Discuss when children might find themselves in a situation where they know the measurement of something, but need to convert that measurement into a different unit to see if they have enough. Use the opportunity to brainstorm real-life scenarios where metric units need to be converted. Ask children to describe which part of the flower bed is 2 metres long. This is an important tool to check that children understand the scenario and realise that 2 m refers to the length of the flower bed and not its perimeter.

PRACTICAL TIPS Provide pairs of children with pieces of string, labelled with their length in millimetres (for example, 900 mm). Set up two tables in the classroom a specific distance apart (for example, 1 m). Challenge children to predict whether their piece of string will reach across the gap. Ask what fact they need to know in order to know whether their piece of string will reach. Use the opportunity to reinforce the concept that 1 m = 1,000 mm.

ANSWERS

Question ❶ a): 1,500 < 2,000, so Ebo does not have enough fencing to go along the flower bed.

Question ❶ b): Alex has put 4·5 litres of water in the watering can.

PUPIL TEXTBOOK 5C PAGE 176

Share

WAYS OF WORKING Whole class teacher led

ASK

- Question ❶ a): *Can you think of any other words that begin with the prefix 'milli'? Are any related to the number 1,000? For example: milligram, $\frac{1}{1,000}$ of a gram; million, 1,000 thousands; millisecond, $\frac{1}{1,000}$ of a second.*
- Question ❶ a): *What does 'one thousandth' mean? Can you think of two ways to write this?*
- Question ❶ a): *How does the bar model help you to convert from metres to millimetres?*
- Question ❶ b): *If 1,000 ml is 1 litre, what is 500 ml in litres?*
- Question ❶ b): *Do you think a watering can of this capacity would usually be marked in millilitres or litres? Why?*

IN FOCUS Both questions ❶ a) and b) use bar models to represent the problem visually. Ask children to explain how the bar models have been used to help solve it.

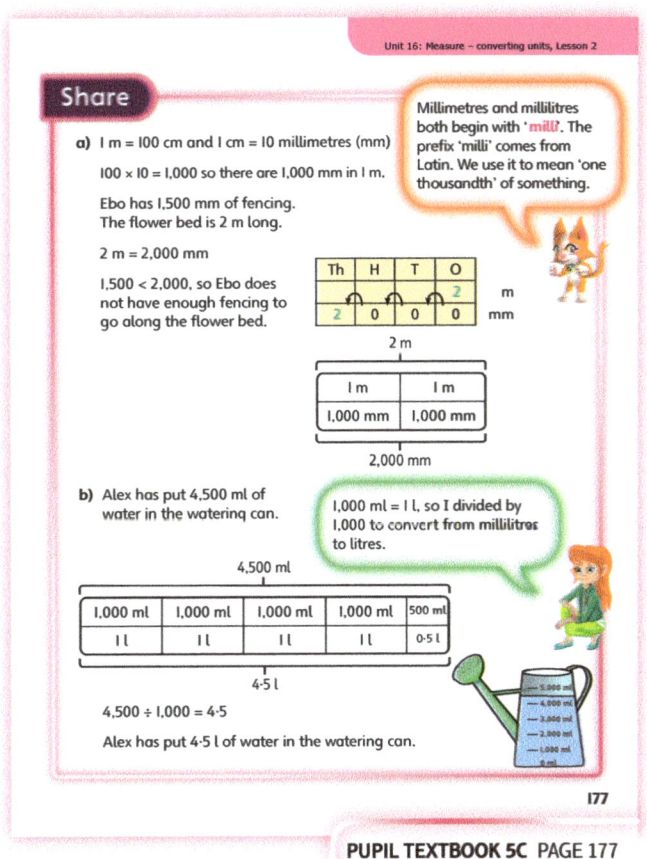

PUPIL TEXTBOOK 5C PAGE 177

213

Unit 16: Measure – converting units, Lesson 2

Think together

WAYS OF WORKING Whole class teacher led (I do, We do, You do)

ASK

- Question ❶: *How many litres of plant food are in the bottle? How can you convert this to millilitres?*
- Question ❷: *In which questions do you need to multiply by 1,000 and in which ones do you need to divide? What patterns do you notice?*
- Question ❸: *What facts do you know about centimetres, millimetres and metres? What is the same and what is different about the measures for length and capacity?*

IN FOCUS Question ❶ provides a context-driven problem, where children have to convert litres to millilitres. Question ❷ offers plenty of fluency practice for children to convert between units. They should notice that each time they are just multiplying or dividing by 1,000. They will start to realise that to move from a smaller unit of measure to a larger one they divide, and vice versa. Question ❸ introduces cm.

STRENGTHEN As any conversion between metric units simply involves a shift of digits, provide children with a place value grid to help them understand. Then challenge children to 'see' the answers without using a place value grid. Ask: *In which direction do the digits shift? Why?*

DEEPEN Question ❸ challenges children to consider the various relationships between units of length and capacity and the similarities between the two. Ensure children recognise that the relationship between a millimetre and a metre is the same as that between a metre and a kilometre. Ask: *How many millimetres are in a kilometre?*

ASSESSMENT CHECKPOINT Use questions ❶ and ❸ to assess whether children can convert between millimetres, centimetres and metres, and between millilitres and litres. They should recognise the relationship between 'milli' units and the concept of one thousandth and apply this correctly.

ANSWERS

Question ❶: The bottle contains 700 ml of plant food.

Question ❷ a): 4 l = 4,000 ml
9 m = 9,000 mm
14 l = 14,000 ml

c): 8·2 l = 8,200 ml
24·5 m = 24,500 mm
0·6 l = 600 ml

b): 4,000 mm = 4 m
19,000 mm = 19 m
185,000 mm = 185 m

d): 6,900 mm = 6·9 m
750 ml = 0·75 litres
26,500 ml = 26·5 l

Question ❸ a): 1 mm = $\frac{1}{1,000}$ of a m   1 ml = $\frac{1}{1,000}$ of a l
1 cm = $\frac{1}{100}$ of a m
1 m = 1,000 mm   1 l = 1,000 ml
1 m = 100 cm
1 mm is 0·001 m or 1 m is 0·001 km
1 ml is 0·001 l

Question ❸ b): The digits are the same in each column of the table.
The units of measurements are different, length is measured in mm, cm and km; capacity is measured in ml and l.

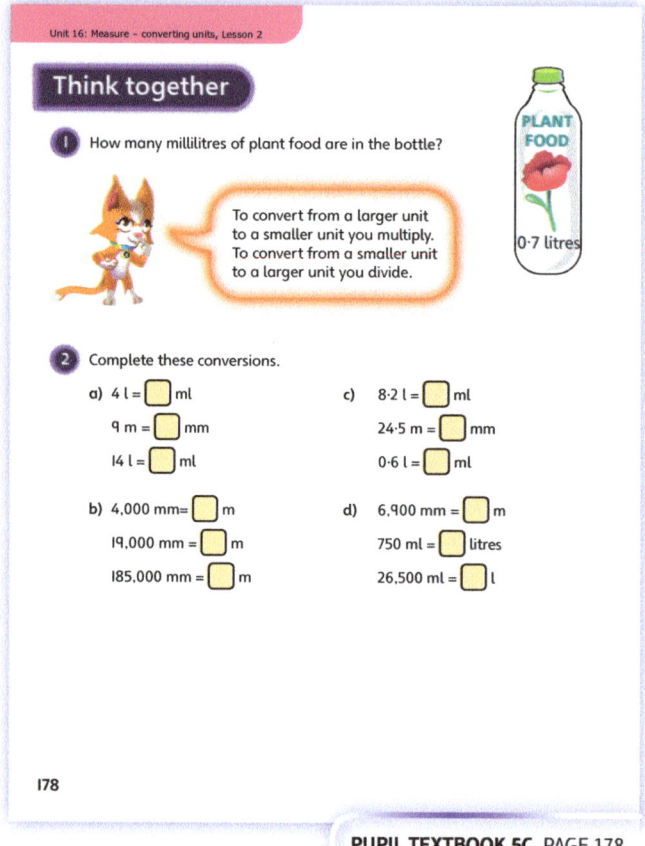

PUPIL TEXTBOOK 5C PAGE 178

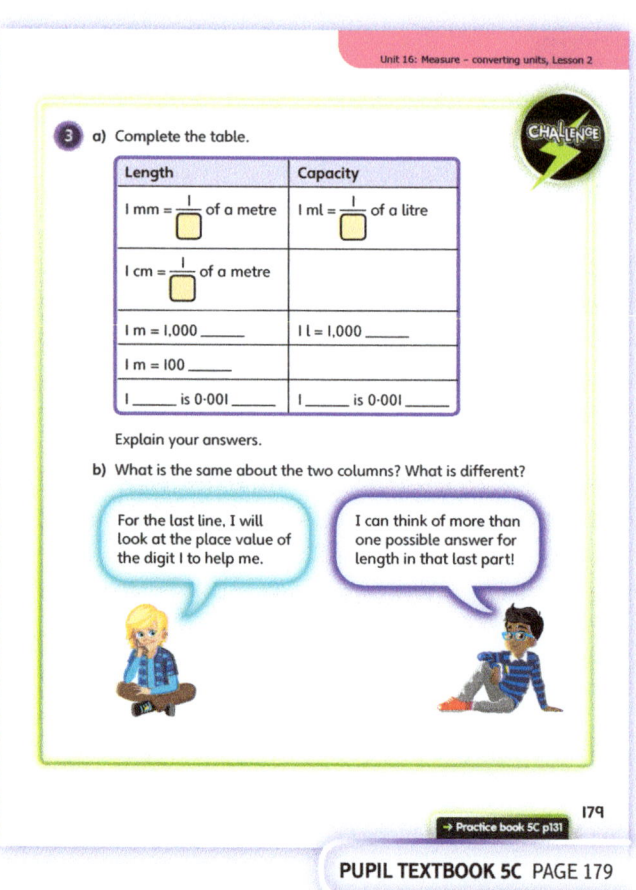

PUPIL TEXTBOOK 5C PAGE 179

Unit 16: Measure – converting units, Lesson 2

Practice

WAYS OF WORKING Independent thinking

IN FOCUS In questions ① and ② children use their understanding from the lesson to decide whether they need to multiply or divide to convert between litres and millilitres and between metres and millimetres. Questions ① and ② are structured so that children can build their confidence looking at different examples. Questions gradually build in complexity to help their understanding. Questions ③ and ④ provide conversions in contexts and children should apply their understanding to these situations. In questions ⑤ e) and f), children see that there are different ways to represent the same measure.

STRENGTHEN Provide children with metre rulers that clearly show millimetres and ask them to label their rulers to help convert between metres and millimetres. For example, the top of the ruler could be labelled in metres (from 0 to 1, with each 0·1 m marked) and the bottom of the ruler in millimetres (from 0 to 1,000, with each 100 mm marked). Children might also find it helpful to mark the equivalence between 1 cm and 10 mm.

DEEPEN In question ⑦, children apply their knowledge of converting between millilitres and litres in a problem-solving context. Three cups each have a capacity measured in millilitres and children need to make totals measured in litres using them. Observe those children who realise that it may be easier to convert the target numbers into millilitres and therefore explore the totals they can make by adding combinations of 100, 75 and 200. Challenge children to find all the possible totals in litres that Amelia could make using either one, two or three of the cups shown.

ASSESSMENT CHECKPOINT Use questions ① to ⑤ to assess whether children are confident when applying their knowledge of metric conversions between metres and millimetres, and between litres and millilitres, in problem-solving contexts and in abstract conversions. They should be able to demonstrate the appropriate methodology and explain why it works. Question ⑥ introduces centimetres. Children should understand that because there are 10 mm in 1 cm, Mo's measurement should be 10 times as big as Lee's measurement.

ANSWERS Answers for the **Practice** part of the lesson can be found in the *Power Maths* online subscription.

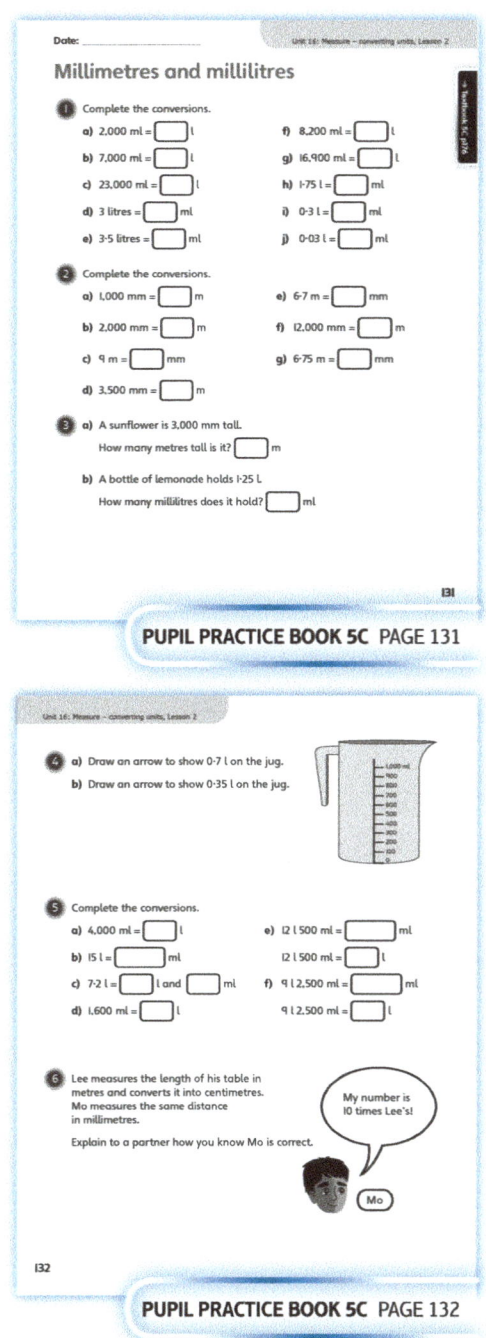

PUPIL PRACTICE BOOK 5C PAGE 131

PUPIL PRACTICE BOOK 5C PAGE 132

Reflect

WAYS OF WORKING Independent thinking

IN FOCUS This **Reflect** question provides an opportunity to assess whether children can apply the concepts learnt during the lesson. While the latter part of the statement is correct (10 mm = 1 cm), the way to convert from centimetres to millimetres is not to divide by 10, but to multiply. Children will need to consider the statement carefully in order to identify the error. Encourage them to convert the measurement in order to check their answers.

ASSESSMENT CHECKPOINT Look for children who can describe how to convert from centimetres into millimetres.

ANSWERS Answers for the **Reflect** part of the lesson can be found in the *Power Maths* online subscription.

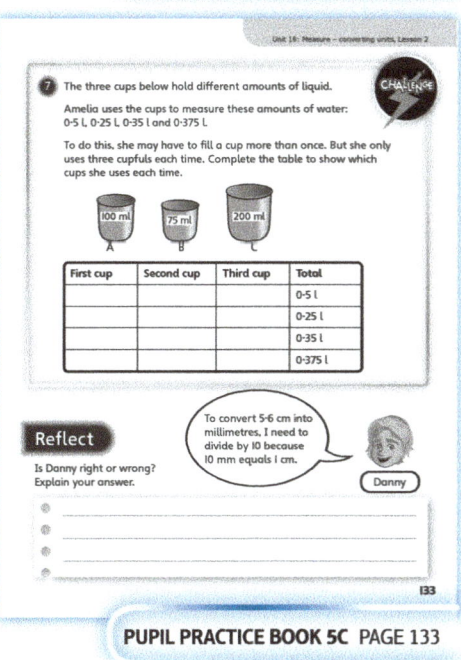

PUPIL PRACTICE BOOK 5C PAGE 133

After the lesson

- Can children recognise when they need to convert units?
- Where did you provide feedback to push children to think more deeply or to learn more about process skills?

215

Unit 16: Measure – converting units, Lesson 3

Convert units of length

Learning focus
In this lesson, children will convert between all combinations of mm, cm, m and km, including splitting conversions into more than one step (for example, mm → m → km). Children will identify the values they need to use for any conversion and apply these in multi-step problem-solving contexts.

Before you teach
- Although the focus of this lesson is on the relationship between metric units of length, how could you apply the rules more generally?
- Can children name the different units of measurement used for measuring length?

NATIONAL CURRICULUM LINKS

Year 5 Measurement

Convert between different units of metric measure (for example, kilometre and metre; centimetre and metre; centimetre and millimetre; gram and kilogram; litre and millilitre).

ASSESSING MASTERY

Children can confidently apply their knowledge of metric units of length to solving problems that involve converting between and calculating with them. They are able to use reasoning to explain their methodology.

COMMON MISCONCEPTIONS

Children may think that they are limited to certain numbers when converting. For example, they may think that all conversions involving kilometres require multiplying and dividing by 1,000. Ask:
- *What facts do you know about the two units you need to convert between? How does this help you know how to convert from one unit to the other?*

Children may think that they are limited to certain units when converting. For example, centimetres to the next smallest unit (millimetres) or the next largest (metres), but not directly to kilometres. This may be because they have less experience of this and the numbers involved are difficult to visualise (for example, 5 cm = 0·00005 km). Ask:
- *Which units of measurement could you convert a distance in millimetres into?*

STRENGTHENING UNDERSTANDING

Encourage children to make flashcard reminders of how to convert between mm and cm, cm and m, and m and km, including simple bar models or number lines as pictorial prompts. To convert between units more than one step removed (for example, cm → km), they can place their flashcards next to each other and talk about the way to find the answer in two jumps (cm → m → km: divide by 100, then divide by 1,000) or in one jump directly to km (divide by 100,000).

GOING DEEPER

Ask children to research distance-based world records that incorporate a variety of units, for example: furthest long jump, furthest journey by tandem bicycle, furthest ocean swim, tallest person, smallest dog, etc. They can use these facts to devise their own questions that require unit conversion for a partner to solve.

KEY LANGUAGE

In lesson: base **metric unit**, millimetre, centimetre, metre, kilometre, length, convert, decimal, unit

STRUCTURES AND REPRESENTATIONS

Bar model, place value grid

RESOURCES

Optional: place value grids, digit cards, a selection of rulers showing different metric units

 In the eTextbook of this lesson, you will find interactive links to a selection of teaching tools.

Quick recap

Practise multiplying and dividing numbers by 10, 100 and 1,000. Numbers could include whole numbers and decimals when multiplying.

Unit 16: Measure – converting units, Lesson 3

Discover

WAYS OF WORKING Pair work

ASK

- Question 1 a): *If you made a similar 1 km chain of coins in a straight line from your school, where do you think you would get as far as? What have you used to help you estimate 1 km?*
- Question 1 a): *'1 km of coins means 1,000 pennies which equals £10·00.' What is wrong with this statement?*
- Question 1 b): *Can you explain why the children will raise more money if they place the coins on their sides? Use plastic counters to show what you mean.*

IN FOCUS Discuss the scenario and talk about whether children have ever seen coins used in this way. As well as chains of coins, children may have seen photos of floors covered with coins as a form of charity fundraising. Ask why they think 1p coins have been chosen for this question. Encourage children to estimate the number of coins they think might stretch to 1 km and the value of this amount. Observe whether children's estimates are reasonable and whether any understand the link between the number of coins and their value (the number of coins divided by 100 gives the value in pounds).

PRACTICAL TIPS Provide children with some 1p coins. Ask them to measure the diameter and depth of each coin and discuss whether 1 cm and 1 mm (respectively) are accurate measurements. Challenge children to suggest the number of coins they would need to cover short distances (for example, 1 cm or 10 cm). Encourage them to translate this into a monetary value. By considering simpler distances, children are building the skills that can then be applied to the longer distances featured in this **Discover**.

ANSWERS

Question 1 a): The children will need 100,000 1 pence coins to make a line 1 km long.
They will have raised £1,000 for charity.

Question 1 b): The children would raise £9,000 more if they placed the coins on their sides.

Share

WAYS OF WORKING Whole class teacher led

ASK

- Question 1 a): *What do you think this statement means: 'The 'base metric unit of length is the metre'? What do you think is the base metric unit of capacity?*
- Question 1 a): *How do the names of each unit give you a clue about what they are worth?*
- Question 1: *What facts will help you to work out the answers quickly? How would the questions be more difficult if the children used a different coin to make a chain from?*

IN FOCUS Both questions have several parts to them and it is important that children can understand the purpose behind each step. Ask children to write a description of each step on a piece of card and arrange them to show the method. For example: find the number of coins in 1 m; calculate the value of 1 m of coins; work out the number of metres in 1 km; calculate the value of 1 km of coins.

PUPIL TEXTBOOK 5C PAGE 180

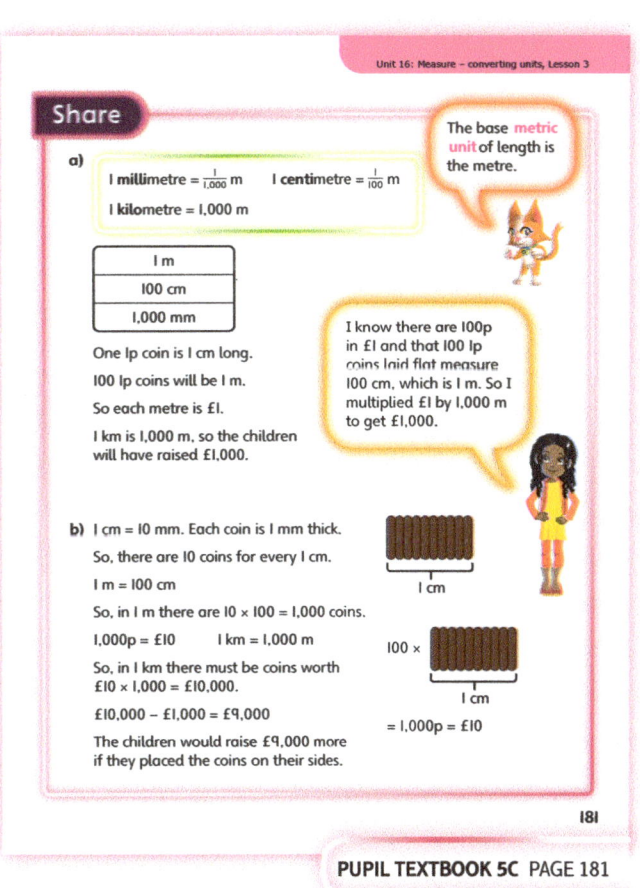

PUPIL TEXTBOOK 5C PAGE 181

Unit 16: Measure – converting units, Lesson 3

Think together

WAYS OF WORKING Whole class teacher led (I do, We do, You do)

ASK

- Question ① a): *How do you convert m to cm? Do you need to multiply 1·6 by 100 or divide it by 100?*
- Question ① b): *How many barrels make up the tower? If the whole tower is 160 cm tall, what do you need to do to find the height of one barrel?*
- Question ②: *How would you convert $\frac{1}{2}$ to a decimal? How does converting $\frac{1}{2}$ to a decimal help? How many centimetres are in 1 m? How many metres are in 1 km?*
- Question ②: *Once you have converted $\frac{1}{2}$ km to a decimal, what unit do you need to change it to before you can change it to cm?*

IN FOCUS In question ① a), because there are 100 cm in every metre, children should realise that they need to multiply 1·6 m by 100 to find the height in cm. For question ① b), children should realise they need to use their answer from ① a).

In question ②, children have to work through the correct order of steps to complete the conversion. Ask children if they can work out what number to multiply by to change from km directly into cm. Elicit that they need to multiply by 1,000 and then multiply by 100, which is the same as multiplying by 1,000 × 100 = 100,000.

In question ③, explain to children that the *standard* units for measuring length are mm, m and km in just the same way that the standard units for measuring capacity are ml and l. For each conversion between mm and m or between m and km, they multiply or divide by 1,000 just as they do with ml and l. For units of length, cm is commonly used (however, 'cl' is not often used).

STRENGTHEN Provide place value grids and digit cards for children to use in questions ① and ②. Children should move the digit cards twice to model converting in two steps.

DEEPEN For question ③, make a blank version of the diagram shown in the answer below (but include the mm, cm, m, and km notation), and challenge children to complete it.

ASSESSMENT CHECKPOINT Use questions ① and ② to assess whether children can switch between different units of length. Can they convert using one or two steps and apply their knowledge to problem-solving contexts?

ANSWERS

Question ① a): 1·6 m = 160 cm

Question ① b): Each barrel is 32 cm tall.

Question ② a): B, C, A

Question ② b): Max walks 50,000 cm.

Question ③: Children should be able to complete a diagram like this.

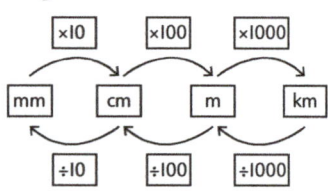

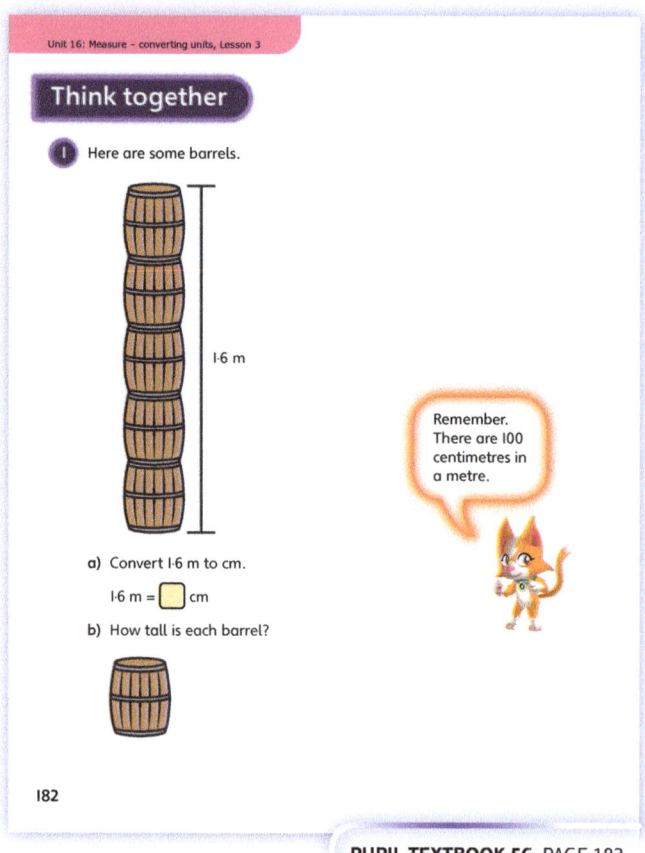

PUPIL TEXTBOOK 5C PAGE 182

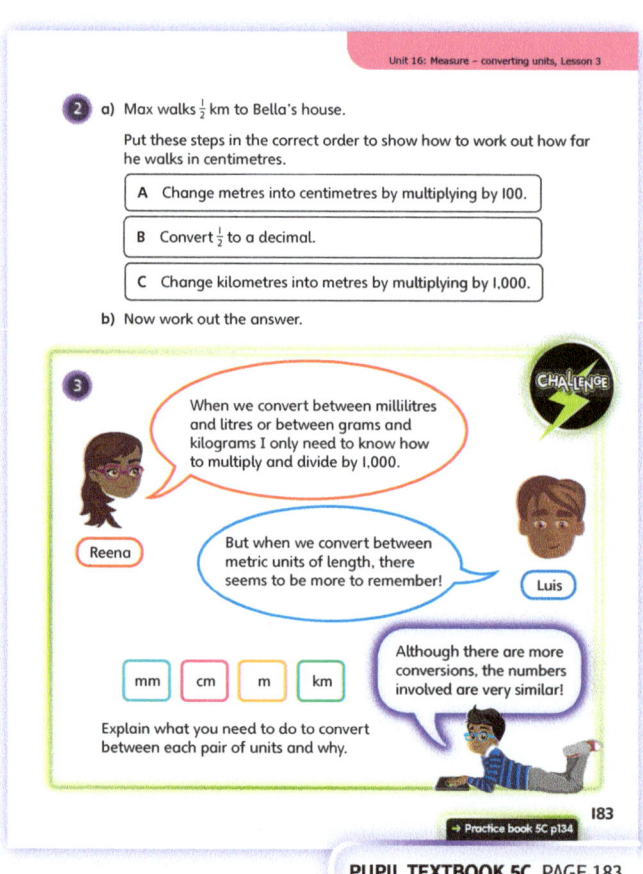

PUPIL TEXTBOOK 5C PAGE 183

218

Unit 16: Measure – converting units, Lesson 3

Practice

WAYS OF WORKING Independent thinking

IN FOCUS In question ❶, children demonstrate their knowledge of different metric conversions. In questions ❷ and ❸, children apply these conversions to work out missing values. Check that children are confident with whether they need to divide or multiply depending on the units that they are moving from and to. Questions ❹ to ❻ provide children with the opportunity to start to apply their conversion of units knowledge to contexts involving measurement.

STRENGTHEN Ask children to model question ❺ by providing them with digit cards and place value grids. Encourage children to convert each measurement into centimetres and to model each shift in digits using the cards. Discuss why it is easier to order the measurements when they are converted to centimetres first. Ask children whether they would get the same answer if they converted the measurements into metres instead.

DEEPEN Ask children to make a poster showing how to convert between different units of length. They could illustrate it with examples of world records from question ❼ or their own research.

THINK DIFFERENTLY Question ❻ requires children to apply their knowledge of conversion to calculating a perimeter. They need to understand that when finding perimeters and areas of shapes, it is very important that all the measurements are in the same unit.

ASSESSMENT CHECKPOINT Use questions ❹ to ❻ to assess whether children can apply their knowledge of metric units of length to solving multi-step problems that involve converting between units and calculating with the resulting numbers. Children should be able to reason with confidence, and explain their methodology for each conversion.

ANSWERS Answers for the **Practice** part of the lesson can be found in the *Power Maths* online subscription.

Reflect

WAYS OF WORKING Independent thinking

IN FOCUS The aim of this **Reflect** question is for children to display their knowledge of metric units in different ways. Although the obvious answer is that there are 10 mm in 1 cm, encourage children to give answers that equal part of a unit and remind them that 10 of one unit does not always have to equal 1 of another (for example, 10 cm in 0·1 m). Prompt them with questions such as: *What is 1 m in terms of kilometres? So, what is 10 m in terms of kilometres?* Encourage children to use their knowledge of conversion to help (for example, converting 10 cm into millimetres).

ASSESSMENT CHECKPOINT Look for children who are able to demonstrate their knowledge of metric units by suggesting appropriate ways in which each sentence can be completed.

ANSWERS Answers for the **Reflect** part of the lesson can be found in the *Power Maths* online subscription.

After the lesson

- The next few lessons deal with imperial and metric units of measurement. Do you feel that children's grasp of metric units is solid enough to apply their knowledge in this way without further consolidation?

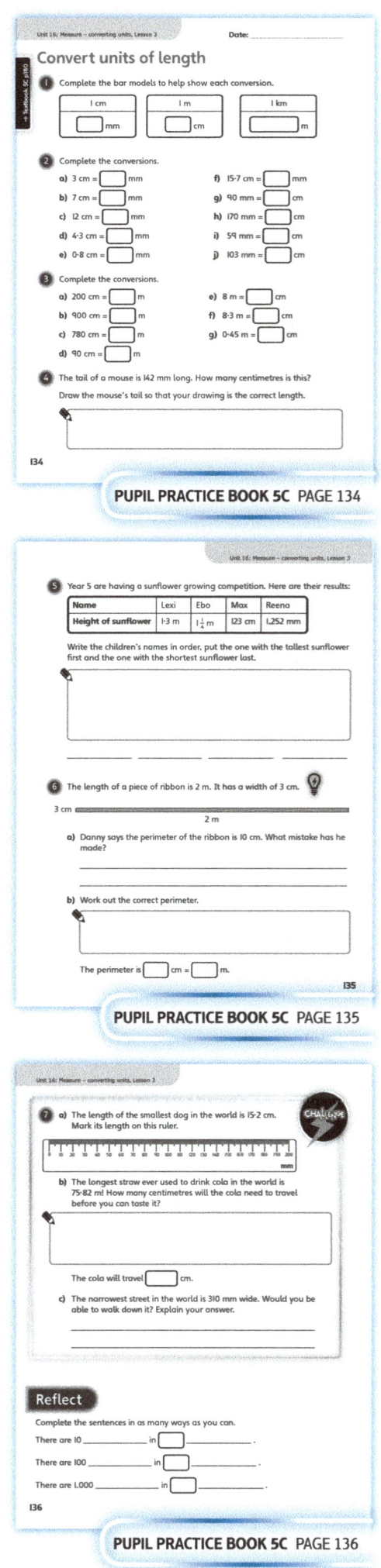

PUPIL PRACTICE BOOK 5C PAGE 134

PUPIL PRACTICE BOOK 5C PAGE 135

PUPIL PRACTICE BOOK 5C PAGE 136

Unit 16: Measure – converting units, Lesson 4

Imperial units of length

Learning focus

In this lesson, children will be introduced to imperial units of length. They will understand the terms inches, feet and yards, convert between these and use approximations to convert from imperial to metric units.

Before you teach

- Based on previous lessons taught in this unit, are there any additional misconceptions you need to consider before this lesson?
- Think about some real-life examples of imperial measurements you can use to reinforce the concepts of the lesson?

NATIONAL CURRICULUM LINKS

Year 5 Measurement

Understand and use approximate equivalences between metric units and common imperial units such as inches, pounds and pints.

ASSESSING MASTERY

Children understand the term 'imperial units'. They can convert between common imperial units of length using given facts (12 inches = 1 foot, 3 feet = 1 yard). They can use given approximations to convert between imperial and metric units (for example, 1 inch is about 2·5 cm). Children are able to apply these skills when solving problems.

COMMON MISCONCEPTIONS

Children may think that it is not possible to convert between imperial and metric units. This may be because so far they have only worked with the metric system. They may think that any unit that does not fit this pattern of base units (the 'metre') cannot be converted. Ask children to show what they think 1 inch looks like and ask:
- *Approximately how many centimetres do you think this is?*

Children may be unsure how to find the answer when more than one conversion is needed to solve a problem (for example, to convert yards → centimetres, first convert yards into inches and then centimetres). Ask:
- *How many inches are there in 1 yard? Approximately how many centimetres are there in 1 inch?*

STRENGTHENING UNDERSTANDING

Give children extensive practical experience of an explorative nature in measuring, comparing and converting imperial units before embarking on problem solving. Give children rulers and tape measures that show inches and centimetres or inches and feet, then ask them to measure different pieces of string that are whole inches long. Children are not expected to recognise each relationship just through measurement, but practical activities will build their understanding before starting on more abstract representations of these concepts.

GOING DEEPER

Encourage children to design their own 'Whodunnit?' problems in the style of the **Discover** exercise, but extended to include a comparison of different units of measurement (for example, four characters with each character's height given in one of inches, feet and inches, centimetres, and metres, the tallest character being the culprit).

KEY LANGUAGE

In lesson: imperial units, metric units, **inch/inches** (in), **foot/feet** (ft), **yard** (yd), centimetre, metre, convert, approximately

STRUCTURES AND REPRESENTATIONS

Bar model

RESOURCES

Optional: rulers and tape measures showing imperial and metric units, sticky labels

 In the eTextbook of this lesson, you will find interactive links to a selection of teaching tools.

Quick recap

Practise multiplication and division facts for the 12 times-table.

Unit 16: Measure – converting units, Lesson 4

Discover

WAYS OF WORKING Pair work

ASK

- Question ❶: *What do you notice about the units of measure shown in the picture?*
- Question ❶: *When have you heard people use inches, feet or yards in real life?*
- Question ❶: *Can you show with your hands how long you think 1 inch is? And 1 foot? 1 yard?*

IN FOCUS Imperial units of measure are introduced for the first time in this picture. It is important that time is spent talking about these new units before attempting to find the solution. Without discussing and exploring imperial units as a concept (particularly in real life), 'inches' and 'feet' do not hold any meaning and the conversion chart could be replaced with nonsense words and children could still calculate the answer (for example, if children knew that 1 zog ≈ 2·5 cm, they could still convert zogs into centimetres!).

PRACTICAL TIPS Provide lots of opportunities for children to explore the imperial measurements used in the picture. A fun idea might be to give children rulers or tape measures marked in inches and ask them to mark the various heights of the birds on a wall in chalk. The experience of measuring in a new unit will help children to both visualise its worth and (if they use tape measures) to spot the equivalence between feet and inches by the markings.

ANSWERS

Question ❶ a): The penguin swallowed the ring.

Question ❶ b): The ostrich is 120 cm tall, which is the same as 1·2 m.

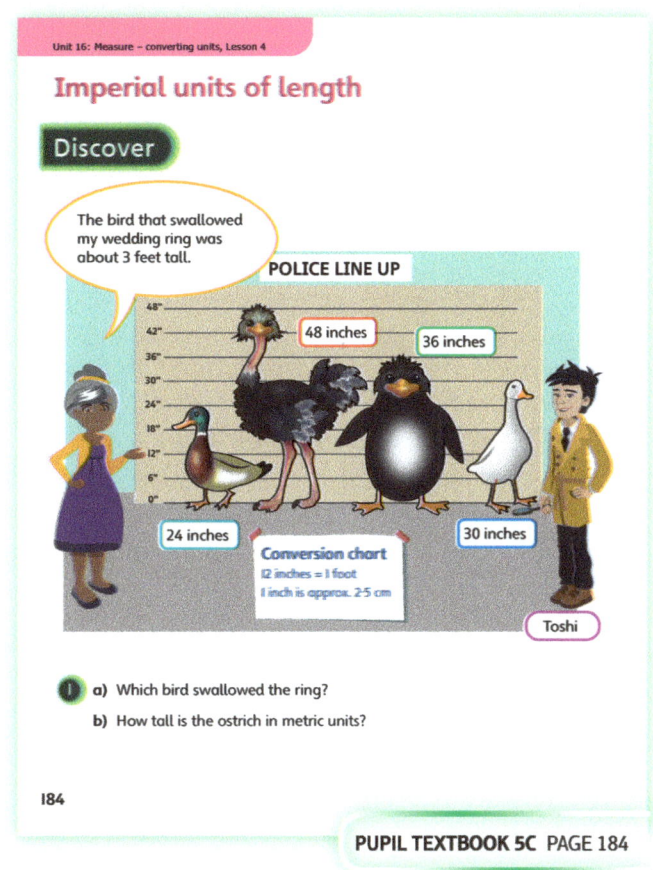

PUPIL TEXTBOOK 5C PAGE 184

Share

WAYS OF WORKING Whole class teacher led

ASK

- Question ❶ a): *Why do you think the metric system might have made the heights easier to work with?*
- Questions ❶ a) and b): *Can you explain how the two bar models show the conversion between the units?*
- Question ❶ b): *Read Dexter's comment. Can you think of another '× 0·5' calculation that you can use to show what Dexter means?*

IN FOCUS There is an interesting difference between questions ❶ a) and b) and it is important that children spend time understanding this. In part a), children are given an exact equivalence (12 in = 1 ft) and in part b), they are given an approximation (1 in ≈ 2·5 cm). Provide rulers marked in inches and centimetres and ask children to draw a line 1 inch long and measure it very carefully in centimetres. They should see that 1 inch is just over $2\frac{1}{2}$ cm (or about 25·5 mm; actual measurement: 25·4 mm or 2·54 cm). Discuss why we approximate this to $2\frac{1}{2}$ cm, and the usefulness of rounding to make calculations easier to work with. Ensure that children understand (and are reminded regularly) that all imperial → imperial conversions will be exact and all imperial ↔ metric conversions will be approximate.

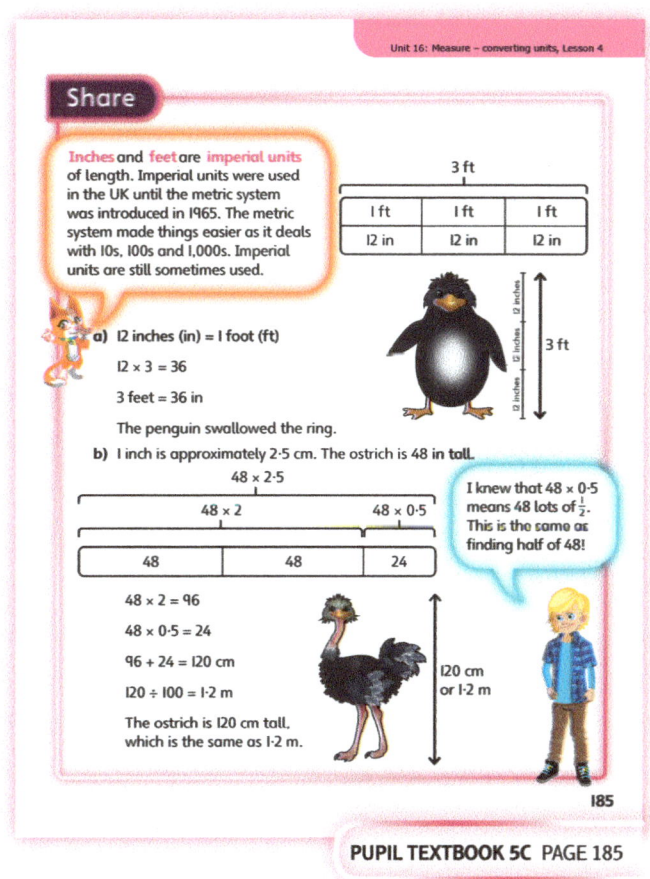

PUPIL TEXTBOOK 5C PAGE 185

Unit 16: Measure – converting units, Lesson 4

Think together

WAYS OF WORKING Whole class teacher led (I do, We do, You do)

ASK

- Question ①: *What strategy can you use to multiply a number by 12 and so convert feet into inches?*
- Question ①: *The emu's height is an amount of whole feet with an amount of inches. Can you explain what to do with this amount of inches when you are converting into inches?*
- Question ②: *Dexter uses two steps. How could you convert yards into inches using only one multiplication?*

IN FOCUS In question ②, the partially completed bar model illustrates how to find the answer by converting from yards to feet (× 3) and then from feet to inches (× 12). Check that children understand these two steps. Ask if it is possible to convert yards to inches in one step. In effect, this is the same as removing the '3 ft' bar of the bar model, so each yard is clearly the same as 36 inches. The answer can therefore be found by multiplying 15 by 36.

STRENGTHEN Cut thick coloured strips of paper 1 yard long (two strips may need to be taped together), 1 foot long and 1 inch long. For question ②, challenge children to begin to make the bar model using the coloured strips, emphasising the equivalences between each unit. Ask how many 1-yard strips they will need to complete the bar model, how many 1-foot strips and how many 1-inch strips.

DEEPEN In question ③, discuss Astrid and Ash's comments and why they might have to convert 1 foot and 1 yard into inches before converting into metric. Children can work in pairs to plan their strategies. Remind them to use various metric units (for example, 1 in is approximately 25 mm and 1 yd is approximately 0·9 m). Give children different measurements for them to practise using their conversions.

ASSESSMENT CHECKPOINT Use questions ① and ② to assess whether children can use given facts to convert between imperial units of length. Use question ③ to assess whether children can convert between imperial and metric units. Children should be growing in confidence when applying these conversions in problem-solving contexts and be able to suggest appropriate strategies.

ANSWERS

Question ①: 5 × 12 = 60   60 + 3 = 63
5 feet 3 inches is equal to 63 inches.
The emu is 63 inches tall.

Question ②: 15 × 3 = 45, so 15 yards = 45 feet
45 × 12 = 540, so 15 yards = 540 inches.
The pond is 540 inches wide.

Question ③: Children's answers will vary. For example:
1 inch ≈ 2·5 cm, 25 mm, 0·025 m
1 foot ≈ 30 cm, 300 mm, 0·3 m (12 × 2·5 = 30)
1 yard ≈ 90 cm, 900 mm, 0·9 m (3 × 30 = 90)

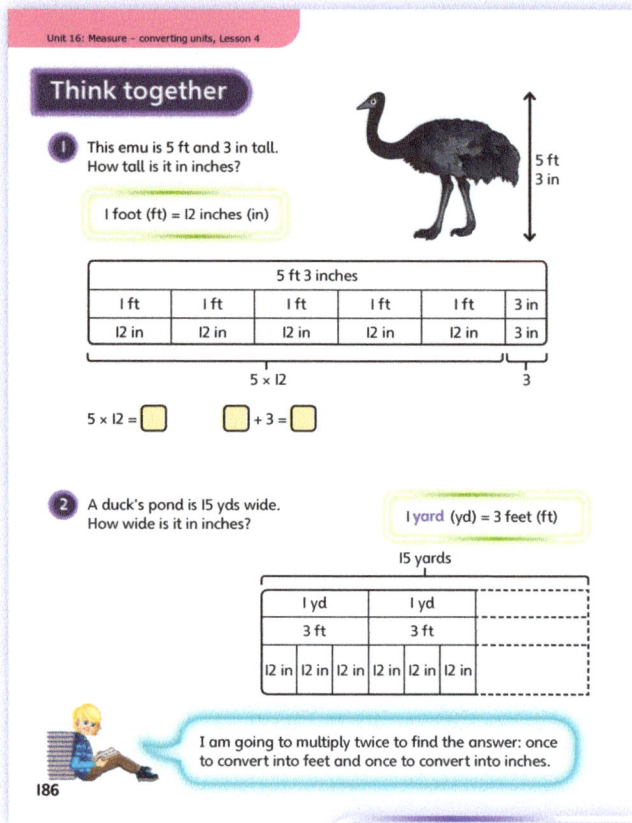

PUPIL TEXTBOOK 5C PAGE 186

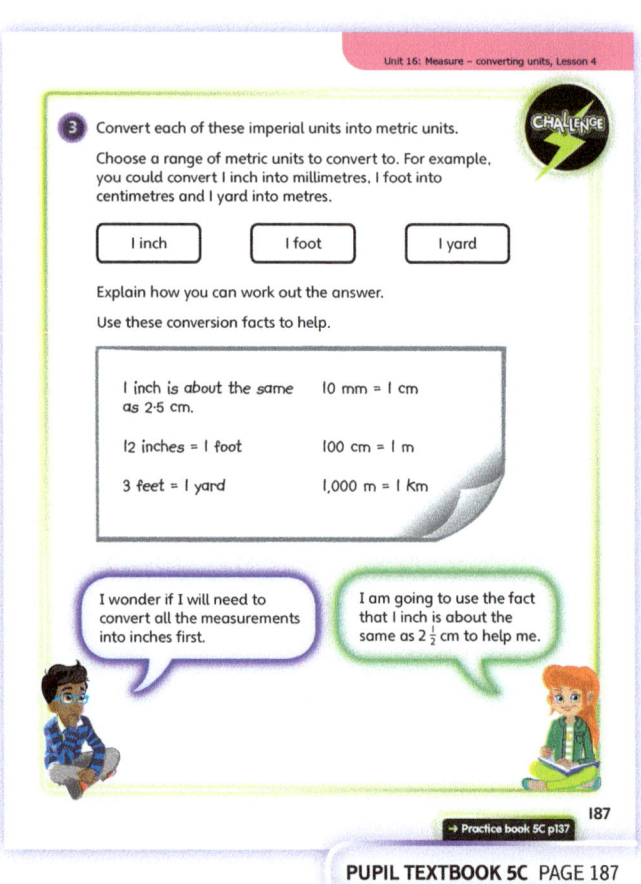

PUPIL TEXTBOOK 5C PAGE 187

Unit 16: Measure – converting units, Lesson 4

Practice

WAYS OF WORKING Independent thinking

IN FOCUS In question ④, children compare pairs of imperial measurements given in different units. From their earlier work comparing metric units children should be aware that they can convert either one of the units. Ask whether there is more than one way to compare the measurements and which they find easier. Discuss whether conversions are easier with metric units and why. As well as the ease of multiplying and dividing by 10, 100, 1,000 and so on, the fact that the digits don't alter in metric conversions means that the answer can often be seen quickly, without the need to do any real calculation.

STRENGTHEN For question ③, children's understanding could be strengthened by providing them with rulers marked in inches and in centimetres, together with sticky labels to label the ruler in the same way as the one in the text.

DEEPEN In question ⑥, ask children to discuss their method with a partner and to decide if they have approached the problem in the most efficient way. For example, to find 100 yds did they go back to the start or use the answer to the 20 yds conversion?

ASSESSMENT CHECKPOINT Use questions ② and ④ to assess whether children can convert between imperial units. Use questions ③ and ⑥ to assess whether children can convert from imperial units to metric units using approximations. Children should be confident when applying their knowledge in problem-solving contexts.

ANSWERS Answers for the **Practice** part of the lesson can be found in the *Power Maths* online subscription.

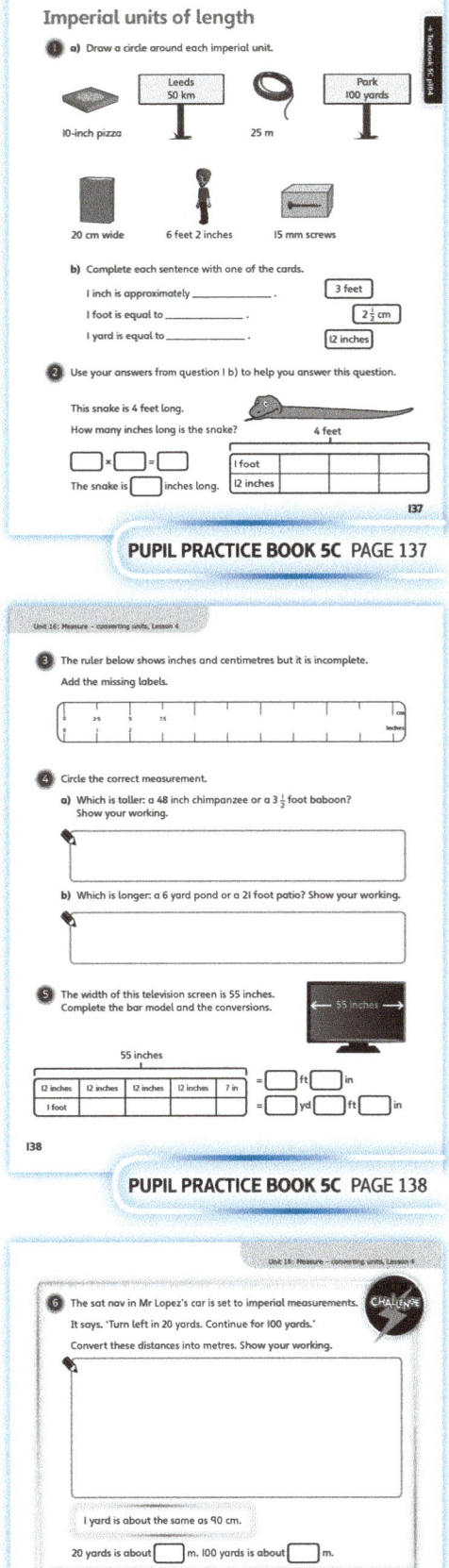

PUPIL PRACTICE BOOK 5C PAGE 137

PUPIL PRACTICE BOOK 5C PAGE 138

PUPIL PRACTICE BOOK 5C PAGE 139

Reflect

WAYS OF WORKING Independent thinking

IN FOCUS This **Reflect** question asks children to identify a mistake or misconception. As part of their explanation, encourage children to give more appropriate answers, correcting the mistake.

ASSESSMENT CHECKPOINT Listen for children giving a clear explanation of Jamie's mistake and correcting it. Look for children who recognise that there may be different explanations for Jamie's mistake. For example, she may have confused feet for yards or metres, and may think (still incorrectly) that a typical human is about 2 m tall.

ANSWERS Answers for the **Reflect** part of the lesson can be found in the *Power Maths* online subscription.

After the lesson

- How well do you feel that children understand the general concept of imperial units of measurement as opposed to metric units?
- What misconceptions and mistakes did they make that you can take into account in the next lesson on imperial measures?

Unit 16: Measure – converting units, Lesson 5

Imperial units of mass

Learning focus
In this lesson, children will be introduced to imperial units of mass. They will understand the terms ounces, pounds and stones, convert between them and use approximations to convert from imperial to metric units.

Before you teach
- How do you plan to include practical experiences of ounces, pounds and stones during the lesson?
- How might you develop and refine children's own representations of problems?

NATIONAL CURRICULUM LINKS

Year 5 Measurement

Understand and use approximate equivalences between metric units and common imperial units such as inches, pounds and pints.

ASSESSING MASTERY

Children can recognise common imperial units of mass and convert between them using given facts (16 ounces = 1 pound, 14 pounds = 1 stone). Chidren can use given approximations to convert between imperial and metric units (for example, 1 oz is about 28 g). Children are able to apply these skills when solving problems.

COMMON MISCONCEPTIONS

Children may be unsure how to find the answer when more than one conversion is needed to solve a problem (for example, to convert stones and pounds → grams, convert stones and pounds → pounds → ounces → grams). Ask:
- *How many pounds are in 1 stone? How many ounces are in 1 pound?*

STRENGTHENING UNDERSTANDING

Provide children with practical experience of measuring, comparing and converting imperial units of mass before embarking on problem solving. Encourage children to investigate and explore as they measure using weighing scales that display both imperial and metric units. For example, ask children to weigh out 1 ounce of sand and to read how many grams this is. They are not expected to recognise each relationship just through measurement, but practical activities will build their understanding.

GOING DEEPER

Ask children to investigate the labelling on food products. They should find that many are labelled with both a metric mass and its imperial equivalent, where the metric value is a round number (for example, 500 g or 1 kg). However, some products (for example, jars of jam and packs of sausages) are labelled with only a mass of 454 g. Discuss why companies have not rounded up to 500 g or down to 450 g. Children may apply their knowledge from the lesson to see that 454 g is close to 448 g and is a way of approximating 1 lb. Even though these products are sold in a country using the metric system, they are still sold by the pound (or by the 454 g). Children could also devise their own conversion problems based on one of the units on a label.

KEY LANGUAGE

In lesson: imperial units, metric units, **pounds** (lb), **ounce** (oz), **stone** (st), gram, kilogram, convert, approximately, mass, long multiplication, estimate

STRUCTURES AND REPRESENTATIONS

Bar model, number line, grid method, long multiplication

RESOURCES

Optional: weighing scales (imperial and metric), examples of masses (for example, 1 oz and 1 lb), two-sided flashcards for conversions, flashcards for imperial and metric units of mass, sticky labels

 In the eTextbook of this lesson, you will find interactive links to a selection of teaching tools.

Quick recap
Ask children to work to calculate 36 × 28 using a grid or area method. What steps do they take? Can they work out the answer in another way? If children need a lot of support, give them more calculations to check that they understand the method.

Unit 16: Measure – converting units, Lesson 5

Discover

WAYS OF WORKING Pair work

ASK

- Question ① a): *When have you heard people use ounces, pounds or stones in real life?*
- Question ①: *How heavy do you think 1 ounce or 1 pound is? Can you estimate the mass of objects in ounces or in pounds?*

IN FOCUS Talk about the scenario. Explain that fruit and vegetables were sold using pounds and ounces until fairly recently, when new laws said traders had to use metric measurements (although they can still also display the imperial equivalent on the label). While units of length and capacity can be understood visually, it is more difficult for children to make links between units of mass and what they can see. When discussing the scenario, provide pre-weighed examples of 32 oz, 16 oz (1 lb), 4 oz and 1 oz to allow children to feel objects that weigh these amounts. This will allow them to make cognitive links between the words on the page and what they represent.

PRACTICAL TIPS Give children plenty of opportunities to explore imperial measures of mass. Provide sets of scales and materials (for example, sand, dried peas or modelling clay). Being able to actually measure amounts in a new unit should have the dual effect of ensuring children can gauge the values of ounces and pounds and (if the weighing scales show both units) spot the equivalence between these imperial units by noting the markings on the scale.

ANSWERS

Question ① a): Alex should ask for 2 lb of apples and $\frac{1}{4}$ lb (or 0·25 lb) of blueberries.

Question ① b): Alex's fruit will weigh about 1·008 kg (1,008 g).

PUPIL TEXTBOOK 5C PAGE 188

Share

WAYS OF WORKING Whole class teacher led

ASK

- Question ①: *In as few words as possible, describe what each part of the question is asking you to do.*
- Question ① a): *How do the bar models help you to convert between units of measurement?*
- Question ① a): *Explain why 4 ounces is the same as $\frac{1}{4}$ of a pound. How could you write this as a decimal?*
- Question ① b): *36 × 28 is shown using the grid method. Can you think of a different method?*

IN FOCUS Challenge children to identify the stages in the calculations where they are converting from one unit to another. Draw arrows between units on the board and ask children to label these with the relevant conversion calculation (for example, ounces → grams: × 28). Ensure that children understand why they need to multiply or divide (larger → smaller unit or smaller → larger, respectively) and how they know what value to multiply or divide by. For question ① b), discuss the different methods children might use to convert 36 oz into kilograms.

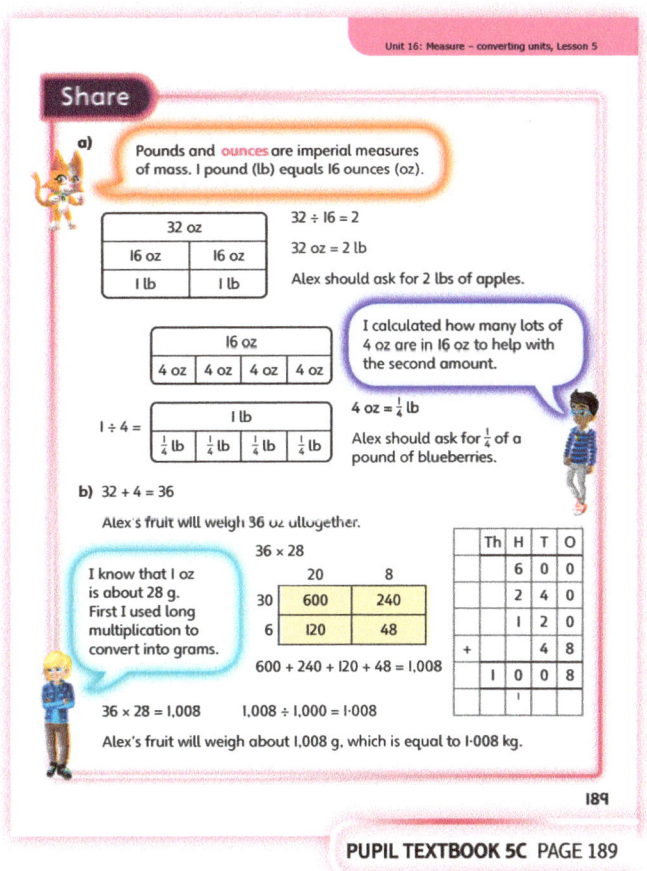

PUPIL TEXTBOOK 5C PAGE 189

Think together

WAYS OF WORKING Whole class teacher led (I do, We do, You do)

ASK

- Question ❶: *What do you know about ounces and grams that will help you find the answer? How would you complete the bar model? Can you explain how the written column method has been used to work out the answer to 28 × 15? Is there a different method you could use to multiply 28 by 15?*
- Question ❷ a): *Which number features in all three calculations? How would you alter the bar model to represent each individual calculation?*
- Question ❷ b): *There is more than one way to find $\frac{3}{4}$ lb in ounces. Can you describe two ways?*

IN FOCUS Although the partially completed bar model in question ❶ has three layers, children should be able to see that the calculation is simply the number of ounces (15) multiplied by the number of grams per ounce (28). After discussing the method shown, ask children what other strategies they might use to find 28 × 15. Ask which method children think is the most efficient and why.

STRENGTHEN To strengthen children's understanding of the connections between units of mass, use flashcards with equivalent measurements on each side (for example, 1 lb and 16 oz). Children can use these flashcards to form their own bar models. For example, in question ❷, they could put four '1 lb' flashcards in a row that they then turn over to show how the mass is converted into four lots of 16 oz, which they can then discuss how to calculate.

DEEPEN Question ❸ challenges children to begin thinking more independently about how to apply their knowledge of imperial units of mass. Ask them to discuss in pairs whether Astrid or Flo is correct and why. Choose different pairs to share their ideas with the rest of the group. Challenge children to work out how to convert from stones to grams and to make up problems using their conversion method.

ASSESSMENT CHECKPOINT Use questions ❶ and ❷ to assess whether children can use given facts to convert between common imperial units of mass, and between imperial and metric units. Children should be growing in confidence when applying these conversions in problem-solving contexts and be able to suggest appropriate strategies.

ANSWERS

Question ❶: 28 × 15 = 420
There are 420 g of raspberries in the container.

Question ❷ a): 4 lb = 16 oz × 4 = 64 oz;
10 lb = 16 oz × 10 = 160 oz;
$\frac{1}{2}$ lb = 16 oz ÷ 2 = 8 oz

Question ❷ b): $\frac{1}{4}$ lb = 16 oz ÷ 4 = 4 oz,
so $\frac{3}{4}$ lb = 4 oz × 3 = 12 oz

Question ❸ a): The second set of scales will show 13·2 lb.

Question ❸ b): The dog weighs 49 lbs, which is about 22 kg (also accept 22·3 kg).

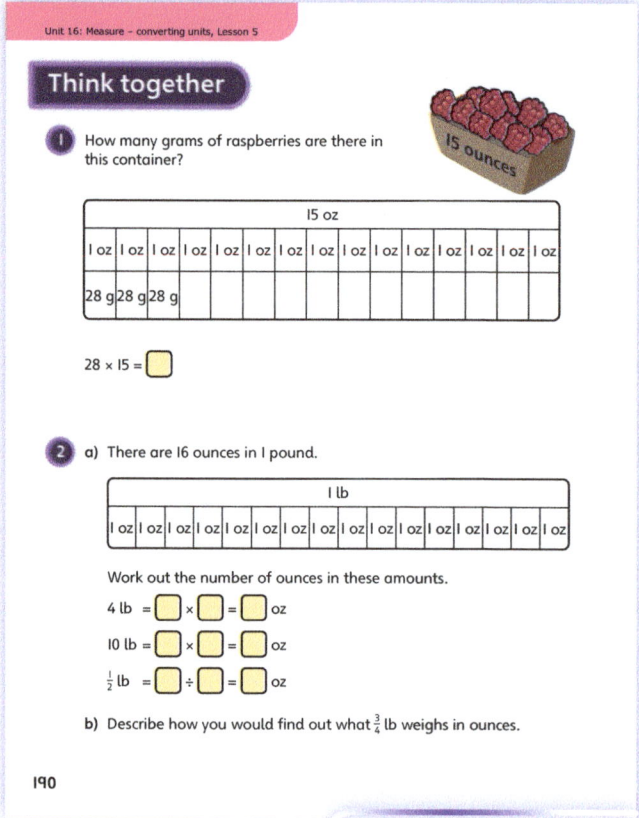

PUPIL TEXTBOOK 5C PAGE 190

PUPIL TEXTBOOK 5C PAGE 191

Unit 16: Measure – converting units, Lesson 5

Practice

WAYS OF WORKING Independent thinking or pair work

IN FOCUS Question ❹ addresses a common misconception. When finding equivalences for mixed numbers, some children will convert the whole number, but then take the fraction and simply add it on the end, writing 3 × 450 = 1,350 and therefore 3½ lb = 1,350½ g. Discuss what the ½ in 3½ lb stands for – half of what? It is crucial that children see that this represents half of *something*.

STRENGTHEN Before children attempt question ❶, give them cards showing different units of mass (both imperial and metric). Ask children to sort these into two groups according to whether they are imperial or metric units. Then draw a large number line for question ❶ for children to label using sticky labels. Encourage children to see a pattern in the multiples of 16. They can use their completed number line to help convert the measurements that follow.

DEEPEN In question ❻, encourage children to share their methods with a partner to help consolidate their understanding. Challenge children to estimate how heavy the giant octopus is in relation to an adult human. If an adult is said to weigh about 13–14 stone, then a giant octopus is about half the mass.

THINK DIFFERENTLY Question ❺ provides children with a code to crack and a new imperial unit of mass to discover: the ton. To convert 10 lb into different units, children should convert from pounds to ounces by multiplying by 16, and use 1 lb ≈ 450 g to convert into grams and then kilograms.

ASSESSMENT CHECKPOINT Use questions ❶ to ❸ to assess whether children can convert between common imperial units, and between imperial and metric units. Children should be confident when applying their knowledge of imperial units of mass.

ANSWERS Answers for the **Practice** part of the lesson can be found in the *Power Maths* online subscription.

Reflect

WAYS OF WORKING Independent thinking

IN FOCUS This question provides an opportunity for children to apply the information they have learnt about pounds. Observe those children who use different strategies (for example, sketching a bar model or a double number line). Encourage children to share their methods and discuss them as a group.

ASSESSMENT CHECKPOINT Look for children who can apply their knowledge of imperial and metric equivalences to express the given mass correctly in either grams or kilograms.

ANSWERS Answers for the **Reflect** part of the lesson can be found in the *Power Maths* online subscription.

After the lesson

- Are children confident in understanding that an ounce is smaller than a pound, which is smaller than a stone?
- Do you feel that children are beginning to be more confident working with imperial measures? How might the final lesson (on capacity) deal with some of their more general misconceptions?

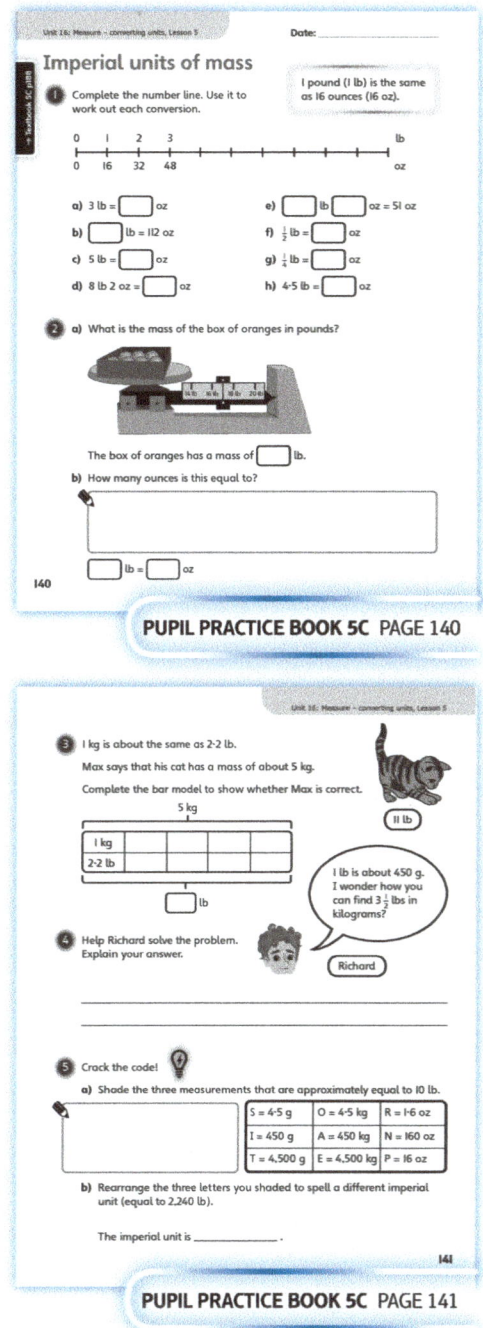

PUPIL PRACTICE BOOK 5C PAGE 140

PUPIL PRACTICE BOOK 5C PAGE 141

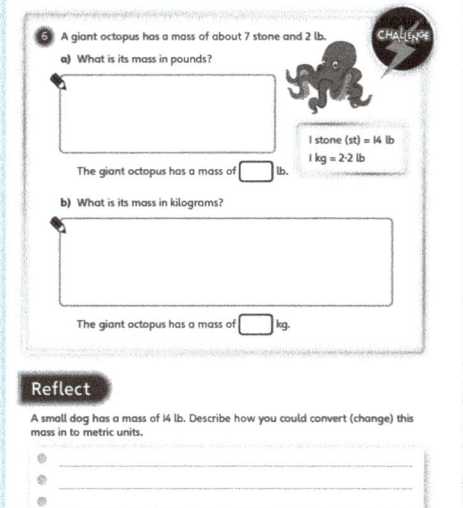

PUPIL PRACTICE BOOK 5C PAGE 142

Unit 16: Measure – converting units, Lesson 6

Imperial units of capacity

> **Learning focus**
>
> In this lesson, children will be introduced to imperial units of capacity. They will understand the terms pints and gallons, convert between them and use approximations to convert from imperial to metric units.

> **Before you teach**
>
> - This lesson is the last in this series on imperial units. Are there any general misconceptions that you feel need to be addressed during the lesson?
> - How might you help children to adapt their strategies when problem constraints are changed?

NATIONAL CURRICULUM LINKS

Year 5 Measurement

Understand and use approximate equivalences between metric units and common imperial units such as inches, pounds and pints.

ASSESSING MASTERY

Children can convert between common imperial units of capacity using given facts (for example, 8 pints = 1 gallon). They can use given approximations to convert between imperial and metric units (for example, 1 pint is about 570 ml). Children are able to apply these skills when solving problems.

COMMON MISCONCEPTIONS

Children may think that it is not possible to convert between units when more than one conversion is needed to solve a problem (for example, to convert gallons → millilitres, convert gallons → pints → millilitres). Ask:
- *What do you know about converting between units? How can you use it to help?*

STRENGTHENING UNDERSTANDING

Provide children with practical experience of measuring, comparing and converting imperial units of capacity before embarking on problem solving. Encourage children to measure exploratively, using measuring jugs that display both imperial and metric units. For example, ask them to measure 1 pint of water and read how many millilitres this is. They can read other equivalences just by looking carefully at the jug.

GOING DEEPER

Give children facts about capacities measured in imperial units (for example, a bath holds 64 gallons, the Niagara Falls has 150,000 gallons of water rushing over it every second, and the daily requirement of water for women is about 3 pints) or encourage them to find their own facts. Challenge them to devise their own conversion problems for a partner to solve (for example: the weekly water requirement for one person in litres; the number of pints of water going over the Niagara Falls every second; the number of litre jugs it would take to fill up a bath).

KEY LANGUAGE

In lesson: imperial units, metric units, **pint**, **gallon**, millilitre, litre, convert, approximately, capacity

STRUCTURES AND REPRESENTATIONS

Bar model, number line

RESOURCES

Optional: measuring jugs (imperial and metric), liquid containers (for example, milk cartons, water bottles), paper rectangles

 In the eTextbook of this lesson, you will find interactive links to a selection of teaching tools.

> **Quick recap**
>
> On the board write the digits 2, 3, 5 and 6. Ask children to make a 3-digit × 1-digit multiplication question and work out the answer. How many different answers can they find? Check that children are confident with short multiplication methods.

Unit 16: Measure – converting units, Lesson 6

Discover

WAYS OF WORKING Pair work

ASK

- Question ❶: *Where might you see imperial units of measurement in a supermarket?*
- Question ❶: *When have you seen people use pints or gallons in real life?*
- Question ❶: *What facts can you see in the image? How can you use these to help convert between units?*

IN FOCUS As with the previous lesson, talk about the fact that, until recently, liquids were sold by imperial measures and also that, although shops now have to label products with metric units, people still use imperial measures (for example, people buy milk in pint cartons and use miles per gallon for fuel consumption). Discuss the facts shown in the picture and ask children to try estimating 1 pint. Encourage children to use reasoning and their prior knowledge of capacity (for example, a 500 ml bottle of water fills about 2 mugs of water, so a pint must contain just over 2 mugs of water).

PRACTICAL TIPS Give children opportunities to explore imperial measures of capacity. Provide the sorts of liquid containers found in supermarkets (milk cartons, water bottles and so on) and encourage children to find their capacity in pints using imperial measuring jugs. Physically measuring amounts in pints should have the dual effect of ensuring children can gauge the value of imperial units of capacity, and (if the measuring jug shows both units) spot the equivalence between imperial and metric units by looking at the scale.

ANSWERS

Question ❶ a): 2·28 l are approximately equal to 4 pints of milk.

Question ❶ b): Mo has 3·42 l of water.

PUPIL TEXTBOOK 5C PAGE 192

Share

WAYS OF WORKING Whole class teacher led

ASK

- Question ❶ a): *Can you describe why the bar model, short multiplication and place value grid are all important in finding the solution?*
- Question ❶ a): *Can you explain how to work out the answer without using a bar model?*
- Question ❶ a): *A pint is just over $\frac{1}{2}$ a litre (500 ml). What would you expect 2 pints to be about the same as? What would you expect 4 pints to be?*
- Question ❶ b): *What is $\frac{1}{4}$ of a gallon? How does this help? Can you think of a different way to solve this? Think of a way that uses subtraction to find $\frac{3}{4}$.*

IN FOCUS Both questions use bar models. Ask children to explain how they have been used to help find the solution. Spend time considering the two bar models used to represent question ❶ b), and how they show how to find $\frac{3}{4}$ by finding $\frac{1}{4}$ and multiplying by 3.

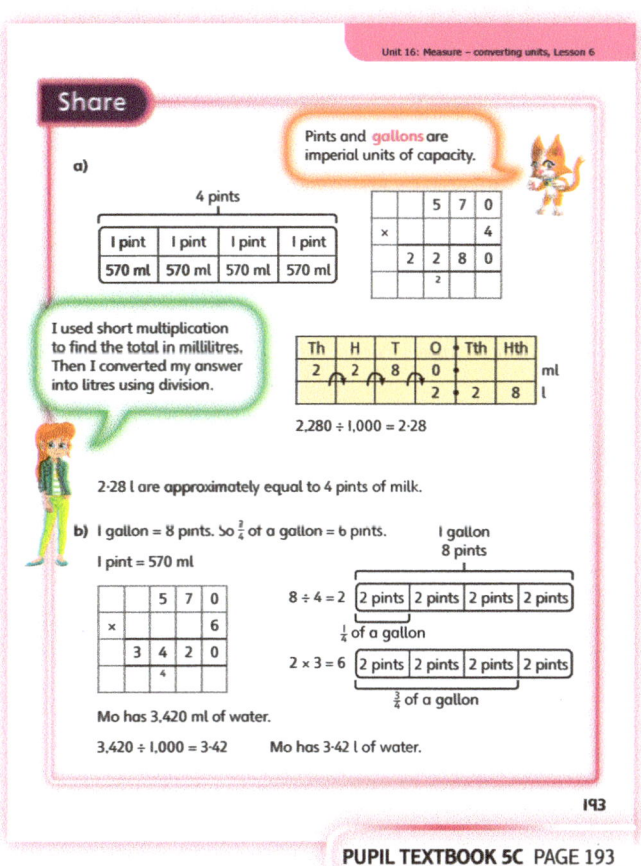

PUPIL TEXTBOOK 5C PAGE 193

229

Think together

WAYS OF WORKING Whole class teacher led (I do, We do, You do)

ASK

• Question ❶ a): *Can you describe how to get from a capacity measured in pints to one measured in litres?*
• Question ❶ a): *Can you think of a way to convert directly from pints to litres?*
• Question ❶ a): *How would you alter the bar model and the method if the water carrier was $\frac{1}{3}$ the size?*
• Question ❶ b): *How has the number line been used to show the difference between the water container's capacity and 1 gallon? How could you alter the number line to include millilitres?*
• Question ❷: *'Is a pint of milk more or less than two cans of lemonade?' Why is answering this the same as answering the given question?*

IN FOCUS Question ❶ scaffolds children's ability to visualise and calculate the conversions needed to find the answer using bar models and number lines. These are removed for questions ❷ and ❸, although children may benefit from drawing their own pictorial representations as part of their working.

STRENGTHEN If children are finding question ❶ a) challenging, provide measuring jugs to help them to find what 1 pint of water is in millilitres. Ask how many pints they need and what units they need to give the answer in. Then ask children to draw a sketch of what the question is asking, after which children should attempt the problem.

DEEPEN Ensure children understand that question ❸ is the sort of conversion problem that occurs in everyday life. Ask children to work in pairs and plan the steps needed to find the answer (for example, convert 1 gallon to pints, pints to ml and ml to l, and then subtract to find the difference). Challenge children to devise new, similar problems for a partner to solve.

ASSESSMENT CHECKPOINT Use questions ❶ and ❷ to assess whether children can use given facts to convert between common imperial units of capacity and between imperial and metric units. Children should be growing in confidence when applying these conversions in problem-solving contexts and be able to suggest appropriate strategies.

ANSWERS

Question ❶ a): 5 × 570 = 2,850
2,850 ÷ 1,000 = 2·85
5 pints are about the same as 2·85 l.

Question ❶ b): 3 pints is the difference.
1,710 millilitres is the difference.
1,710 ÷ 1,000 = 1·71
1·71 litres is the difference.

Question ❷: 1 pint is approximately 570 ml.
Half a pint is approximately 285 ml.
So half a pint of milk is less than a 330 ml can of lemonade.

Question ❸: Yes, the bucket can be filled and there will be 0·26 litres (or 260 ml) left over.

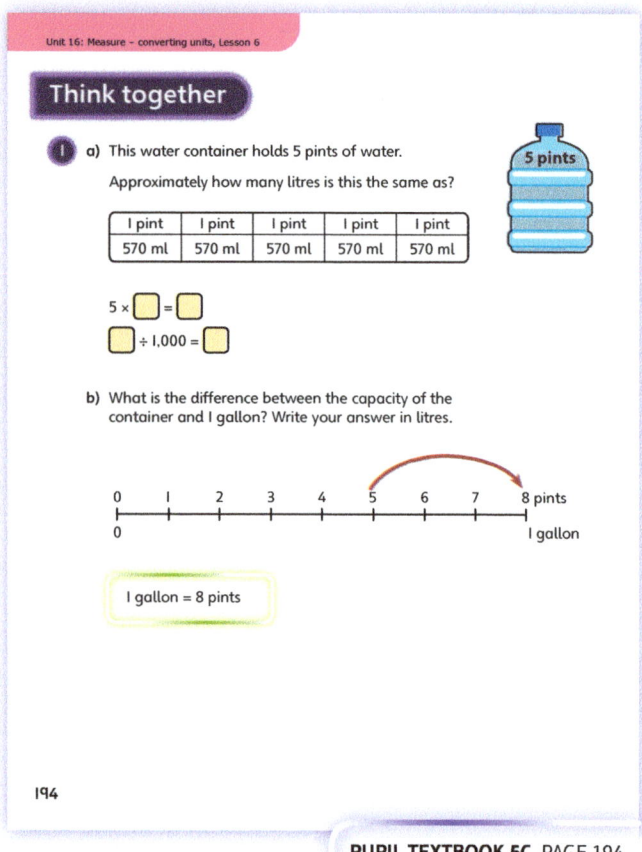

PUPIL TEXTBOOK 5C PAGE 194

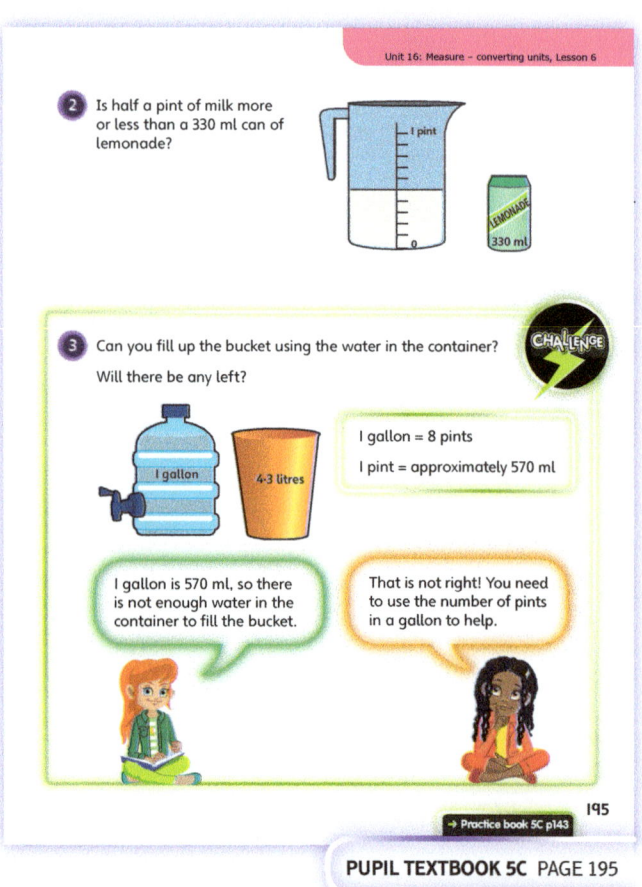

PUPIL TEXTBOOK 5C PAGE 195

Unit 16: Measure – converting units, Lesson 6

Practice

WAYS OF WORKING Independent thinking

IN FOCUS In question ③, children are expected to make the important link between gallons and litres. They can use the bar model from question ① to find the equivalence of 8 pints in terms of litres. The answer, 4,560 ml, rounded to the nearest half litre is 4·5 l (or $4\frac{1}{2}$ l). As an extension, ask how many litres are nearest to $\frac{1}{2}$ a gallon.

STRENGTHEN For question ④, encourage children to label the notches on the milk jug. Establish which two amounts the milk is between and thus that there are 3·5 pints. Provide rectangles of paper for children to create a bar model, writing that each whole pint is worth 570 ml. Ask them to cut one rectangle in half to represent half a pint and ask how many millilitres this is. Once children have calculated the answer in millilitres, provide place value grids, where appropriate, to support conversion from millilitres into litres.

DEEPEN Ask children to write a series of steps that will convert from gallons to millilitres. How many different ways can they can find to do this? What do they need to know in order to help them with the conversion?

THINK DIFFERENTLY In question ⑤, the only calculation that children need to do is 3 × 200 ml. They will then need to recognise that 600 ml is just over a pint.

ASSESSMENT CHECKPOINT Use questions ① to ④ to assess whether children can convert between common imperial units and between imperial and metric units. They should be confident when applying their knowledge of imperial units of capacity.

ANSWERS Answers for the **Practice** part of the lesson can be found in the *Power Maths* online subscription.

Reflect

WAYS OF WORKING Independent thinking

IN FOCUS This activity provides an opportunity for children to apply what they have learnt in a real-life context, using prior knowledge of the equivalence between pints and millilitres. Children's explanations may go on to justify the number of litres they would decide to buy. An answer of 1·14 l shows that they are able to convert correctly. Answers of either $1\frac{1}{2}$ l or 2 l shows that children are considering the real-life scenario and have recognised that milk may possibly be sold only in half or whole litres, so they have rounded their answers up.

ASSESSMENT CHECKPOINT Look for children who are able to convert 2 pints into litres by using known facts (1 pint ≈ 570 ml). Observe those children who round their answer up to the next half or whole litre.

ANSWERS Answers for the **Reflect** part of the lesson can be found in the *Power Maths* online subscription.

After the lesson

- Do you feel that children's misconceptions were addressed during the lesson? What reasons did children have for making these misconceptions and what learning points did they show?
- How well do you think that children have understood and applied the new concept? Do you feel that they are ready to move on?

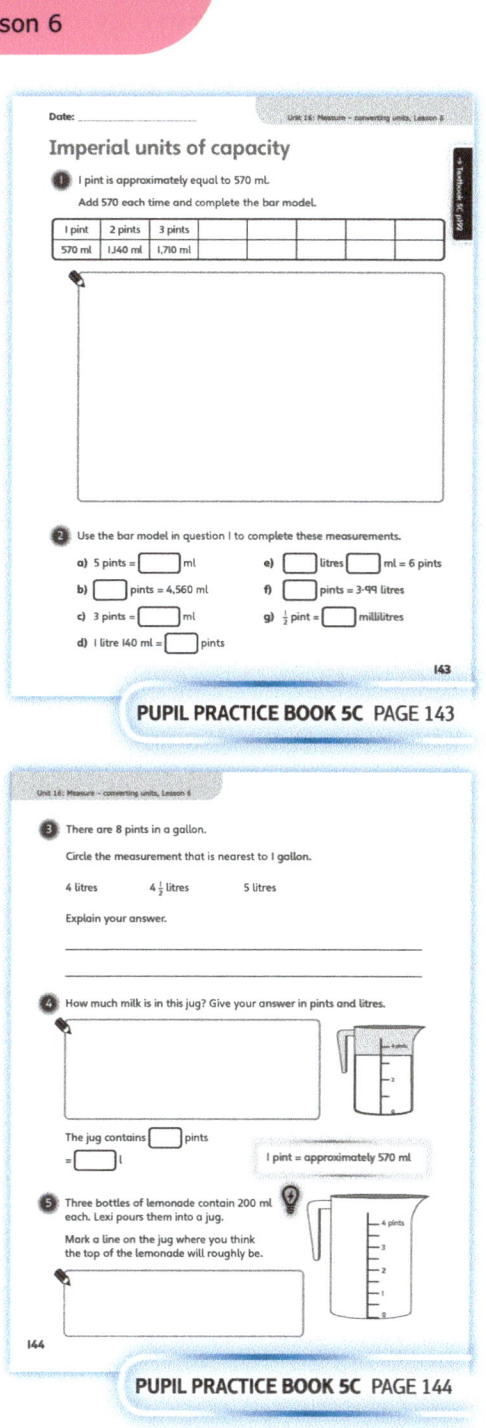

PUPIL PRACTICE BOOK 5C PAGE 143

PUPIL PRACTICE BOOK 5C PAGE 144

PUPIL PRACTICE BOOK 5C PAGE 145

Unit 16: Measure – converting units, Lesson 7

Convert units of time

Learning focus

In this lesson, children will solve problems where they have to convert between units of time, including those where there is a remainder.

Before you teach

- This lesson applies the concepts of converting between units of time. How confident are children with these concepts?
- How can you improve the teaching of problem solving and reasoning through this lesson?

NATIONAL CURRICULUM LINKS

Year 5 Measurement

Solve problems involving converting between units of time.

ASSESSING MASTERY

Children can solve time-based problems confidently, where they are required to convert between units of time. They can make appropriate decisions about how to present their answers if there is a remainder (for example, 150 seconds = 2 minutes 30 seconds or 2·5 minutes).

COMMON MISCONCEPTIONS

Children may think that they can express remainders as decimals, as with metric measures (for example, 2 kg 500 g = 2·5 kg). However, 2 weeks 5 days cannot be expressed as 2·5 weeks and 3·25 hours is not 3 hours 25 minutes. Ask:
- *What proportion of a week is 5 days? Can you write this as a decimal?*
- *What does the decimal 0·25 in 3·25 hours represent? What fraction of an hour is the same as 0·25?*

STRENGTHENING UNDERSTANDING

To support children before the lesson, provide examples showing different units of time (for example, analogue and digital clocks and calendars), together with representations, to consolidate their knowledge of equivalences. Ask questions about equivalences. For example, when looking at calendars ask how they show how many days there are in two weeks. Do the same for minutes, seconds and hours using analogue clock faces.

GOING DEEPER

Give children problems where they need to interpret decimals and fractions of different units of time. For example, ask: *A book order is going to be delivered in $3\frac{1}{2}$ weeks. How many days is this? How many hours is the same as 4·5 days?* Ask children to explain what the fraction or decimal part represents and what mistakes people might make. Extend the challenge to consider further examples (for example, 0·25, 0·75, $\frac{1}{4}$, $\frac{3}{4}$) where finding these parts of the unit is possible (for example, 0·25 hours in minutes or $2\frac{3}{4}$ days in hours).

KEY LANGUAGE

In lesson: remainder, multiple, second, minute, hour, day, week, month, year

STRUCTURES AND REPRESENTATIONS

Calendar, bar model, number line

RESOURCES

Optional: analogue and digital clocks, calendars, year planners, interlinking cubes

 In the eTextbook of this lesson, you will find interactive links to a selection of teaching tools.

Quick recap

Ask: *What facts about time do you know or remember?* Children may give examples such as 'There are 365 days in a year, except leap years' or 'There are 7 days in a week'. Challenge the class to write down 10 facts that include seconds, minutes, hours, days, weeks, years and so on.

232

Unit 16: Measure – converting units, Lesson 7

Discover

WAYS OF WORKING Pair work

ASK

- Question ❶: *What different units of time can you see?*
- Question ❶ a): *How much longer than 1 month has Toshi had his phone? Why is it difficult to say for sure?*
- Question ❶ b): *How long does the whole battery on Amal's phone take to charge fully?*
- Question ❶ b): *How many minutes has Amal's phone been charging for?*

IN FOCUS Use the picture to discuss children's wider experience of this real-life scenario. For example, ask questions about how long they think they would normally have to return an electrical item if it is not working, and whether they think that 5 hours is a quick length of time to charge up a battery. Discuss when else they might think about units of time in relation to a mobile phone (for example, length of a call, the number of free minutes, setting an alarm, etc.).

PRACTICAL TIPS Prior to the lesson, give children experience of everyday durations in the context of electronic devices. For example, children could time how long it takes for their class laptop to charge, explore advertisements for mobile phone deals that talk about free minutes, or investigate the time it takes to download files on a computer. For each example, ask children to practise converting into alternative units of measurement.

ANSWERS

Question ❶ a): Toshi has had his phone for 5 weeks and 4 days.

Question ❶ b): 4 bars of Amal's battery should be charged fully. There are 15 minutes left until the next bar is charged.

PUPIL TEXTBOOK 5C PAGE 196

Share

WAYS OF WORKING Whole class teacher led

ASK

- Question ❶ a): *What do you think Flo means by 'I predicted that there would be a remainder'?*
- Question ❶ a): *How can you predict whether there will be a remainder when converting units of time?*
- Question ❶ a): *What does the 5 represent in the answer 5 remainder 4? What does the 4 represent?*
- Question ❶ b): *Can you give some amounts of minutes that are the same as a whole number of hours? Which two multiples of 60 does 285 minutes come between?*
- Question ❶ b): *Explain how the bar model represents the problem.*

IN FOCUS Both questions ❶ a) and b) use bar models to represent the problem. Ask children to explain how the bar models have been used. In particular, challenge them to explain the relevance of the remainder at the end of the second bar. Check that children recognise that this leftover amount is less than a whole week or an hour (depending on the question).

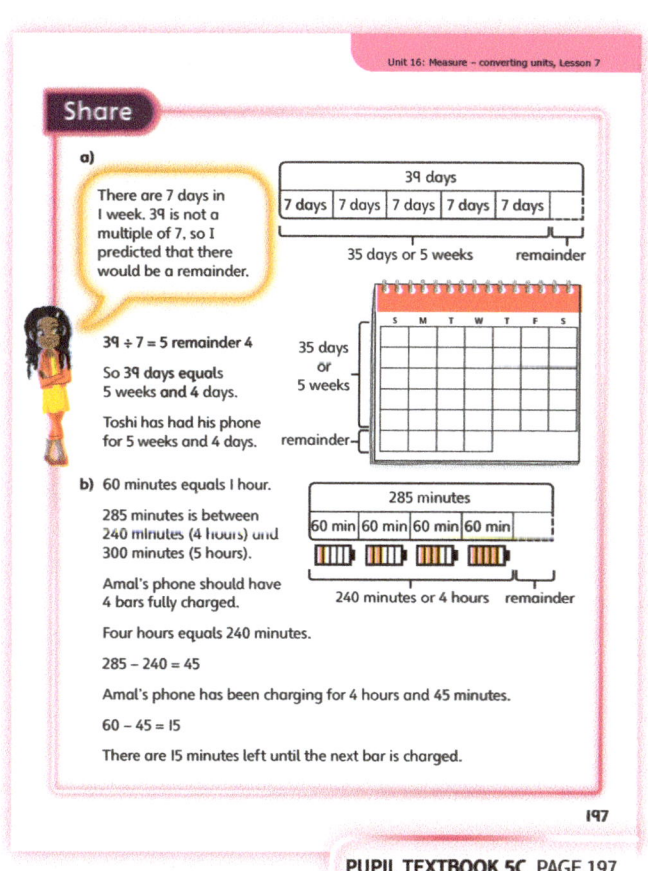

PUPIL TEXTBOOK 5C PAGE 197

233

Unit 16: Measure – converting units, Lesson 7

Think together

WAYS OF WORKING Whole class teacher led (I do, We do, You do)

ASK

- Question ①: *What unit of time is given in the question? What unit is needed for the answer?*
- Question ①: *What could you do to find out whether Amal's phone has been downloading for a whole number of minutes or for a whole number of minutes and a part of a minute?*
- Question ①: *What two whole-minute values is 378 seconds in-between?*
- Question ①: *Can you explain how to work out the remainder in seconds?*
- Question ②: *How is this question the same as question ①? How are the questions different?*

IN FOCUS The scaffolding for question ② involves calculating 5 weeks in terms of days, and then subtracting the number of days that have already gone. Ask children whether they can think of a different way to find the answer. They may be able to recognise that they could convert the number of days that have gone by into weeks and days, and then use this to find the difference.

STRENGTHEN If children need representations to help answer question ②, refer back to the use of a number line in question ①. Ask: *What else could you use that shows weeks and days clearly?* Provide calendars (or × 7 grids) to help children count in multiples of 7 and so quickly identify that 5 weeks are the same as 35 days. Discuss how they can use that fact to work out the number of days left.

DEEPEN Question ③ deepens children's understanding by giving them a series of durations in hours and asking them to translate these into clock times. As the durations are mostly greater than 24 hours, children will need to convert into days and hours (for example, 93 hours = 3 days 21 hours) and identify the effect that this will have on the day of the week as well as the time. Ask children to create a similar problem based on their school week.

ASSESSMENT CHECKPOINT Use questions ① and ② to assess whether children can give solutions involving conversion and comparison. Ensure children are able to explain clearly why and how they are using bar models to help solve each of the problems.

ANSWERS

Question ①: 378 is between 360 and 420.
So there are 6 minutes and there will be a remainder of seconds.
378 − 360 = 18
Amal's phone has been downloading updates for 6 minutes and 18 seconds.

Question ②: 5 weeks = 5 × 7 = 35 days
35 − 22 = 13
There are 13 days until the sale ends.

Question ③ a): On the ferry: 13:00 Tuesday; arrived: 19:00 Tuesday; visiting auntie: 13:00 Thursday; theme park: 10:00 Friday.

Question ③ b): Jen has 154 hours until she returns home.

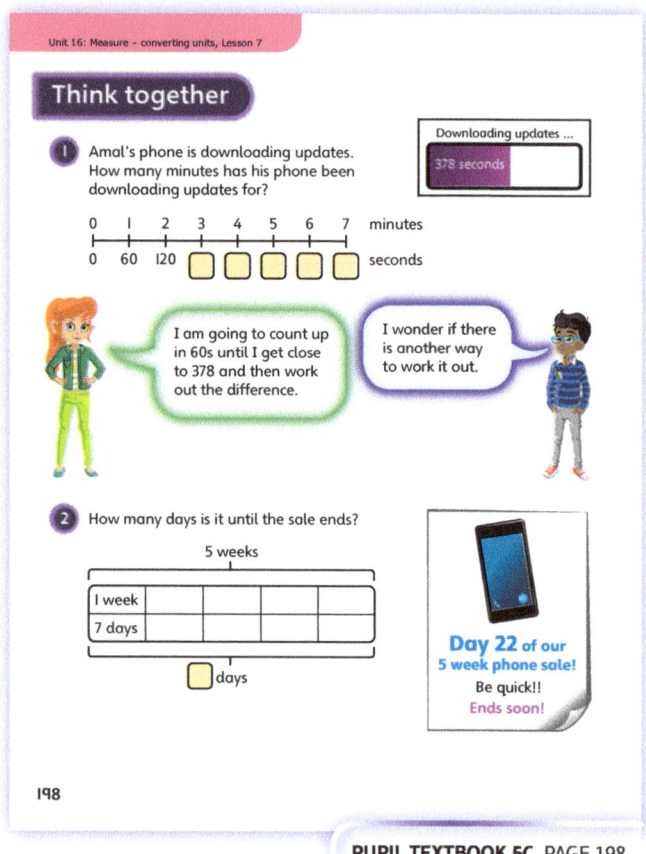

PUPIL TEXTBOOK 5C PAGE 198

PUPIL TEXTBOOK 5C PAGE 199

Unit 16: Measure – converting units, Lesson 7

Practice

WAYS OF WORKING Independent thinking

IN FOCUS Question ③ addresses the common misconception that units of time behave in a similar way to metric units, with a remainder appearing to the right of the decimal point. It is essential that children make the link between the decimal and the fraction of an hour as their understanding depends on them recognising that 0·25 of an hour = $\frac{1}{4}$ of an hour (= 15 minutes), so 4·25 hours is the same as $4\frac{1}{4}$ hours or 4 hours 15 minutes.

STRENGTHEN For question ⑤, provide children who need further support with scaffolding in the form of a list for them to fill in. Children should write this out and fill in the gaps: '1 week = ☐ days; 1 day = ☐ hours; 1 hour = ☐ minutes; 1 minute = ☐ seconds.' Then ask children how they would convert each of these (for example, how would they work out the number of weeks there are if there are 21 days, and how would they work out the number of days in 3 weeks).

DEEPEN Question ⑤ deepens children's understanding by challenging them to consider the combinations of operations needed to convert across several units. Encourage children to apply their knowledge of larger and smaller units (larger → smaller unit requires multiplication) to work out the final units. Ask children to draw sets of function machines for two conversions, one 'larger → smaller unit' and one 'smaller → larger unit'.

THINK DIFFERENTLY In question ④, children are presented with four durations of summer holidays expressed using different units. Ask why the different durations are tricky to compare and what children need to do to make the comparison easier.

ASSESSMENT CHECKPOINT Use questions ① to ③ to assess whether children are confident when solving problems where they need to convert between units of time that are not whole amounts. They should display reasoning skills, explain appropriate methods and think with confidence.

ANSWERS Answers for the **Practice** part of the lesson can be found in the *Power Maths* online subscription.

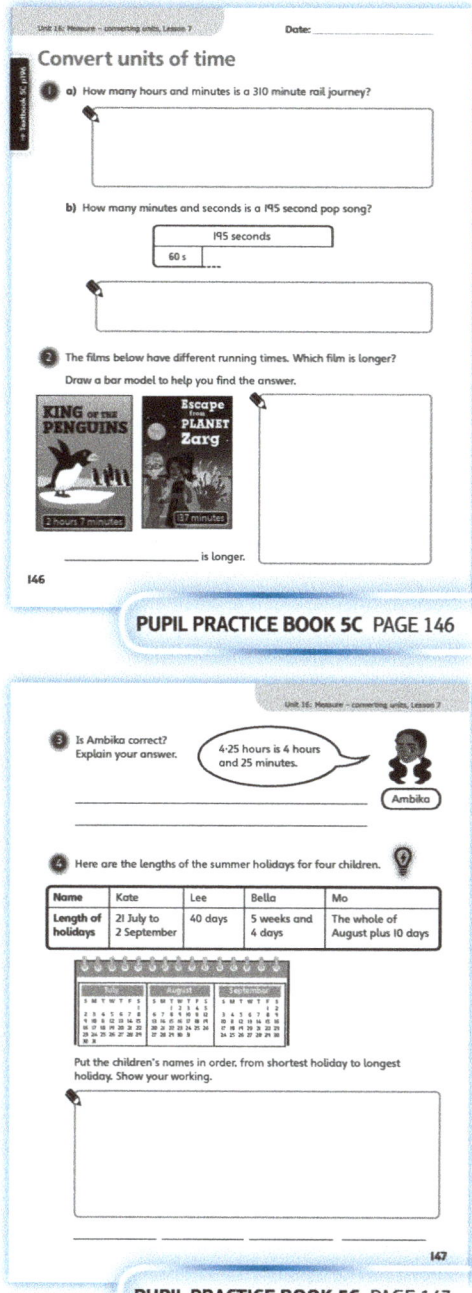

PUPIL PRACTICE BOOK 5C PAGE 146

PUPIL PRACTICE BOOK 5C PAGE 147

Reflect

WAYS OF WORKING Independent thinking followed by pair work

IN FOCUS This **Reflect** question provides an opportunity to check children's methodology. Initially, children should think individually and decide on a method, which they should then explain to a partner. They should mention that 1 year = 12 months and that 30 is not a multiple of 12, so the answer is not a whole number of years. Children should describe how they would use this information – for example: 'I would count in 12s until I got near to 30. This would show the years, and then the remainder would be the number of extra months.' Encourage children to compare their methods.

ASSESSMENT CHECKPOINT Look for children who describe clearly and accurately how to convert 30 months into years and months.

ANSWERS Answers for the **Reflect** part of the lesson can be found in the *Power Maths* online subscription.

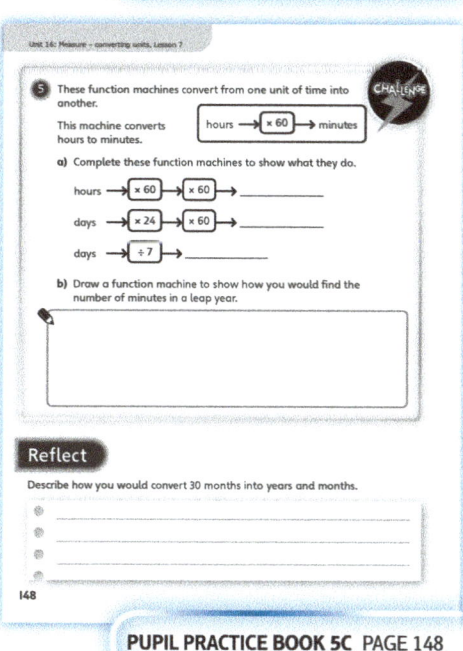

PUPIL PRACTICE BOOK 5C PAGE 148

After the lesson ⏸

- How did children respond mathematically to the problems and how did their mathematical understanding develop during the lesson?
- Did you provide feedback to push them to think more deeply or to learn more about process skills?

235

Unit 16: Measure – converting units, Lesson 8

Timetables – calculating

Learning focus

In this lesson, children will learn to use timetables, applying their knowledge of 24-hour times to read arrival and departure times and calculate durations.

Before you teach

- This lesson requires a prior knowledge of 24-hour times. How well do you feel children understand this way of representing times? Does this skill need consolidating first?
- How can you link this lesson to other lessons or to the use of timetables in everyday life?

NATIONAL CURRICULUM LINKS

Year 5 Measurement

Solve problems involving converting between units of time.

ASSESSING MASTERY

Children can read information from timetables accurately. They can apply their knowledge of timetables to solve time-based problems confidently, including where they need to convert between units of time.

COMMON MISCONCEPTIONS

Children may think that a journey is read from left to right, across the row of the timetable, instead of down a column. Ask:
- *Use your finger to trace the journey that one single bus/train makes. How many times does it stop? What time does it finally arrive? What do you think a row of the timetable shows?*

Children may think that timetables should show every time mentioned in a problem. For example, with 'Olivia arrives at Boston at 10 o'clock and catches the next train …', children become confused because they cannot locate 10:00 on the timetable. Ask:
- *What is the first train after Olivia arrives? What time does it leave?*

STRENGTHENING UNDERSTANDING

It is essential that children's use of timetables is made as relevant as possible to their own lives. Where possible, take photos of local bus stops and ask children to match them up with a timetable. Children can document their journeys on buses and compare the times the buses stopped with the scheduled times. For children who do not use public transport regularly, start by using timetables that apply to them (a school timetable or cinema showing times). These will not be as complex as bus or train timetables and so will provide a useful introduction.

GOING DEEPER

Challenge children to devise their own bus timetables with several columns, showing at least five buses, at least five stops, at least one bus not stopping so regularly (and is therefore a faster bus), and with consistent times between stops. They can then create their own problems for a partner to solve, based on their timetables.

KEY LANGUAGE

In lesson: timetable, 24-hour, digital, duration, hour, minute, departs/departure, arrives/arrival

STRUCTURES AND REPRESENTATIONS

Timetable, analogue clock face, number line

RESOURCES

Optional: different kinds of timetables, analogue clock faces with movable hands, mini-whiteboards, paper strips

 In the eTextbook of this lesson, you will find interactive links to a selection of teaching tools.

Quick recap

On the board, show the digital times 09:32, 3:15 pm and 18:40. Ask children to tell you what time each clock shows. Discuss what each clock tells them and the similarities and differences. They should be able to read the times confidently and tell you what part of the day it is.

Unit 16: Measure – converting units, Lesson 8

Discover

WAYS OF WORKING Pair work

ASK

- Question ①: *Where have you seen a timetable like this before? Have you ever needed to use one?*
- Question ①: *Put your finger at the top of a column and move it downwards. What does this column show?*
- Question ①: *Put your finger at the start of a row and move it from left to right. What does this row show?*
- Question ①: *How many buses are shown on this timetable? How often do they leave?*

IN FOCUS Use the picture as a way to introduce children to the concept of timetables. Talk about how each column has been labelled with a bus name. Tell children to put their finger at the top of a bus route and follow its journey. Ask questions about the bus journey.

PRACTICAL TIPS Provide pairs of children with timetables showing local bus routes or film times at a local cinema. Many children will never have used a timetable and the more relevant it is to your school's locality, the better. Ask them to look at the way the times are written (layout, 24-hour) and what each column and row represents. Encourage pairs to share their findings.

ANSWERS

Question ① a): Emma catches Bus A. Emma arrives at school at 08:05 (five minutes past 8).

Question ① b): Bus C arrives at school at 08:35.

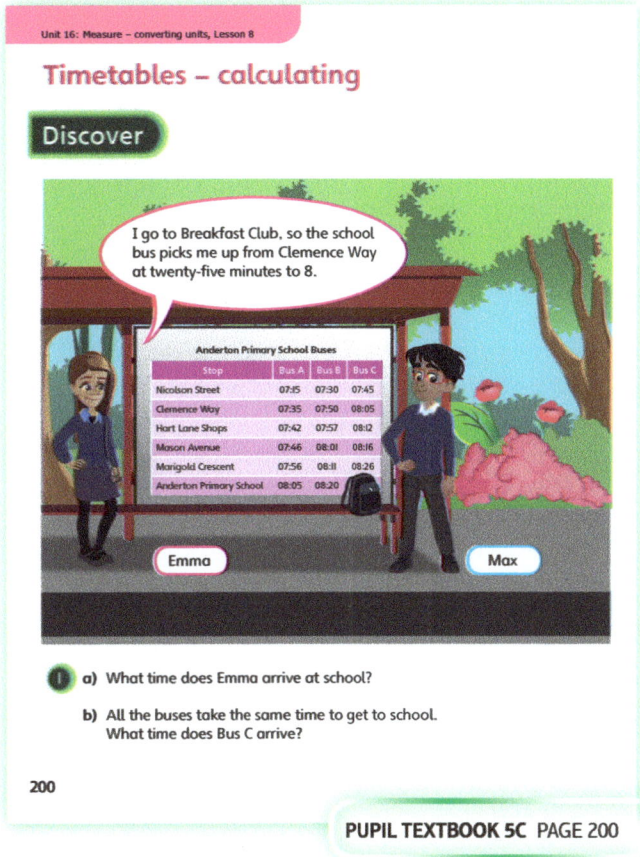

PUPIL TEXTBOOK 5C PAGE 200

Share

WAYS OF WORKING Whole class teacher led

ASK

- Question ① a): *Why do you think timetables are usually written in 24-hour digital time?*
- Question ① a): *Emma leaves at twenty-five to 8: what time will this look like on the timetable? Which bus does Emma catch?*
- Question ① a): *How do you know what time Emma will arrive at school?*
- Question ① b): *How does the number line show how long each bus takes?*
- Question ① b): *Why do you think the clock faces show 15 minutes being added up to 08:00, and then the remaining 35 minutes separately?*

IN FOCUS Question ① b) employs a method children will be familiar with from previous work when bridging an o'clock time. Both the number line and the clock faces show time being added up to an o'clock time and then from the o'clock time to the final time. When used in timetables, the 24-hour clock reduces confusion between am and pm times and makes it easier to calculate duration over several hours, although children have not had to use this strategy here. For example, 8:30 am to 1:30 pm is 5 hours – this is easier to see when written as 08:30 to 13:30 because 8 + 5 = 13.

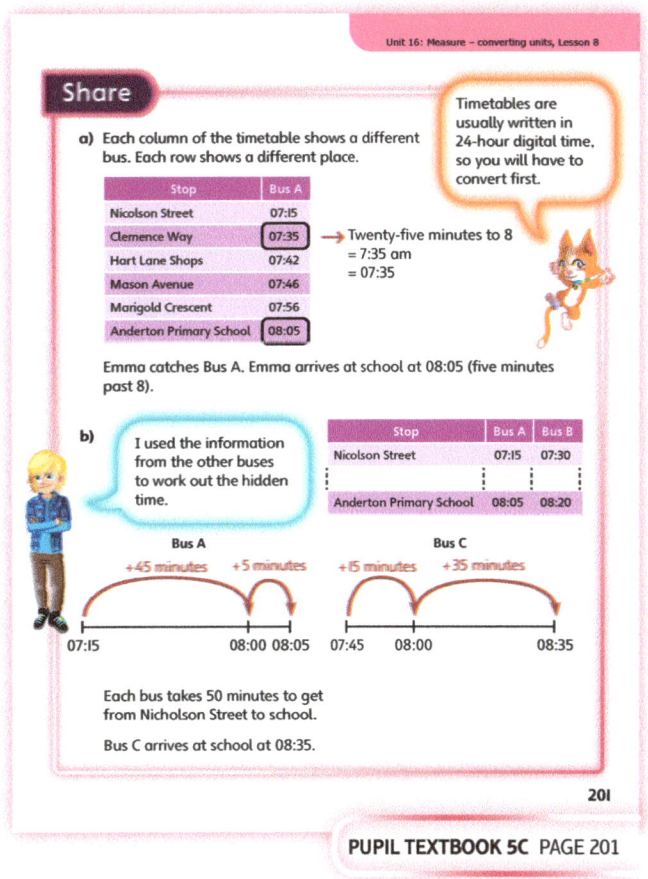

PUPIL TEXTBOOK 5C PAGE 201

237

Unit 16: Measure – converting units, Lesson 8

Think together

WAYS OF WORKING Whole class teacher led (I do, We do, You do)

ASK

- Question ❶: *Compare this timetable to the timetable in* **Discover**. *What is the same? What is different?*
- Question ❶: *What does each column show? Each row?*
- Question ❶ a): *How can you find which train Lexi gets on?*
- Question ❶ a): *Describe Lexi's arrival time in two ways.*
- Question ❶ b): *How can you find which train Andy gets on?*
- Question ❶ b): *How will you find the duration of the journey? How does the representation of the clocks help?*
- Question ❷: *Can you find this part of the timetable? What time does the train leave? What time does it arrive? Why does the number line show two separate jumps? How does this help? All the trains take the same time to get from Birchfield to Ashtown Parkway. Can you prove this by using a number line and working out the time using different trains?*

IN FOCUS Encourage children to use the large timetable as much as possible and to locate the relevant times and trains within the whole timetable, rather than using the small timetable representations. For question ❷, you may find it useful to initially ask the question separately, without getting children to consider the supporting image of part of the timetable. Ask whether it matters which train they choose. Then go on through the methodology shown.

STRENGTHEN To strengthen children's understanding of the relationship between a train's journey and the timetable, provide analogue clock faces. Encourage children to draw a line on a mini-whiteboard and then talk through the journey of one of the trains. At every station, children should make the new time on their analogue clock and label the time and station on their time line. This should help translate the columns of the timetable into something that resembles a journey from A to B.

DEEPEN Use question ❸ to discuss when a train might *not* stop at particular stations. Ask why children would expect the express train to be quicker than the next train, without looking at the times. If children find a duration in hours and minutes, encourage them to convert before comparing and calculating. Ask children to discuss Flo's comment in pairs and to find the answer using her method. They can compare the 14:13 and 15:13 trains: the 15:13 leaves one hour later but arrives only 40 minutes later, so it is 20 minutes quicker.

ASSESSMENT CHECKPOINT Use questions ❶ to ❸ to assess whether children are growing in confidence when reading timetables. They should be able to identify departure and arrival times and calculate durations, using representations to help them. They should understand how to apply these skills in problem-solving contexts.

ANSWERS

Question ❶ a): Lexi arrives in Ashtown Central at 15:50.

Question ❶ b): It takes Andy 24 minutes to get to Birchfield.

Question ❷: It takes 32 minutes to get from Birchfield to Ashtown Parkway.

Question ❸: The express train is 20 minutes quicker.

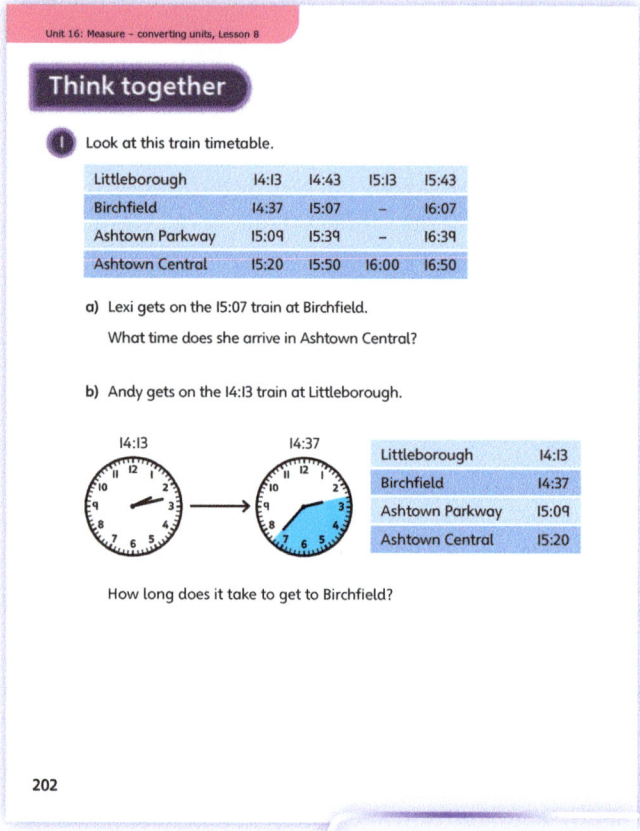

PUPIL TEXTBOOK 5C PAGE 202

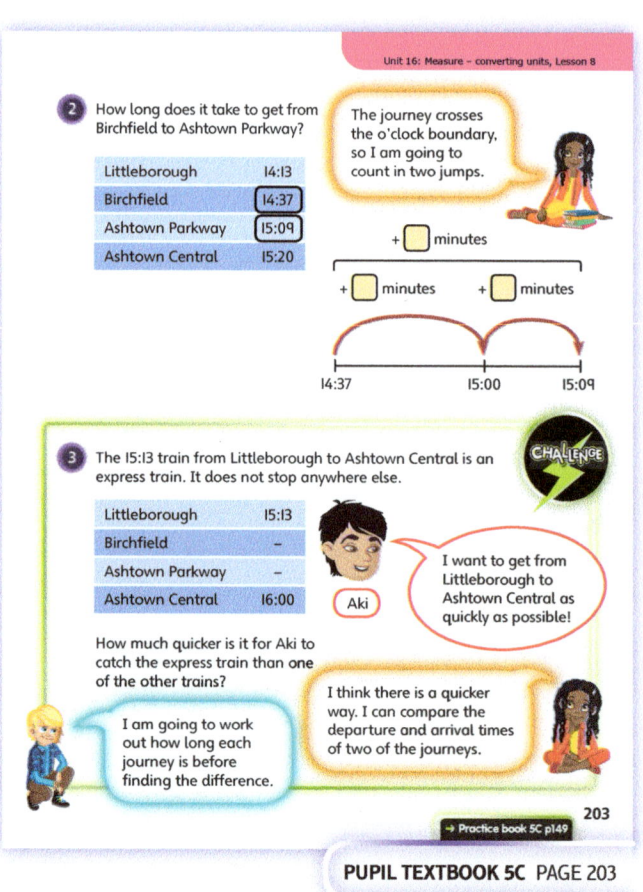

PUPIL TEXTBOOK 5C PAGE 203

Unit 16: Measure – converting units, Lesson 8

Practice

WAYS OF WORKING Independent thinking

IN FOCUS Question ❷ shows a timetable that seems to have times missing. Discuss what children think these gaps mean. Ask which station would be the next stop if they caught the 06:31 train from Grantham. Children may spot similarities in the different train journeys. For example, the first and the fourth trains have a similar journey (no stop at Rauceby), as do the second and third trains (no stops at Thorpe Culvert or Havenhouse) and the fifth and sixth trains (no stops at any of these three stations).

STRENGTHEN Encourage children to focus on the relevant part of the timetable by providing them with two pieces of paper to cover up the timetable either side of the particular column or row that they need to look at. They may also benefit from drawing time-line representations of journeys on mini-whiteboards as described in **Think together**.

DEEPEN Question ❸ challenges children to fill in the bus timetable given certain parameters. Establish that they can write in two times immediately from the information they are given and ensure they understand how to find the other times. Ask children to create a similar set of instructions for producing a timetable for a bus to their school.

ASSESSMENT CHECKPOINT Use questions ❶ and ❷ to assess whether children can read timetables with confidence, identifying departure and arrival (or start and end) times and calculating durations accurately. They should be able to apply these skills confidently in problem-solving contexts.

ANSWERS Answers for the **Practice** part of the lesson can be found in the *Power Maths* online subscription.

Reflect

WAYS OF WORKING Independent thinking

IN FOCUS This **Reflect** question encourages children to consider why the 24-hour clock is used in timetables. Possible answers may include the following: it is not possible to get confused between whether a time is am or pm; it is easier to calculate the duration of something with 24-hour clock times, particularly when it goes across 12:00; the lack of 'am' and 'pm' means that the times take less space on the timetable.

ASSESSMENT CHECKPOINT Look for children who are able to explain why timetables use 24-hour clock times by referring to some of the benefits of using 24-hour over 12-hour times.

ANSWERS Answers for the **Reflect** part of the lesson can be found in the *Power Maths* online subscription.

After the lesson ⏸

- To what extent were children's various responses the ones that were anticipated? Review their ideas and questions, the obstacles they encountered, and their misunderstandings and mistakes.
- What percentage of children do you feel mastered the lesson? How confident do you think they will be when using timetables in everyday situations?

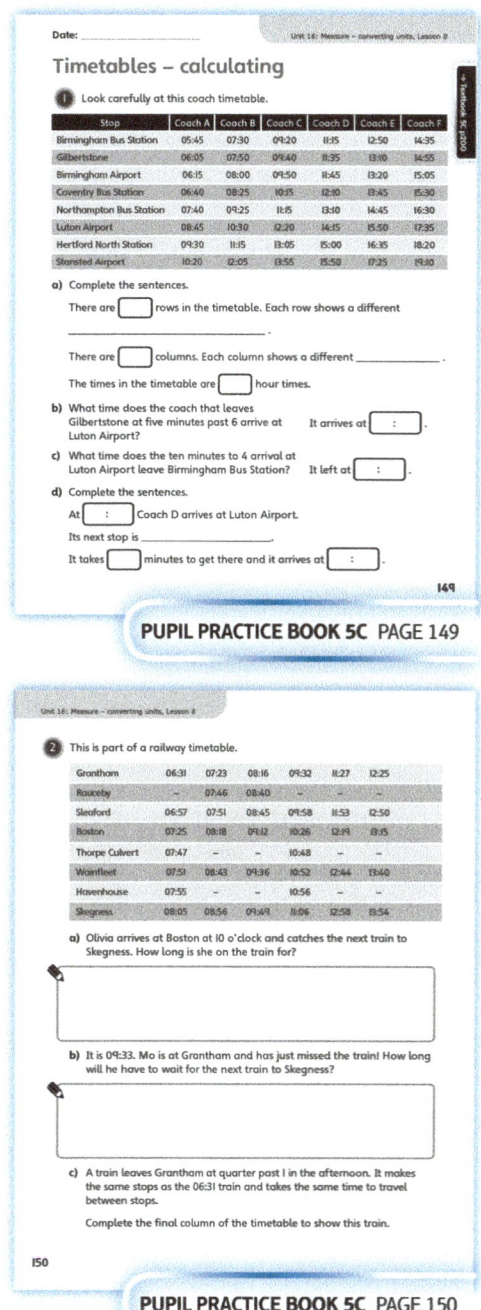

PUPIL PRACTICE BOOK 5C PAGE 149

PUPIL PRACTICE BOOK 5C PAGE 150

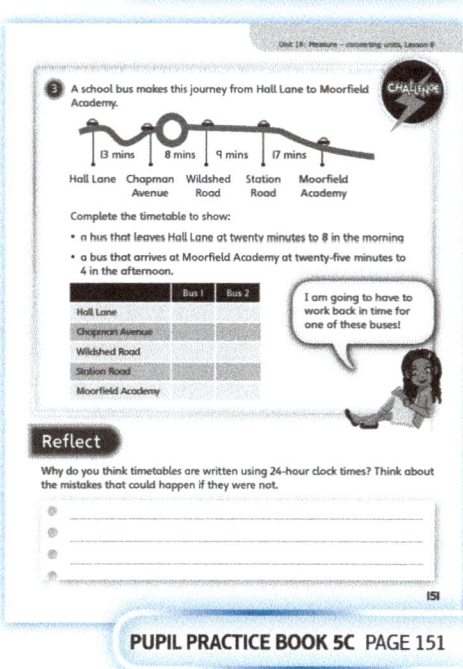

PUPIL PRACTICE BOOK 5C PAGE 151

239

Unit 16: Measure – converting units, Lesson 9

Problem solving – units of measure ①

Learning focus

In this lesson, children will apply their knowledge of metric units of length, mass and capacity to problems that require calculating with different units. They will work with measurements that have different numbers of decimal places and fractions of units.

Before you teach

- How will you help to refine children's own representations for different problems?
- How can you improve the teaching of problem solving and reasoning through this lesson?
- Are children confident naming which units of measurement are used for weighing and which for measuring distances and capacity?

NATIONAL CURRICULUM LINKS

Year 5 Measurement

Use all four operations to solve problems involving measure [for example, length, mass, volume, money] using decimal notation, including scaling.

ASSESSING MASTERY

Children are able to solve problems that require converting between metric units of measurement. They can apply their knowledge of fractions to work out fractions of larger units in terms of smaller units.

COMMON MISCONCEPTIONS

Children may think that they can solve a problem without converting any units (just using the numbers from each measurement regardless of units). Ask them to point to all units in the question. Ask:
- Do you need to convert to find the answer? Which measurement(s) will you need to change?

When converting fractions of units of measurement (specifically mixed numbers), children may think that the fraction can be simply transferred across to the smaller unit (for example, 1 $\frac{1}{2}$ litres to 1,000·5 ml). Draw a bar model showing the equivalence of litres and ml, similar to the one shown here. Ask:
- What is $\frac{1}{2}$ a litre the same as? So how would you write 2 $\frac{1}{2}$ litres as millilitres?

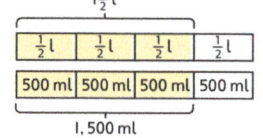

STRENGTHENING UNDERSTANDING

Ensure that children use the structures and representations used in the previous lessons. Ask them to underline the parts of the problem that are most important. Provide blank bar models, number lines and place value grids with digit cards to help children model each problem and convert between metric units.

GOING DEEPER

Challenge children to create their own problems involving metric units of measurement. Begin by ensuring children recognise some of the common characteristics of conversion problems (information given in one or two units and the answer required in a different unit). Children can share their problems and explain how they think their problems ought to be solved.

KEY LANGUAGE

In lesson: convert, smaller, larger, unit, metre, centimetre, millimetre, litre, millilitre, kilogram, gram, heaviest, lightest

Other language to be used by the teacher: greater, more, less, fraction, decimal

STRUCTURES AND REPRESENTATIONS

Number line, place value grid

RESOURCES

Optional: blank bar models, number lines, place value grids, digit cards, flashcards

 In the eTextbook of this lesson, you will find interactive links to a selection of teaching tools.

Quick recap

On the board or with flashcards, show children the measures 4,000 ml, 600 g, 1.7 m and 750 cm. Ask children to convert each measure to a different unit of measure. Ask: How many different answers can you find for each measurement? What rules do you use to find their answers?

Unit 16: Measure – converting units, Lesson 9

Discover

WAYS OF WORKING Pair work

ASK

- Question 1: *Where can you see different units of measurement in the picture?*
- Question 1: *What facts do you know about these units of length and capacity? How many centimetres are in 1 metre? How many millilitres are in 1 litre?*
- Question 1 a): *To be allowed on the roller coaster, does Isla's height need to be less than or greater than 1·45 m?*
- Question 1 b): *Why is it tricky to add the fizzy drinks as they are? What could you do to help?*

IN FOCUS Discuss the sorts of height restriction signs that children have seen. These may include those at a theme park, bridge or barrier heights for traffic, and so on. Discuss why height restrictions exist. Talk about the ruler and ask children to describe how they might work out the answer as the ruler shows a different unit of measurement.

PRACTICAL TIPS Provide tape measures marked in millimetres for children to measure each other. Give them different height restrictions in metres for them to work out whether they are shorter or taller than each height.

ANSWERS

Question 1 a): 1·40 m < 1·45 m so Isla is not tall enough to go on the roller coaster.

Question 1 b): Aki is buying 750 ml of fizzy pop altogether.

PUPIL TEXTBOOK 5C PAGE 204

Share

WAYS OF WORKING Whole class teacher led

ASK

- Question 1 a): *How does the number line help you to convert between units?*
- Question 1 a): *Can you think of another way to compare the two heights (instead of converting 140 cm into metres)?*
- Question 1 b): *0·25 l is the same as $\frac{1}{4}$ of a litre. 500 ml is the same as $\frac{1}{2}$ of a litre. Can you explain why? How can you say the total (750 ml) as a fraction of a litre?*

IN FOCUS The number line used in question 1 a) is a useful way of representing the equivalences between two different units of measurement. It is a model that is used in later lessons on imperial and metric units and so it is important that children recognise its importance at this stage. Give children time to practise using the line as a conversion tool by calling out different measurements and asking them to respond by using the number line to convert to the other unit of measurement.

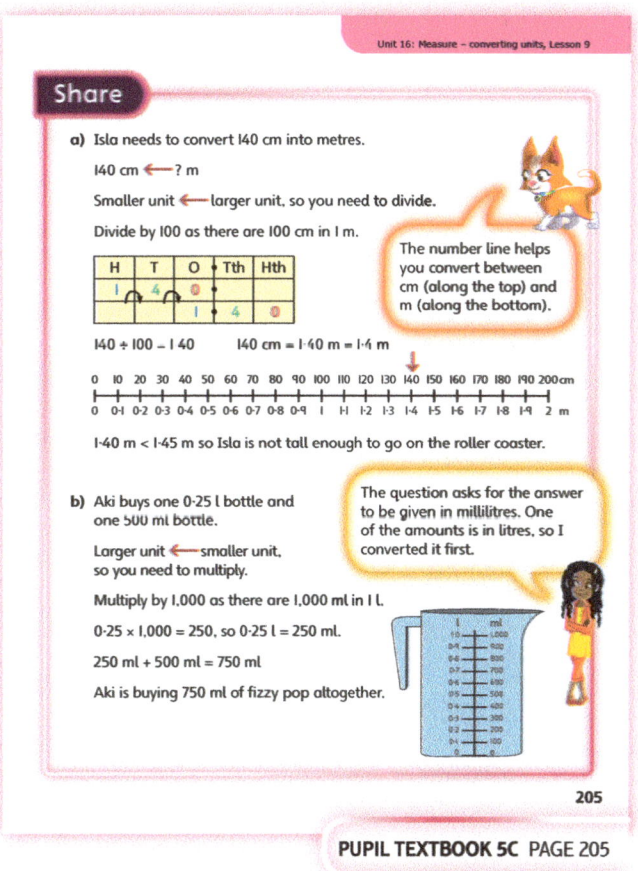

PUPIL TEXTBOOK 5C PAGE 205

Think together

WAYS OF WORKING Whole class teacher led (I do, We do, You do)

ASK

- Question ❶: *How do you think Ambika has estimated the masses of the cakes? What strategies might she have used?*
- Question ❶: *What does Ambika need to calculate?*
- Question ❶: *How could you find the same answer, but only convert once?*
- Question ❷: *Can you think of a mistake you could make when answering this question? What could you do to avoid it?*
- Question ❷: *How would you convert 300 cm into metres?*

IN FOCUS Questions ❶ and ❷ both incorporate an element of problem solving that children will need to familiarise themselves with – the presentation of information in one unit and the requirement to give the answer in a different unit. It is important that children are able to identify this in a question. Encourage children to discuss potential errors (including giving the answer in the wrong unit) and to suggest ways of preventing them.

STRENGTHEN To strengthen children's understanding of the connections between metric units, use flashcards with equivalent measurements on each side (for example, 1 kg on the front and 1,000 g on the back). Once children are familiar with these equivalences, extend the activity to include the amounts in each problem, with children using the cards to form bar models. For example, in question ❶, children could model the first measurement (0·9 kg) by placing 9 '0·1 kg' cards in a row. By turning over the cards, they show how the measurement is converted into 9 lots of 100 g.

DEEPEN Question ❸ extends children's knowledge of metric units in context by introducing comparison and order into a problem that already involves reading scales. The character comments highlight an easy mistake to make when answering this question and what needs to be done to avoid this mistake. Discuss each character's statement with the class. Challenge children to produce a similar problem on ordering lengths for a partner to answer.

ASSESSMENT CHECKPOINT Children should be able to apply with confidence their knowledge of metric units to solve problems that involve conversion. Check that children can describe the strategies they are using to find each answer. Check how and when they are converting between units.

ANSWERS

Question ❶: 0·9 kg = 900 g   0·3 kg = 300 g
0·9 kg + 0·3 kg = 900 g + 300 g = 1,200 g
Ambika should guess a total of 1,200 g.

Question ❷: The roller coaster is now 597 metres long.

Question ❸ a): A, D, B, C, E

Question ❸ b): Yes, you would get the same order.
Children's explanations should mention that the mass of each parcel is still the same, whether it is expressed in grams or kilograms.
Converting to kilograms and then comparing or converting to grams and then comparing will both give the same order.

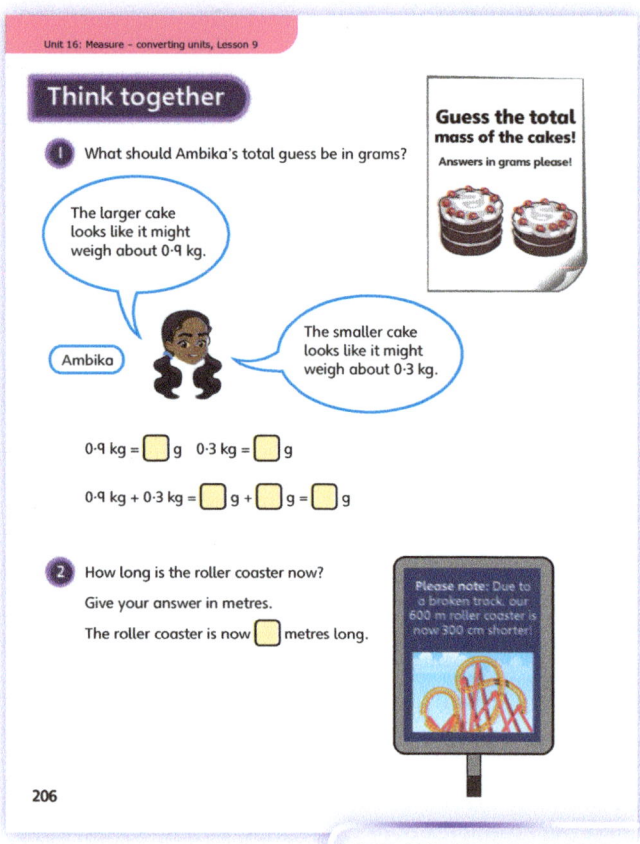

PUPIL TEXTBOOK 5C PAGE 206

PUPIL TEXTBOOK 5C PAGE 207

Unit 16: Measure – converting units, Lesson 9

Practice

WAYS OF WORKING Independent thinking

IN FOCUS Question 3 explores how pairs of measurements can be converted in two different ways and how the same answer can be expressed differently. Discuss the difference between the calculations on the left and those on the right. Check that children recognise that they are essentially the same calculation, but that the answer has been worked out in a different unit of measurement, and that the answers are therefore equal.

STRENGTHEN Ask children to identify the different units used in each problem, highlighting them in some way. Ask questions about the problem, such as the unit asked for in the answer, the unit(s) given in the question, and when children need to convert. Provide children with digit cards and encourage them to draw their own place value grids, shifting the digits physically. Children may also find it useful to cross out numbers in each problem once they have converted them and replace them with their equivalents.

DEEPEN Question 5 a) requires children to work through a series of steps to solve the problem: finding $\frac{1}{10}$ of a litre, adding it to 1 litre, then subtracting 300 ml. Challenge children to write the problem as a single number sentence (brackets will not be needed to express the order of operations: 1,000 ÷ 10 + 1,000 − 300). Ask how the number sentence would be different if the question asked 'How many litres of squash does he have left?'

THINK DIFFERENTLY Question 4 first requires children to read scales showing masses in different ways. In some cases they need to work out the interval between each mark and hence the mass that is being shown. They then need to convert the masses so that they can order them.

ASSESSMENT CHECKPOINT Use questions 1 to 4 to assess whether children are confident when applying their knowledge of metric conversions in problem-solving contexts.

ANSWERS Answers for the **Practice** part of the lesson can be found in the *Power Maths* online subscription.

Reflect

WAYS OF WORKING Independent thinking

IN FOCUS This question has been designed to assess children's ability to add two different units of measurement. Look for children identifying which is the smallest unit and whether they recognise that they could convert to either kilometres or metres.

ASSESSMENT CHECKPOINT Look for children who can apply their knowledge of metric units and can explain the strategies they used.

ANSWERS Answers for the **Reflect** part of the lesson can be found in the *Power Maths* online subscription.

After the lesson

- To what extent were children's various responses the ones that were anticipated. Review children's ideas and questions, the obstacles they encountered, and their misunderstandings and mistakes.
- How well do you feel children were able to express the strategies they chose to use? Do you feel that children have a broad knowledge of mental strategies to draw on?

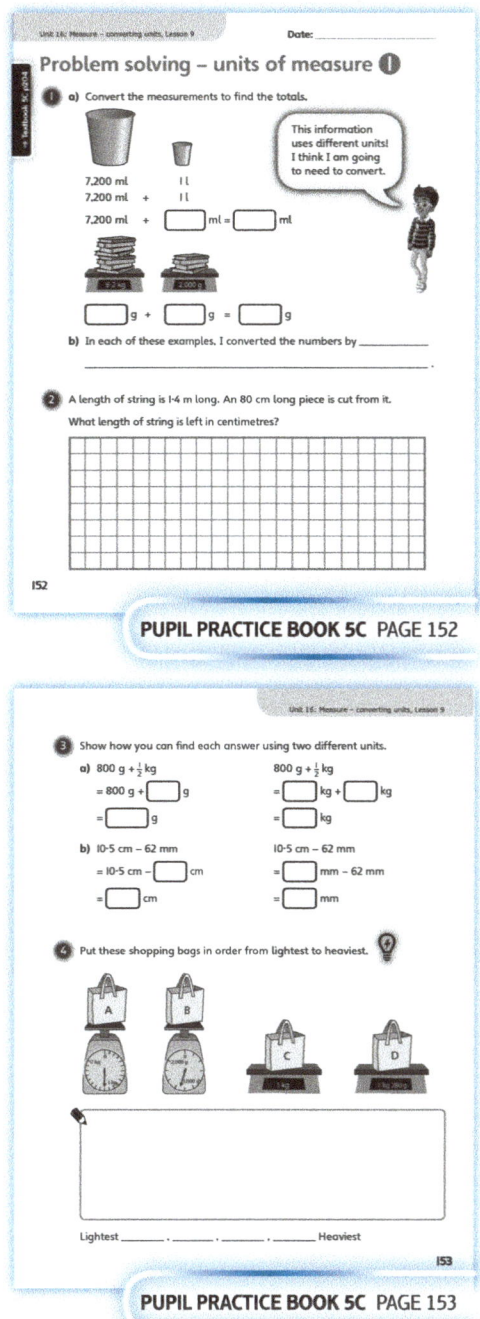

Unit 16: Measure – converting units, Lesson 10

Problem solving – units of measure ❷

Learning focus

In this lesson, children will apply their knowledge of converting units to solving problems. These will include a range of problem types, including those that involve applying multiplication and/or division facts to solve scaling problems.

Before you teach

- This lesson draws together everything children have learnt in Unit 16. What misconceptions and mistakes will you look out for?
- Where might you give children more opportunities to develop their reasoning skills?

NATIONAL CURRICULUM LINKS

Year 5 Measurement

Use all four operations to solve problems involving measure [for example, length, mass, volume, money] using decimal notation, including scaling.

ASSESSING MASTERY

Children can apply their knowledge of converting between units of measurement (metric → metric and metric ↔ imperial, using approximate equivalences) in the process of solving a range of problems. Children can apply their knowledge of multiplication and division in particular when converting and solving scaling problems.

COMMON MISCONCEPTIONS

Children may think that they can compare or calculate using amounts without converting when it is necessary, failing to check that two amounts are in the same unit first. Ask:
- *What unit of measurement will your answer be in? What will you need to do to find the answer?*

STRENGTHENING UNDERSTANDING

Prepare children for this lesson by encouraging them to look back through the previous nine lessons and the different conversions they have learnt, including the equivalences between imperial and metric units. Show children flashcards with different conversions written on them (for example, mm → cm). Encourage children to match these up with a list of operations written on the board (mm → cm: ÷ 10). Ask children to justify their decisions and use the opportunity to refresh their knowledge of what each unit is equivalent to.

GOING DEEPER

Find examples of simple recipes that serve 2, 5 or 10 people (numbers that are fairly easy to divide a measurement by). Challenge children to develop their knowledge of scaling by exploring the ways they can alter a recipe to make different amounts. For example, this makes 10 pancakes: 100 g plain flour, 1 pinch of salt, 2 large eggs, 300 ml milk. Children should alter the ingredients to make 12 pancakes. They will need to think carefully about how to alter a pinch of salt and 2 eggs.

KEY LANGUAGE

In lesson: unit, measure, quantity, convert, centimetre, metre, millilitre, litre, gram, kilogram, ounce, pound, pint, inch, yard, per

Other language to be used by the teacher: approximately, kilometre, millimetre, gallon, foot, for every, scaling, value

STRUCTURES AND REPRESENTATIONS

Number line, bar model

RESOURCES

Optional: measuring tape (cm), set of scales (g), string, dried rice and pasta, analogue clock faces with movable hands, strips of paper, rulers, flashcards

 In the eTextbook of this lesson, you will find interactive links to a selection of teaching tools.

Quick recap

As a class, practise simple 2- and 3-digit by 1-digit multiplications. Check that children have an efficient method to work out the answers.

Unit 16: Measure – converting units, Lesson 10

Discover

WAYS OF WORKING Pair work

ASK

- Question ❶: *List these categories of measurement (capacity, mass, length and time) in order from most to least important in the kitchen. Why did you choose that order?*
- Question ❶: *Why is it important for Reena and Lee to solve their problems? What might be the result if they do not?*

IN FOCUS Talk about the different types of units of measurement used when cooking. Length may be the least obvious one; examples include the diameter of a pizza (a recipe for a 10-inch pizza), the size of a cake tin or baking tray, or the depth of an ingredient ('Roll out the icing until it is about half a centimetre thick'). Discuss when children might need to convert between measurements when baking.

PRACTICAL TIPS Set up very simple activities where these sorts of problems are explored practically. For example: provide a measuring tape that only shows centimetres and ask children to cut a piece of string that measures about 5 inches; give children 90 g of pasta without telling them how heavy it is and explain that this is enough for 3 people – ask them to find out how much pasta is needed for 1 person or 4 people.

ANSWERS

Question ❶ a): Reena needs to convert ounces into grams:
  2 oz = 56 g oats
  4 oz = 112 g brown sugar
  4 oz = 112 g butter
  5 oz = 140 g plain flour

Question ❶ b): 5 cooking apples, 70 g oats, 140 g brown sugar, 140 g butter, 175 g plain flour

Share

WAYS OF WORKING Whole class teacher led

ASK

- Question ❶ a): *How do you convert from ounces into grams? How do you know this?*
- Question ❶ a): *How did Dexter use doubling to help convert the different amounts of ounces?*
- Question ❶ a): *Can you think of two methods to convert 5 oz – one using addition and one using subtraction?*
- Question ❶ b): *If a recipe is for 1 person, how can you work out the quantities for 5 people? Can you explain what Flo is suggesting?*
- Question ❶ b): *How does the bar model show the method? Can you describe what to do?*

IN FOCUS Explain that converting units and altering recipes to make different amounts are common real-life problems that anyone who cooks or bakes has to solve regularly. For question ❶ b), encourage children to think about how the recipe might be written differently to be more helpful to Lee. Introduce the term 'scaling' as a way of describing how to make something larger or smaller whilst keeping everything in proportion.

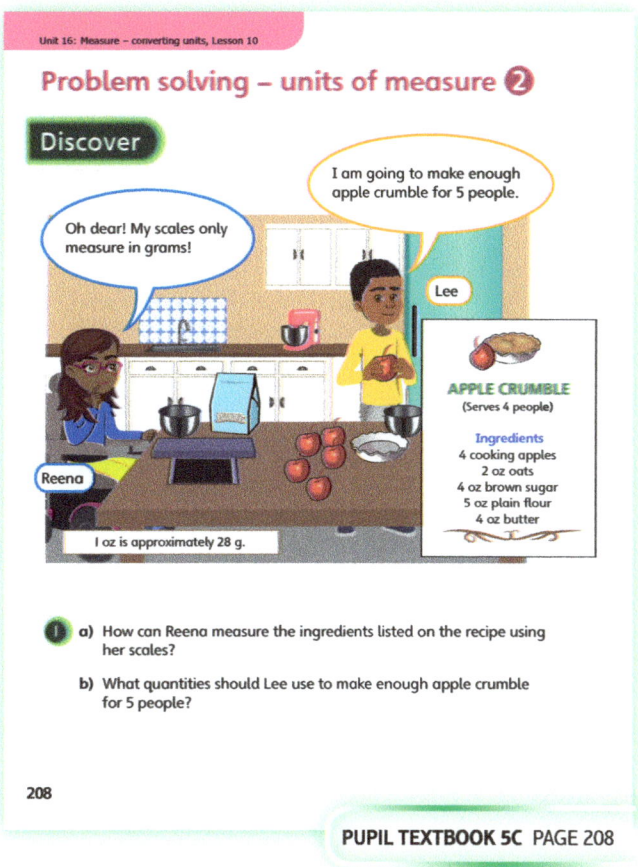

PUPIL TEXTBOOK 5C PAGE 208

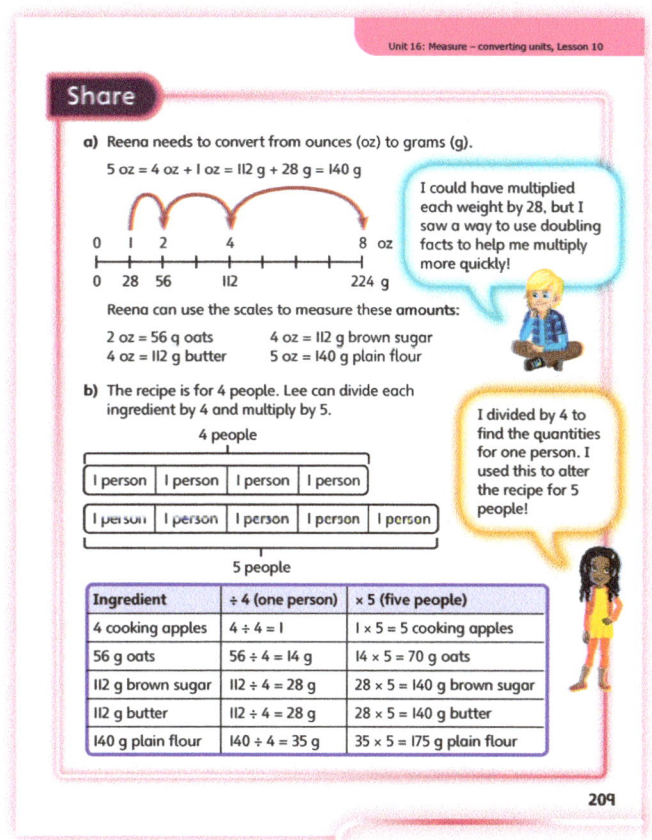

PUPIL TEXTBOOK 5C PAGE 209

245

Unit 16: Measure – converting units, Lesson 10

Think together

WAYS OF WORKING Whole class teacher led (I do, We do, You do)

ASK

- Question 1 a): *Can you predict the number of cartons Jamie needs without calculating anything?*
- Question 1 a): *How is the bar model useful in helping to solve the problem? How would you complete it?*
- Question 2: *Is 17:10 the start or the end time of Danny's baking? Do you need to work backwards or forwards to find the answer?*
- Question 2: *What could you do to represent this problem visually? How might it help you to find the answer?*
- Question 3: *What does 'best value' mean?*

IN FOCUS When answering question 1 a), encourage children to consider how they might predict the answer using mental methods. For example, if they know that 2 l = 2,000 ml and that a pint of milk is just over 500 ml, they can predict that they will need to open four pints of milk to get the 2,000 ml that they need. They should then use the bar model and calculate whether their prediction is correct.

STRENGTHEN Support children when working backwards in question 2 by providing them with analogue clocks with movable hands. Encourage children to make the end time (17:10) on their clocks and then count 55 minutes backwards to represent the time in the oven, and a further 25 minutes backwards to represent the preparation time. Discuss how they can check whether the start time is correct by counting forwards from 15:50.

DEEPEN When considering question 3, establish why the bags of sugar are difficult to compare (they are all the same price, they all look the same, but the units of measurement are all different). Ask whether the best value bag is going to be the one that weighs the least or the most. Discuss why Dexter suggests that the gram is the best unit of measurement to convert them into and whether there any other units they could be converted into. Give children similar problems to solve using length or capacity.

ASSESSMENT CHECKPOINT Use questions 1 and 2 to assess whether children are displaying good confidence when applying their knowledge of unit conversion in problem-solving contexts. They should know how to use given representations and models to help support their working.

ANSWERS

Question 1 a): 2 litres = 2,000 ml
Jamie needs to open 4 cartons of milk.

Question 1 b): Jamie will have 280 ml left over.

Question 2: The latest time Danny should start preparing is 15:50 (ten minutes to 4).

Question 3: A: 1·4 kg = 1,400 g, B: 10 oz = 280 g,
C: 1,250 g, D: 2 lb = 900 g
Bag A is the best value.

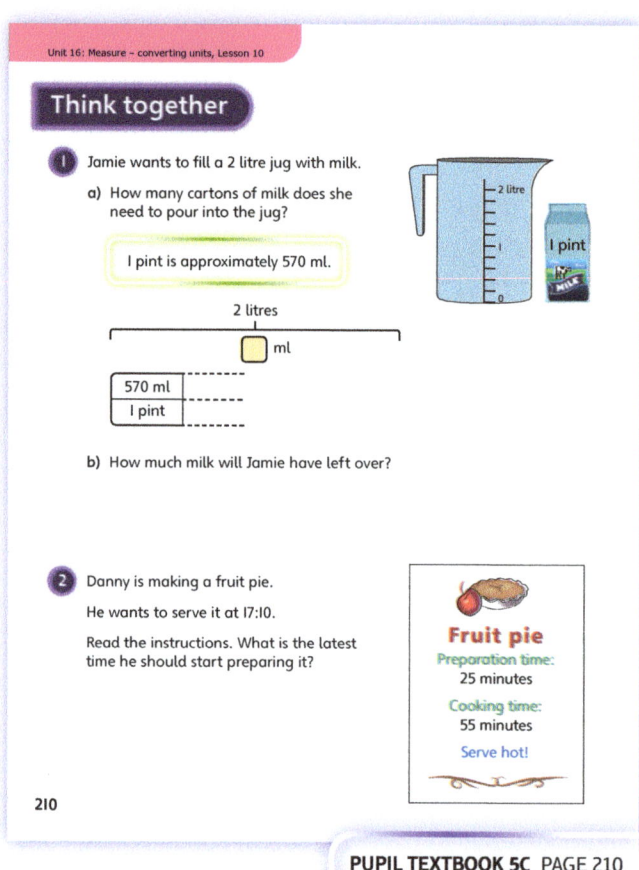

PUPIL TEXTBOOK 5C PAGE 210

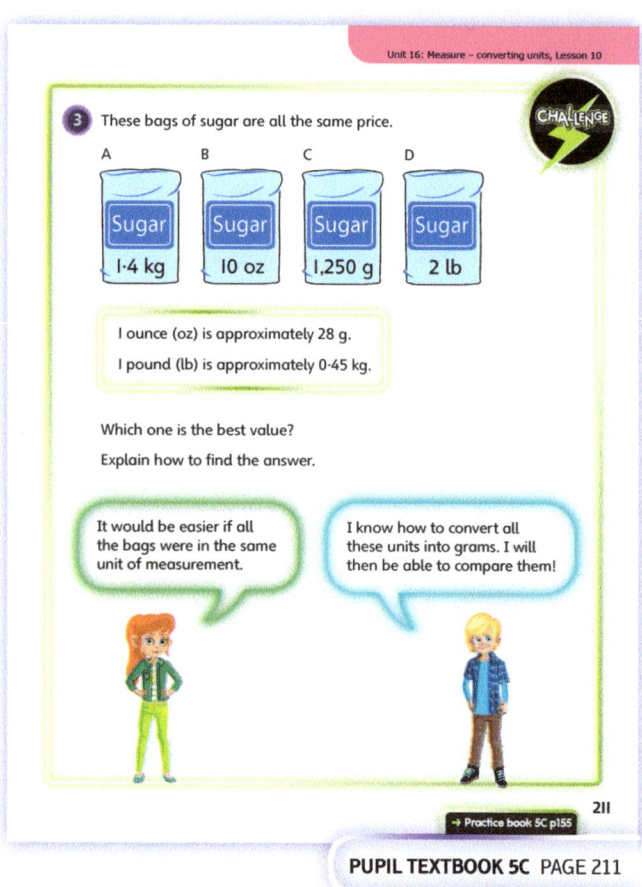

PUPIL TEXTBOOK 5C PAGE 211

Unit 16: Measure – converting units, Lesson 10

Practice

WAYS OF WORKING Independent thinking

IN FOCUS Encourage children to consider at which point in each question they need to convert between units. In some instances, children can work out the answer just as easily by converting at the beginning or at the end. For example, in question ❹, calculate 1·6 ÷ 10 = 0·16 kg and then convert to give 160 g, or convert 1·6 kg into 1,600 g and then calculate 1,600 ÷ 10 = 160 g. Discuss whether it makes a difference which method children use.

STRENGTHEN For question ❺, encourage children to make actual-size bar models, measuring and cutting strips of paper 1 m and 90 cm (≈ 1 yd) long. Give children time to consider the problem in pairs, using these concrete representations to help. They may choose different ways to find the cheaper option, ranging from visual reasoning to calculating the answer.

DEEPEN Question ❼ can be used to deepen children's problem-solving skills as it contains several calculation steps (find the total of four frames, subtract from the total length, divide by 3 for the size of one gap) and also incorporates the need to convert between metric units. Discuss which unit it is easier to work in: centimetres, because 1·25 – 0·8 is more difficult to calculate than 125 – 80. Say that there are now 5 frames and ask how this changes the calculations.

ASSESSMENT CHECKPOINT Use questions ❶ to ❻ to assess whether children can apply their knowledge of unit conversions confidently. Children should be using representations and models to help support their working where necessary. They should identify times when they need to convert between units and recognise how to do this. They should display reasoning skills, explaining appropriate methods and thinking with confidence.

ANSWERS Answers for the **Practice** part of the lesson can be found in the *Power Maths* online subscription.

Reflect

WAYS OF WORKING Independent thinking followed by pair work

IN FOCUS In this **Reflect** question, children explain how they would apply what they have learnt about converting between units, and about the values of metres and inches in terms of centimetres. Initially, children should think individually and decide on a method, which they can then explain to a partner. Children should include the equivalence of 1 m with 100 cm, and that 1 inch is approximately the same as 2·5 cm. Encourage children to compare their methods, identifying what is the same and what is different.

ASSESSMENT CHECKPOINT Look for children who describe how to convert 1 m into centimetres and then use their knowledge of the approximate length of an inch in centimetres.

ANSWERS Answers for the **Reflect** part of the lesson can be found in the *Power Maths* online subscription.

After the lesson

- Were children able to clearly explain their method to convert 1 m into inches to a partner? Can they work as a pair to generalise the steps for converting from metres to inches?
- How much do you feel that the success of this lesson reflects the success of the unit as a whole?

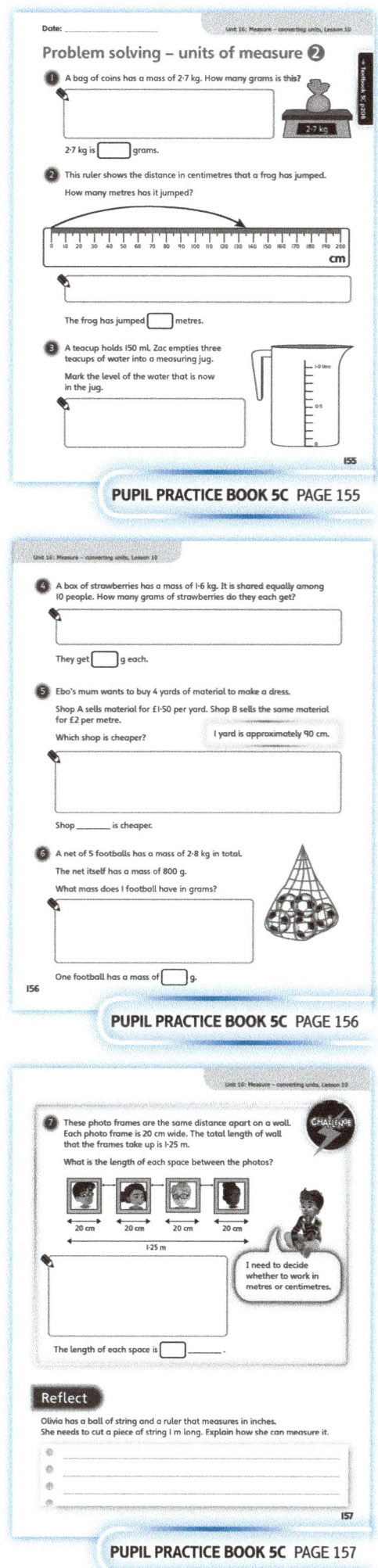

PUPIL PRACTICE BOOK 5C PAGE 155

PUPIL PRACTICE BOOK 5C PAGE 156

PUPIL PRACTICE BOOK 5C PAGE 157

End of unit check

Don't forget the unit assessment grid in the *Power Maths* online subscription.

WAYS OF WORKING Group work adult led

IN FOCUS

- Questions 1, 2, 4 and 6 are all designed to assess children's confidence with converting units. Questions 1 and 4 require children to convert between metric units and question 6 between imperial and metric units, while question 2 assesses their understanding of the method of conversion.
- Questions 3 and 5 are both concerned with time. Question 3 assesses children's ability to convert from a time in hours and minutes into minutes, and question 5 requires children to identify information from a timetable.
- Question 7 is a SATs-style question where children need to convert between units of time to order durations.

ANSWERS AND COMMENTARY

Children who have mastered this unit will be able to convert between units of mass, length and capacity (metric → metric, imperial → imperial and metric ↔ imperial) as well as between units of time, including where there is a remainder, and confidently apply this knowledge to solve problems. Children can use reasoning to explain their methodology. Children can read information from timetables and use it to solve time-based problems.

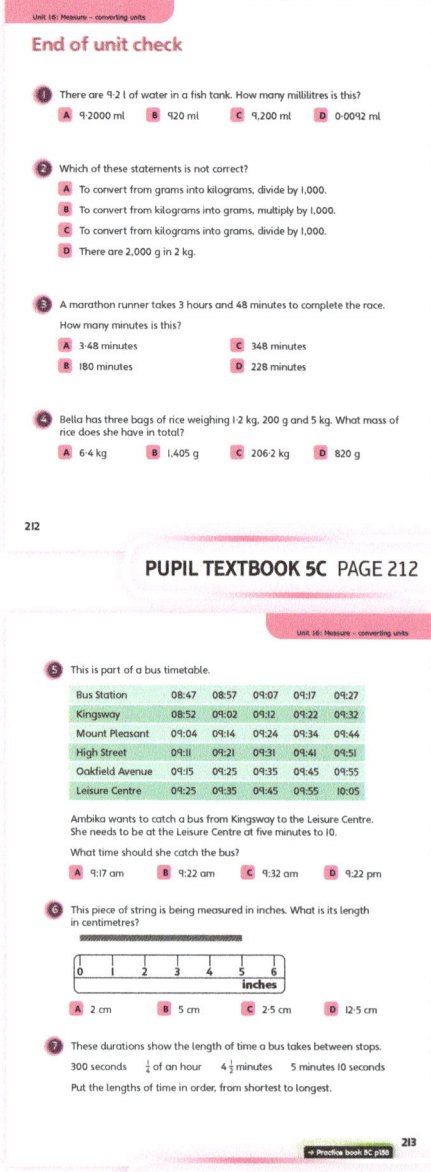

PUPIL TEXTBOOK 5C PAGE 212

PUPIL TEXTBOOK 5C PAGE 213

Q	A	WRONG ANSWERS AND MISCONCEPTIONS	STRENGTHENING UNDERSTANDING
1	C	D shows that the child has divided instead of multiplied.	Play matching pairs games using flashcards where children are given measures expressed in one unit (for example, 4·5 litres) and must find the matching partner (for example, 4,500 ml). Provide opportunities to measure simple lengths, masses and capacities using different units. Where possible, provide equipment that shows different units on its scale.
2	C	Not choosing C suggests that the child is unsure of the correct operation required to convert g ↔ kg.	
3	D	A and C suggest that the child has an incorrect understanding of how to convert a time into minutes.	
4	A	B and C suggest that the child has not converted all the measurements into the same unit.	
5	B	C suggests an incorrect identification of 09:55.	
6	D	A suggests an incorrect understanding of the equivalence between inches and cm (using 1 inch = 2·5 cm).	
7	$4\frac{1}{2}$ min, 300 s, 5 min 10 s, $\frac{1}{4}$ hr Some children may assume that 300 seconds is the longest duration because it involves the highest number, and that $\frac{1}{4}$ hour is the smallest because it is the only one less than 1.		

Unit 16: Measure – converting units

My journal

WAYS OF WORKING Independent thinking

ANSWERS AND COMMENTARY Explanations may vary, but should refer to the equivalence between the units.

a) 1·2 litres = **1,200** millilitres. I know this because there are 1,000 ml in 1 l. To convert, multiply by 1,000.

b) 490 minutes = **8** hours **10** minutes. I know this because there are 60 minutes in 1 hour. To convert, divide by 60 and write the remainder as minutes.

c) 60 inches = **1·5** metres. I know this because there are 2·5 cm in 1 inch and 100 cm in 1 m. To convert inches to centimetres, multiply by 2·5 and then to convert centimetres to metres, divide by 100.

If children are finding working out how to answer the questions challenging, discuss what they know about the units in the question. Change the number in the question to an easier number and ask children to describe how they would work out the answer. Then encourage children to use this method to answer the actual question.

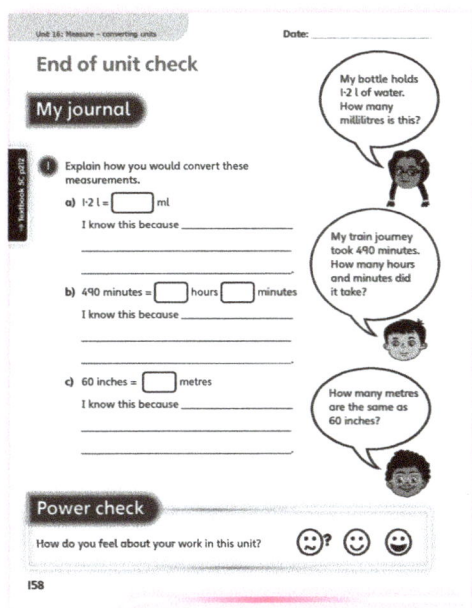

PUPIL PRACTICE BOOK 5C PAGE 158

Power check

WAYS OF WORKING Independent thinking

ASK
- Do you think you would be able to convert between two metric units on your own?
- If you were reminded of the approximate value, do you think you would be able to convert between a metric measurement and an imperial measurement?
- What would you say is the difference between metric and imperial units?

Power play

WAYS OF WORKING Pair work

IN FOCUS The purpose of this **Power play** is to continue to consolidate children's familiarity with reading data from timetables in a game context. The activity challenges children to consider the various ways that the timetable can be read – vertically (considering one train's journey), horizontally (considering the trains that visit one station) and calculating durations.

ANSWERS AND COMMENTARY Look for children demonstrating fluency with 24-hour times, using timetables and adding or subtracting with time.

Facts might include information such as:
- a correct reading of the time in 24-hour vocabulary
- a reference to another time in the same column
- a reference to another time in the same row
- a reference to the duration.

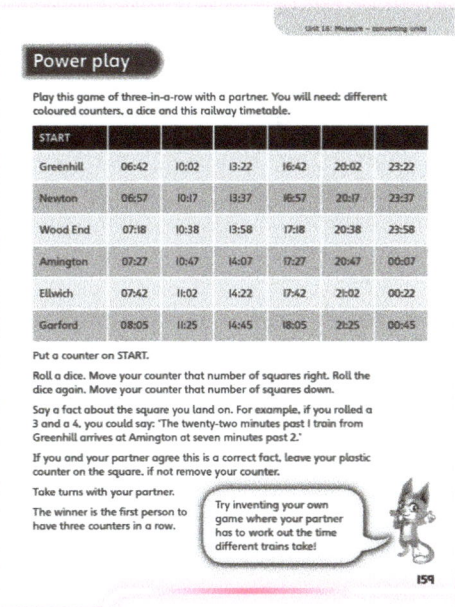

PUPIL PRACTICE BOOK 5C PAGE 159

After the unit

- Are children familiar with the different metric and imperial units of measurement?
- Can children read timetables confidently?
- How do you feel that the unit assessment went?

Strengthen and **Deepen** activities for this unit can be found in the *Power Maths* online subscription.

Unit 17
Measure – volume

Mastery Expert tip! 'When I taught this unit, I encouraged children to work together in small groups to make storage boxes by recycling card boxes, shoe boxes or old containers. I asked children to estimate the volume of the storage boxes they made. We then used the boxes to store some of our maths resources and gave the rest to the local nursery.'

Don't forget to watch the Unit 17 video!

WHY THIS UNIT IS IMPORTANT

This unit introduces the concept of volume, giving children a tangible way to measure and compare a shape's size. Until now, children will have been able to say whether a shape is longer or shorter, wider or narrower, and will have been able to measure a shape's length, width and area. However, this unit now provides them with the tools to measure the amount of space taken up by an object.

Volume is introduced through children using non-standard units and seeing how many of these units will fit within a shape. Children will learn to measure volume by counting the number of cubes that fit within a shape. Although the relationship between length, width and height is not expounded in this unit, children may still recognise informal links between the three concepts from their activities when counting cubes. These links will provide a foundation for further development of the concept of volume in Year 6.

WHERE THIS UNIT FITS

→ Unit 16: Measure – converting units
→ **Unit 17: Measure – volume**

This unit builds on children's understanding of the properties of cubes, cuboids and different solids. It extends children's basic comprehension of how to measure and calculate the area of a shape to estimating the amount of space taken up by an object and the amount a container can hold.

Before they start this unit, it is expected that children:
- understand what is meant by a 3D shape and are able to identify the space inside it
- understand simple properties of cubes and cuboids.

ASSESSING MASTERY

Children who have mastered this unit will understand that the volume of a 3D shape is the amount of space taken up by an object. Children will be able to confidently estimate the volume of 3D shapes. They can do this by counting the number of cubes in each layer, then adding the results. Children are able to recognise that this is an estimation of volume and can explain why this is the case.

COMMON MISCONCEPTIONS	STRENGTHENING UNDERSTANDING	GOING DEEPER
Children may assume that longer or taller shapes have more volume than shorter shapes, simply because they are longer.	Provide children with two containers: one that is long and thin (such as a measuring cylinder) and another that is shorter but has a greater capacity (such as a mug). Ask children to fill each with water and pour each one separately into a measuring jug. Establish that the volume of water in the mug was greater than the volume of water in the measuring cylinder, even though the measuring cylinder is taller.	Provide children with an array of different beakers or containers. Ask them to estimate which has the greatest and which the least volume, and to put the containers in order from smallest volume to greatest volume. Test children's estimations by filling the containers with water.
Some children may view the estimation process as a form of calculation, providing the actual volume, because it involves counting.	It is important children understand that when estimating, the answers can vary. Discuss with children why the answers may vary and what an acceptable answer is.	Challenge children to predict the volumes of different shapes.

Unit 17: Measure – volume

UNIT STARTER PAGES

Use these pages to introduce the concept of the volume of 3D shapes to the whole class, checking their understanding of the importance of estimating volume.

STRUCTURES AND REPRESENTATIONS

3D shapes made of cm³ cubes: Models like this allow children to count the number of cubes in each solid in order to measure volume.

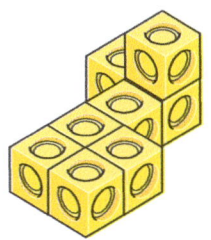

KEY LANGUAGE

There is some key language that children will need to know as part of the learning in this unit:

- volume, capacity, solid
- cube, cuboid, triangular, prism
- 3D shapes, objects
- calculate, estimate, compare, count, accurately, order, amount, irregular, prediction, exact
- cm³ cubes, units of measurement, measure
- less, more, less than (<), more than (>), largest, smallest, least, greatest, equal
- space inside
- height, length, width, size, tall
- layer, slice
- multiple, total, take away, whole, part, almost half, identical

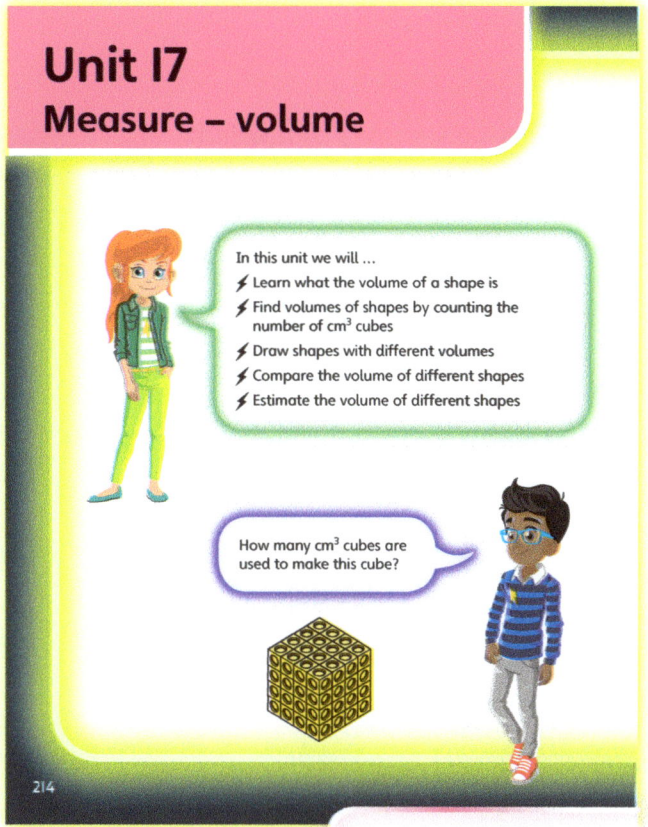

PUPIL TEXTBOOK 5C PAGE 214

PUPIL TEXTBOOK 5C PAGE 215

Unit 17: Measure – volume, Lesson 1

Cubic centimetres

Learning focus

In this lesson, children will be introduced to the concept of the volume of a 3D shape. They will measure this by counting the unit cubes used to make each shape.

Before you teach

- Can you think of any misconceptions that children may have when measuring volume?
- How can you use your school environment to introduce and reinforce the concept of volume?

NATIONAL CURRICULUM LINKS

Year 5 Measurement

Estimate volume [for example, using 1 cm^3 blocks to build cuboids (including cubes)] and capacity [for example, using water].

ASSESSING MASTERY

Children can confidently explain what volume is. Children can use cm^3 cubes to make solids of different volumes and draw the shapes they make on an isometric grid.

COMMON MISCONCEPTIONS

Children should recognise the word 'volume', but may not fully understand it. Some children link volume to loudness and are unable to understand what volume represents. Ask:
- *What do you understand by volume? How does it help to tell us how much space something takes up?*

Children may assume that longer or taller shapes have more volume than shorter shapes, simply because they are longer. Provide them with two containers – one long and thin; another shorter but with a greater volume. Ask:
- *Can you fill both containers with water and see which one holds the most?*

STRENGTHENING UNDERSTANDING

To strengthen understanding, give children simple tasks that involve exploring volume more generally as the space that a 3D shape takes up. Provide small boxes of differing sizes and cm^3 cubes or multilink cubes to support discussion. Ask: *What is the maximum volume this box can hold? How many cubes are used to make this solid? What is the volume of this solid?*

GOING DEEPER

Give children sheets of isometric paper and ask them to draw composite 3D shapes made of cm^3 cubes. They could make the shapes using cubes and then they could draw them. If children work with a partner, they could swap papers and try to work out the volume of each other's shapes without the use of physical cubes to help them. They should pay particular attention to the cubes that may be hidden behind another cube in their partner's diagram.

KEY LANGUAGE

In lesson: volume, 3D shapes, **cm^3 cubes**, amount, space, object, measure, cuboid, identical

Other language used by the teacher: capacity, height, length, width, space inside, units of measurement, layer

RESOURCES

Mandatory: cm^3 cubes, isometric paper

Optional: a variety of measuring items, different-sized bottles and containers

 In the eTextbook of this lesson, you will find interactive links to a selection of teaching tools.

Quick recap

Play 'Square and Cube Number Bingo'. Ask children to write down five square numbers (up to 12 × 12) and one cube number from 0, 1, 8, 27, 125 or 1,000. Call out examples such as 3^2; if children have the number 9 written down, they cross it out. The first player to cross out all six numbers wins.

Unit 17: Measure – volume, Lesson 1

Discover

WAYS OF WORKING Pair work

ASK

• Question 1 a): *What size cubes are Zac and Reena using to make their shapes? Is there a way you could measure the volume of each shape?*
• Question 1 a): *Should you add all the cubes in the shape or just the ones that are visible?*
• Question 1 b): *If the shapes have the same volume, why do you think they look different?*

IN FOCUS Question 1 a) encourages children to begin talking about what volume is. Challenge children by asking: *How can two shapes have equal volumes? If you turn the shape around, would the volume change? What happens if you move one of the cubes from the second layer and place it on the first layer?* Draw out the concept of the space that a shape takes up and that the volume is the amount of 3D space the shape occupies.

PRACTICAL TIPS Give children cubes to build the solids shown in the **Discover** picture. Examining how the solids are constructed by building them for themselves will help children develop their understanding of volume. Ask them to build other shapes using the same cubes and describe what they see.

ANSWERS

Question 1 a): Zac is correct. Both shapes have a volume of 6 cm³.
Volume is the amount of space that an object fills.

Question 1 b):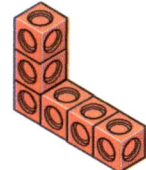

Share

WAYS OF WORKING Whole class teacher led

ASK

• Question 1 a): *What does the word 'volume' mean? When have you heard it used before?* [Children should be familiar with the term 'volume' in terms of level of sounds.]
• Question 1 b): *What do you notice about the way the cubes have been arranged? Are there other ways that the cubes could be arranged?*
• Question 1 b): *Would the volume increase if you put the cubes one above the other to make a tall tower?*

IN FOCUS Questions 1 a) and b) help children understand how to use cm³ cubes as measures to compare the volumes of two shapes. It is necessary that children make the connection that the more cubes used to make a shape, the greater the volume of that shape. It is important for children to make the solids used in this task themselves so that they can see what the volume of 6 cm³ cubes looks like.

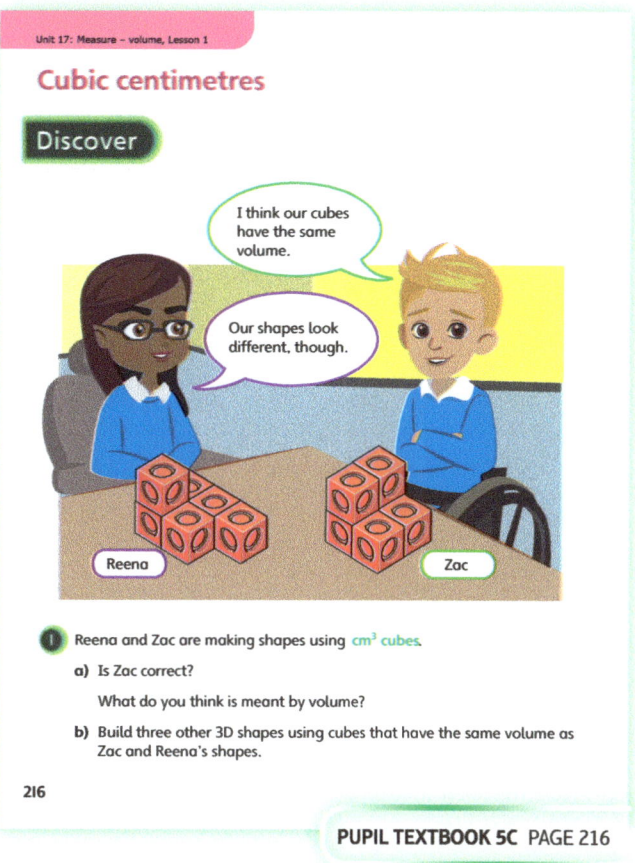

PUPIL TEXTBOOK 5C PAGE 216

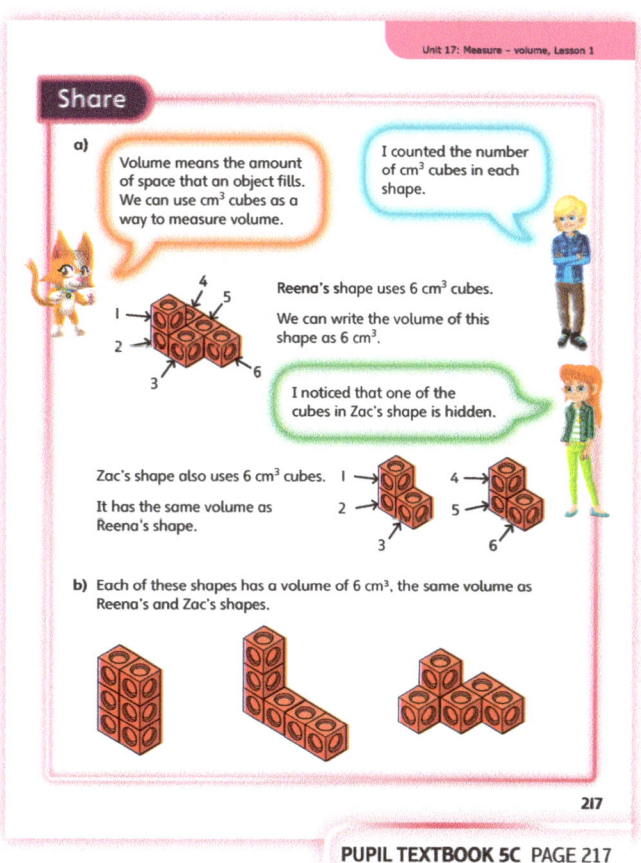

PUPIL TEXTBOOK 5C PAGE 217

Unit 17: Measure – volume, Lesson 1

Think together

WAYS OF WORKING Whole class teacher led (I do, We do, You do)

ASK

- Question ❶: *Can you describe the shapes that you see? How many layers are there? How many cubes are there in each layer?*
- Question ❷: *Can you make a cuboid? Can you make a different cuboid? Will the volume change? Is it possible to make a cube?*
- Question ❸: *How easy is it to draw shapes using isometric paper? What do you have to remember?*

IN FOCUS Question ❶ leads children through calculating the volume of a solid when not all the cubes are visible. Encourage children to discuss their methods of counting the cubes with each other. Allow children to share their most efficient ways of calculating the volume of a solid with the whole class.

STRENGTHEN For question ❶, encourage children to make the shapes shown in the pictures. This way, they will realise that some drawings of shapes have hidden cubes. For question ❷, give children opportunities to discuss their method of making shapes using a certain number of cm³ cubes. Encourage children to make a cuboid as well as other shapes. Ask: *What is the maximum number of layers the shape could have? How do you know?*

DEEPEN In question ❸ b), ask children to make the different solids using the 10 cubes used in the question. Can they make two cubes and a cuboid? Encourage them to discuss the difference between the shapes they make. Ask: *Can you draw the shapes on isometric paper?*

ASSESSMENT CHECKPOINT Children should recognise what volume refers to and understand that this is a measure of the amount of space that an object fills. Children should be able to confidently suggest ways to measure volume using cm³ cubes.

ANSWERS

Question ❶ a): 8 cm³

Question ❶ b): 8 cm³

Question ❶ c): 20 cm³

Question ❶ d): 20 cm³

Question ❷: Children's shapes will vary, but should all include 12 cubes to make 12 cm³.

Question ❸ a):

Question ❸ b): Left-hand shape: 3 cubes
Middle shape: 3 cubes
Right-hand shape: 4 cubes
(There is a hidden cube underneath the top cube.)

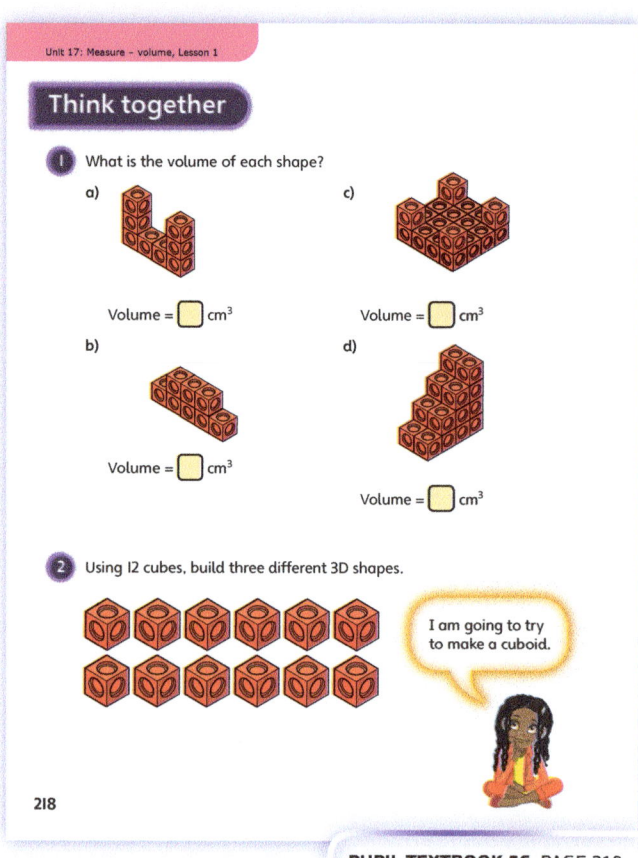

PUPIL TEXTBOOK 5C PAGE 218

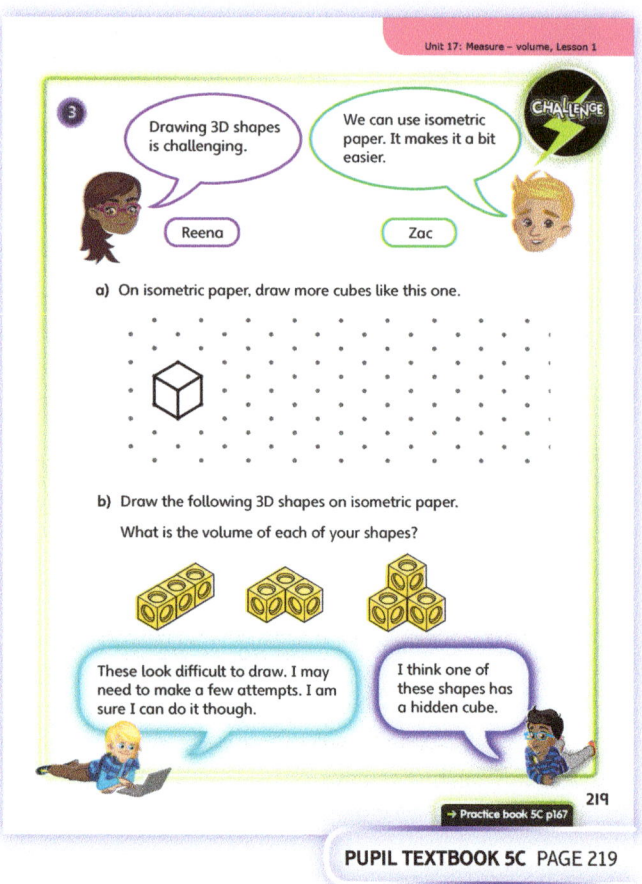

PUPIL TEXTBOOK 5C PAGE 219

254

Unit 17: Measure – volume, Lesson 1

Practice

WAYS OF WORKING Independent thinking

IN FOCUS Questions 1 and 2 allow children to consolidate their understanding of measuring the volumes of shapes using cm³ cubes. Questions 3 and 4 are designed to help children understand that they may not always be able to see all the cubes that make up a shape. However, the cubes they cannot see will still contribute to the volume of the shape. Ask: *How can you find the volume when you cannot see all the cubes? How can you ensure you have not forgotten any?* The more practice children have with making different solids with a given number of cubes, the easier it will be for them to visualise the shapes.

STRENGTHEN In questions 3 and 4, ask children to make the different shapes using multilink cubes. Then ask them to place the shapes in different positions, so that they can see the cubes that were hidden in the illustrations shown in the Practice Book. Ask: *Can you point to the cubes that you cannot see in the picture? What is the total number of cubes that make up this shape?*

DEEPEN For questions 5 and 6, ask children to calculate the volume of each shape. Encourage children to consider the relationship between a shape's dimensions and its volume. To extend understanding even further, children could play the game 'What am I?'. One child describes a shape, while another makes it. Emphasise the importance of using the correct mathematical language.

ASSESSMENT CHECKPOINT At this point in the lesson, children should be confident in finding the volumes of different solids. Children should be able to make different shapes with a given volume and draw cubes and cuboids on isometric paper.

ANSWERS Answers to the **Practice** part of the lesson can be found in the *Power Maths* online subscription.

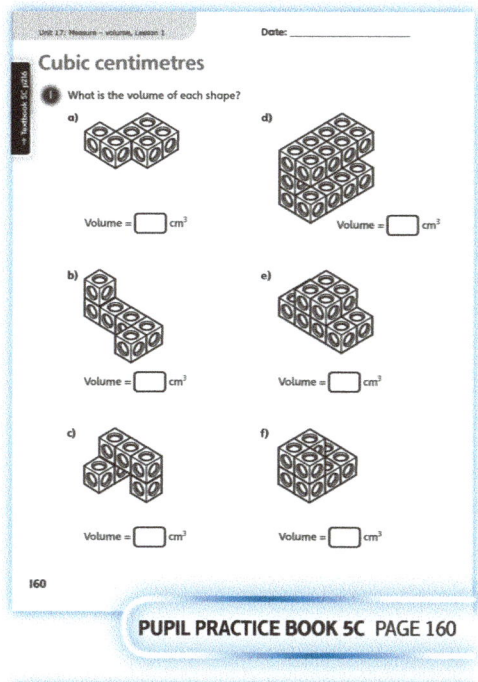

PUPIL PRACTICE BOOK 5C PAGE 160

PUPIL PRACTICE BOOK 5C PAGE 161

Reflect

WAYS OF WORKING Independent thinking

IN FOCUS This **Reflect** question provides an opportunity to check children's understanding of volume and the methodology they will use in calculating volume. The aim of the question is that children are not necessarily asked to calculate the volume of a specific shape but to explain clearly what the volume represents and how they would find it. To extend understanding, ask: *Can different shapes have the same volume?*

ASSESSMENT CHECKPOINT Are children able to define and measure the volume of a shape correctly?

ANSWERS Answers for the **Reflect** part of the lesson can be found in the *Power Maths* online subscription.

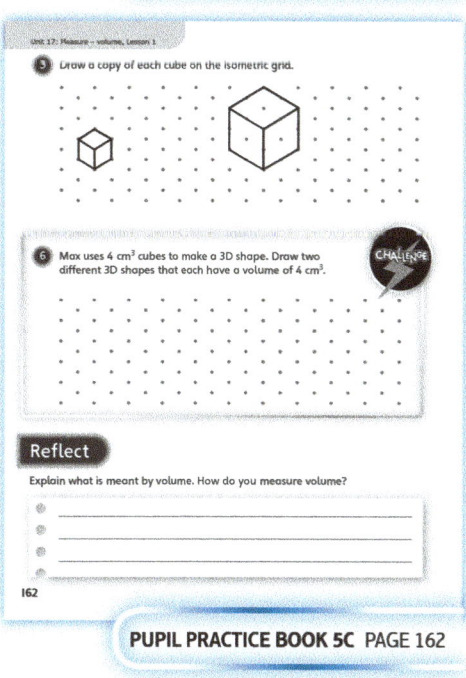

PUPIL PRACTICE BOOK 5C PAGE 162

After the lesson

- Have children mastered what volume is?
- Could more opportunities be given for children to explain the concept of volume?
- What opportunities can be provided for children to calculate volume outside of this lesson?

Unit 17: Measure – volume, Lesson 2

Compare volumes

Learning focus
In this lesson, children will learn how to compare shapes according to their volumes.

Before you teach
- Are children confident in making 3D shapes?
- Do they understand what volume is?

NATIONAL CURRICULUM LINKS

Year 5 Measurement

Estimate volume [for example, using 1 cm^3 blocks to build cuboids (including cubes)] and capacity [for example, using water].

ASSESSING MASTERY

Children can confidently compare the volume of two 3D shapes, measuring their volume accurately by counting cubes, then comparing both values to see which is larger. Children can apply this skill to order several shapes according to their volumes.

COMMON MISCONCEPTIONS

Children may compare the volume of shapes visually by the way they look, rather than by counting cubes and comparing volumes. They may ignore all three dimensions of 3D shapes and compare volumes by comparing the height of the shapes, rather than considering all aspects. Ask:
- *What do you need to do to work out which shape has the largest volume? What is the length, width and height of the shapes?*

STRENGTHENING UNDERSTANDING

Give children the same number of cubes and ask them to build different cuboids and 3D shapes. Place the shapes in different positions and orientations and ask children to compare their volumes. Ask: *If you were to order the shapes, how would you do this?* Allow children to touch the shapes and see for themselves that they have the same number of cubes. If children assumed that the tallest shapes had the largest volume, ask them to look at the shapes again and discuss this misconception with the whole class. Allow time for children to explore the shapes and repeat the activity with a different number of cubes.

GOING DEEPER

Challenge children by asking: *Is it always, sometimes or never true that 3D shapes with the same height have equal volumes?* Encourage children to reason and support their answer with examples, rather than simply saying yes or no.

KEY LANGUAGE

In lesson: compare, volume, 3D shapes, cube, cuboid, irregular, cm^3 cubes, layer, greatest, more, smallest, order, least, greater than (>), prediction

Other language to be used by the teacher: less than (<), count, solid, measure, accurately, multiply, total, take away, size

STRUCTURES AND REPRESENTATIONS

3D shapes

RESOURCES

Mandatory: cm^3 cubes

Optional: ruler, squared dotted paper, isometric paper

 In the eTextbook of this lesson, you will find interactive links to a selection of teaching tools.

Quick recap
Give children six cm^3 cubes each. Ask them as a group to make as many different 3D shapes with a volume of 6 cm^3 as they can.

Discover

WAYS OF WORKING Pair work

ASK

- Question 1 a): *Can you predict who has built the 3D shape with the greatest volume just by looking? Why do you think this shape has the greatest volume?*
- Question 1 a): *How are the shapes the same? How are they different?*
- Question 1 b): *How many cubes does Isla's shape have? How many cubes does Emma's shape have?*

IN FOCUS Talk about the **Discover** picture and encourage children to describe what they see. They will already have plenty of experience of looking at two shapes and choosing the one that looks larger. Now they are required to find the volume of each shape by counting cubes and then comparing both values to find the shape with the greatest volume.

PRACTICAL TIPS Provide children with activities similar to the one shown in the **Discover** picture. Give children two or three shapes. Ask them to compare the volumes. Ask: *What do you have to do to make the volumes of the shapes equal? How many more cubes do you need to add? Why? Show me. Is there another way to do it?* Listen for children who realise that they can add or remove cubes to make the volumes equal.

ANSWERS

Question 1 a): 9 < 10 < 12
Emma has built the shape with the greatest volume.

Question 1 b): Isla needs to add 3 more cubes, for example:

Share

WAYS OF WORKING Whole class teacher led

ASK

- Question 1 a): *What information do you need to know to be able to compare the volumes? Can you tell which shape has the greatest volume just by looking? How could you check which shape has the greater volume?*
- Question 1 b): *How many more cubes does Isla need to add? What would the shape look like now? Is there only one way to do this?*
- Question 1 b): *How would you describe the shapes you made? How many layers does the shape have? How many cubes are required to make each layer? Can you see all the cubes used to make each shape?*

IN FOCUS Question 1 a) consolidates children's understanding of measuring volume by counting cubes. Question 1 b) develops children's understanding of how adding or removing cm³ cubes changes the volume of a shape. The question is important as it allows children to see how shapes with the same volume can look different. Encourage children to explore the different shapes that can be made in question 1 b).

257

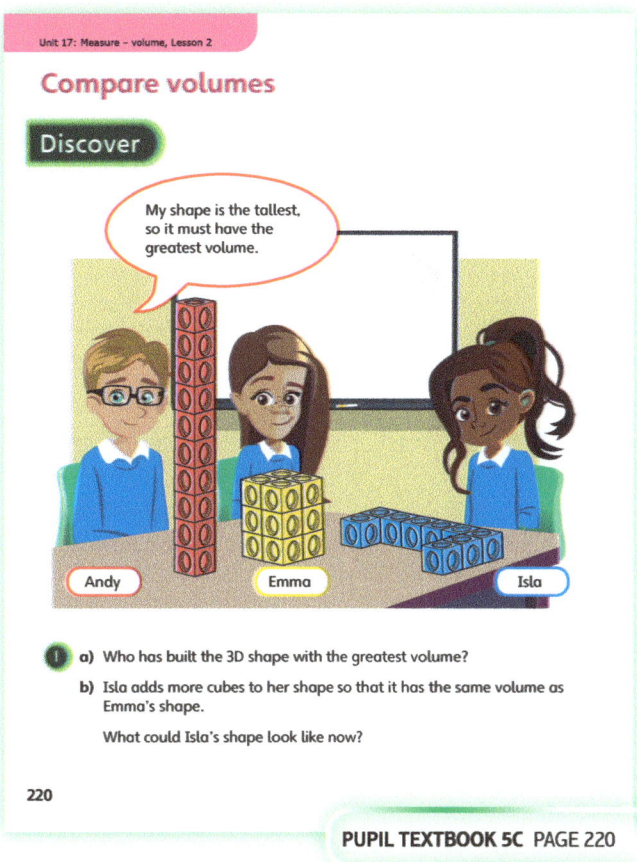

PUPIL TEXTBOOK 5C PAGE 220

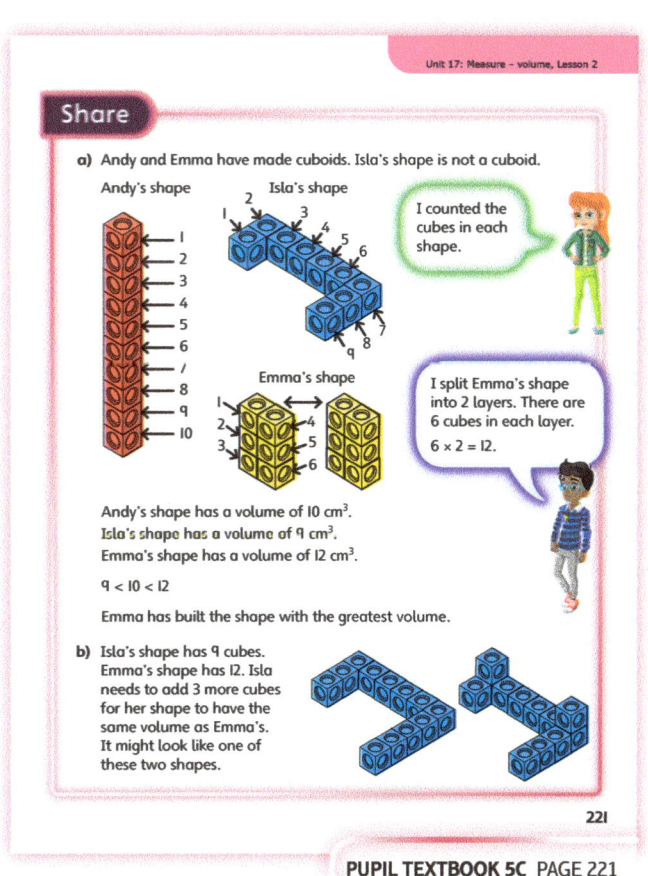

PUPIL TEXTBOOK 5C PAGE 221

Think together

WAYS OF WORKING Whole class teacher led (I do, We do, You do)

ASK
- Question ①: *What method can you use to compare the volumes accurately?*
- Question ②: *Can you see all the cubes in each shape?*
- Question ②: *What steps do you need to take to put these shapes into size order?*
- Question ③ b): *What is different about Isla's shape?*

IN FOCUS Questions ① and ② take children through the steps needed to compare the volume of 3D shapes. First, children should find the volumes by counting cubes. They should then compare the values to see which is smaller or larger. For question ②, children are expected to use their comparisons to write each volume in order. Question ③ a) requires children to factor in hidden cubes when measuring and comparing volumes.

STRENGTHEN For questions ① and ②, give children cubes so that they can make the shapes themselves. When working on question ②, discuss with children how they can find the volume of a shape with 'hidden cubes'. The more practice children have in making different shapes, the better their visualisations of the 3D shapes will be. Discuss how multiplication can be used instead of addition to find the number of cubes in each layer.

DEEPEN In question ③, discuss the importance of using the same sized cubes when comparing shapes. To deepen children's understanding further, discuss the volume of a shape made from plastic building blocks and a shape made from real-life bricks. Show children pictures or provide them with examples of shapes made from different size cubes or blocks. Ask: *What do you need to consider when comparing volumes? Does a shape made from bigger cubes have a larger volume?* Allow children time to explore and generalise.

ASSESSMENT CHECKPOINT Children should understand that to compare volumes, they need to first find their values. They need to consider the number of cubes used to make each shape. They also need to ensure that the sizes of the cubes used are equal.

ANSWERS

Question ①: Emma's shape has the smallest volume. The volume is 12 cubes. Isla's shape has a volume of 13 cubes.

Question ②: A < C < B (A = 16 cubes, B = 20 cubes, C = 18 cubes)

Question ③ a): Emma is correct. Andy's shape has a volume of 8 cubes and Emma's shape has a volume of 8 cubes (including a hidden cube at the back).

Question ③ b): Isla's shape will have a larger volume. She has the same number of cubes as Andy but since the cubes are larger, the total volume will also be larger.

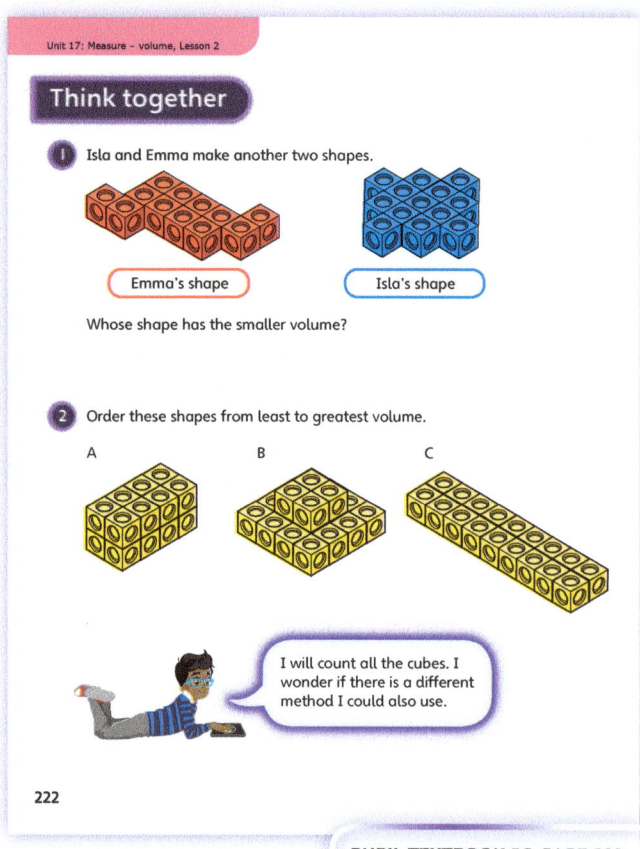

PUPIL TEXTBOOK 5C PAGE 222

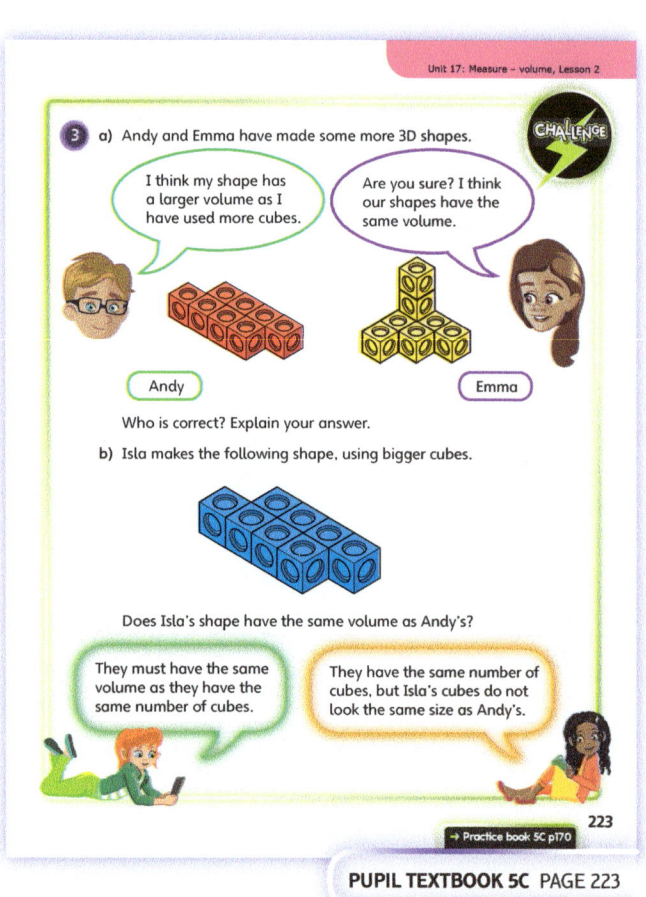

PUPIL TEXTBOOK 5C PAGE 223

Unit 17: Measure – volume, Lesson 2

Practice

WAYS OF WORKING Independent thinking

IN FOCUS Question ③ encourages children to think carefully about what volume represents. They need to read the clues given and identify the shapes that belong to each child. It may be beneficial to revisit what 'a third of', 'half' and 'more than' mean. Ask: *What is the question asking? What do you know? How can you find the volume of each shape? Which clues are you using? Is there another way to find the volume?* Encourage children to give clear reasons for their answers. Ensure that they do not just give numbers when calculating the volumes, but use the correct unit for the volumes too; for example, they should say 'the volume is 10 cm^3 cubes', rather than 'the volume is 10'.

STRENGTHEN If children need support to accurately count hidden cubes in shapes, provide them with cubes and ask them to model each of the shapes in questions ① to ③ concretely. Encourage children to be systematic in their approach. Ask: *How can you record the information to ensure you have not missed any cubes?* Children could count and write the totals of each layer of cubes as they go, then add the results up to find the overall volume of each shape.

DEEPEN Question ⑤ involves reasoning about the relationship between the length, width and height of a shape and its volume. This is another opportunity to clarify the misconception that the taller the shape is, the larger its volume will be. Provide children with cubes so that they can make the shapes themselves and allow plenty of opportunities for them to discuss their estimates and results.

ASSESSMENT CHECKPOINT Children should be working confidently to find volumes by counting cubes, and then comparing and ordering those volumes. Children should be able to use reasoning to explain their answers and make generalisations.

ANSWERS Answers to the **Practice** part of the lesson can be found in the *Power Maths* online subscription.

Reflect

WAYS OF WORKING Independent thinking

IN FOCUS This **Reflect** question asks children to explain how shapes that are made up of the same number of cubes, and therefore have the same volume, can still be different shapes. Their answers should include an explanation along the lines of 'volume being the amount of 3D space a shape occupies'.

ASSESSMENT CHECKPOINT Look for children who can explain the importance of using the same unit of measurement when comparing the volumes of shapes. Children who have mastered this unit can demonstrate that if two shapes are made from the same number of cubes, then they share the same volume, no matter the shape.

ANSWERS Answers for the **Reflect** part of the lesson can be found in the *Power Maths* online subscription.

After the lesson

- How confident are children in comparing the volumes of shapes?
- Are children now less reliant on comparison by sight? Do they understand why it is important to measure before comparing the volume of shapes?

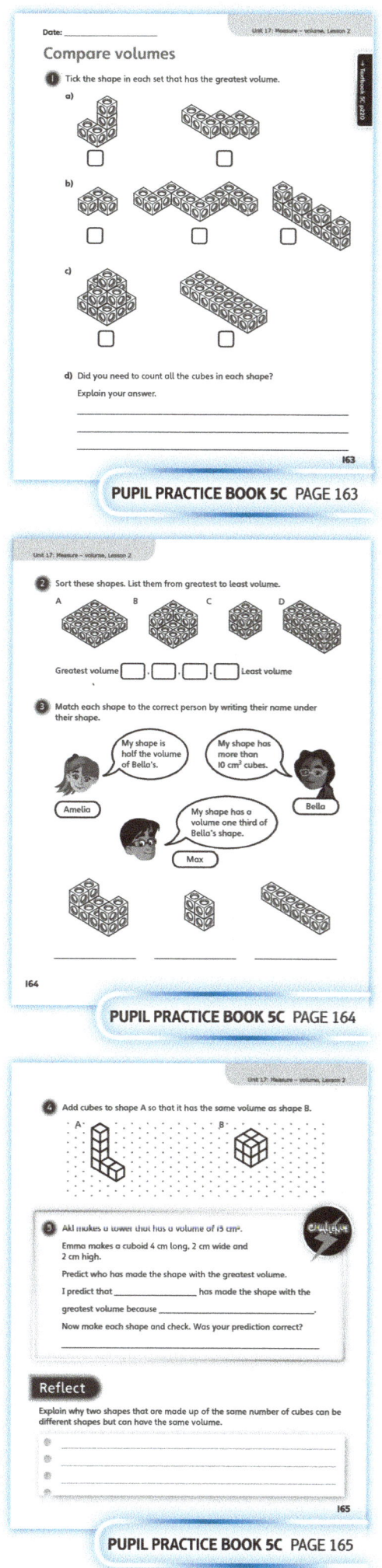

PUPIL PRACTICE BOOK 5C PAGE 163

PUPIL PRACTICE BOOK 5C PAGE 164

PUPIL PRACTICE BOOK 5C PAGE 165

Unit 17: Measure – volume, Lesson 3

Estimate volume

Learning focus

In this lesson, children will apply their knowledge of volume to estimate the volume of 3D shapes and consider how accurate their estimates are.

Before you teach

- How confident are children with the concept of estimation generally?
- How will you link their prior knowledge to the concept of estimation of volume?

NATIONAL CURRICULUM LINKS

Year 5 Measurement

Estimate volume [for example, using 1 cm³ blocks to build cuboids (including cubes)] and capacity [for example, using water].

ASSESSING MASTERY

Children can estimate the volume of 3D shapes confidently. They can do this by modelling the shape using cm³ cubes, counting the number of cubes in each layer, and then adding the results. Children are able to recognise that this is an estimate of volume and can explain why their answer is an estimate and assess how accurate it is.

COMMON MISCONCEPTIONS

Children may view the estimation process as a form of calculation, thinking that it tells them the actual volume, because it involves counting. Other children may assume that the answer or the method of calculation doesn't matter, as it is only an estimate. Ask:

- *What is the difference between estimating and calculating an exact value? Why is estimating important?*

Children may think that if their estimate is slightly different from a partner's estimate, then one of them has to be wrong. Explain to children that there is no one correct answer when estimating; the important thing is to find an answer that is reasonably close to the the true volume.

It is important to use the correct vocabulary in the lesson ('estimate', 'approximately', 'about') and for children to understand that estimating is an important skill, first and foremost because it is a way of determining how reasonable an answer is.

STRENGTHENING UNDERSTANDING

It is important that children understand that, when estimating, the answer can vary and that there is no one correct answer. Discuss with children why their answers may vary and what an acceptable answer is. For example, if a solid is similar in volume to 12 cm³ cubes, but children estimate the volume to be 4 cubes (because they have counted the cubes on only one of the faces), then you may suggest that children check the answer again. Ask: *How many layers does your shape have? How many cubes are there in each layer? How do you know that you have included all the parts in your estimation?*

GOING DEEPER

Challenge children to predict the volumes of different shapes. Ask: *What would be a fair answer? What would not be a realistic answer? How do you know?* Provide children with the same solid shape shown in different orientations. Ask: *Does your method of estimation change? Will your method affect your answer?*

KEY LANGUAGE

In lesson: estimate, volume, 3D shapes, triangular, prism, unit, cubes, layer, slice, exact, compare, accurate, sphere

Other language to be used by the teacher: whole, part, almost half, less than (<), more than (>)

STRUCTURES AND REPRESENTATIONS

3D shapes

RESOURCES

Mandatory: cm³ cubes

Optional: 3D shapes drawn on square dotted paper, modelling clay

In the eTextbook of this lesson, you will find interactive links to a selection of teaching tools.

Quick recap

Ask children to draw as many different rectangles as possible, each with an area of 24 cm². Have them repeat the activity for different shapes.

Unit 17: Measure – volume, Lesson 3

Discover

WAYS OF WORKING Pair work

ASK

- Question 1 a): *Can you describe the shape Jamilla has estimated the volume of? How can you estimate its volume?*
- Question 1 a): *How can you split the shape to make your calculations easier?*
- Question 1 b): *Why do you think you are being asked to estimate, rather than calculate, the volume of the shape? When might it be beneficial to estimate and not calculate?*

IN FOCUS Ask children to suggest how the shapes shown in the **Discover** picture differ from those they have been using so far. Children should appreciate all the shapes shown here (apart from the cube) cannot be made by only using cm³ cubes. They have slanted faces or curved faces, which you cannot model with cubes. Establish that children will use cubes to *approximate* the volume of each shape. Ask: *When might someone need to estimate the volume of a shape?* For example, it might be needed when deciding whether a toy will fit into a suitcase or box. Look at the **Discover** picture again and ask: *Which of the shapes will it be easiest or most difficult to estimate the volume of?*

PRACTICAL TIPS Play estimation games where children are split into two teams. Display one 3D shape made from cubes and ask each team to predict the volume of the shape, without counting cubes. Begin with simple shapes like cubes and cuboids, then gradually make them more complex. As the shapes grow in complexity, discuss different methods with the whole class.

ANSWERS

Question 1 a): Jamilla estimated the volume of the triangular prism.
Each cube has a volume of 1 cm³. There are 90 cubes. An estimate of the triangular prism is, therefore, 90 cm³.

Question 1 b): The volume is an estimate because it is not exact, there would still be spaces left in the triangular prism if it were filled with cubes.

Share

WAYS OF WORKING Whole class teacher led

ASK

- Question 1 a): *Why do you think it is important to find the cubes in each layer first, then add them together? How does Ash's method work? What does a 'slice' mean?*
- Question 1 b): *How is this method an 'estimation'? Is it just a way to 'calculate' volume?*

IN FOCUS Question 1 a) requires children to consider various methods of estimating the volume of a shape. Remind children of the aim of the task. Ask: *Why is 90 cm³ the volume of the 3D shape, but only an estimation of the volume of the prism?* For question 1 b), ensure that children understand that the estimate is less than the actual volume of the shape as the cubes could fit inside the prism with space to spare.

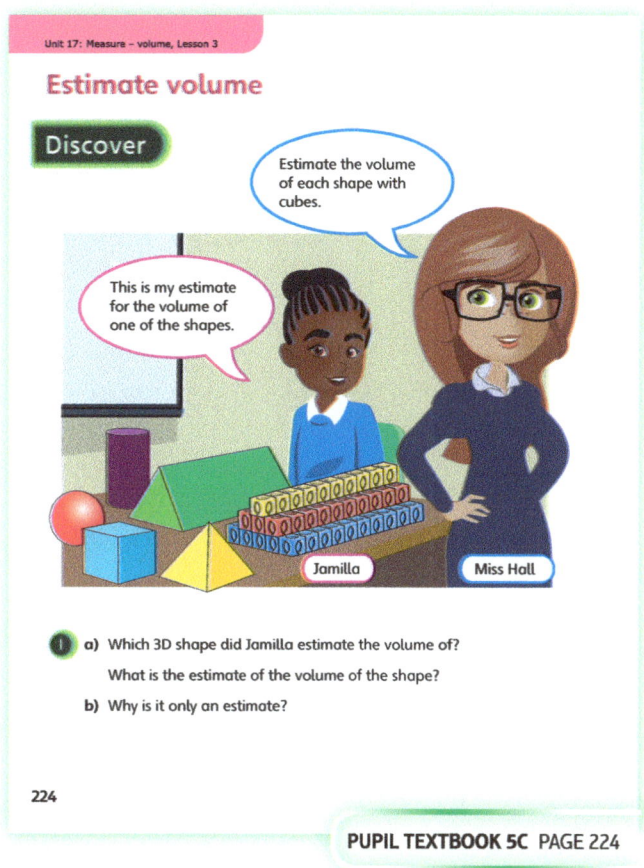

PUPIL TEXTBOOK 5C PAGE 224

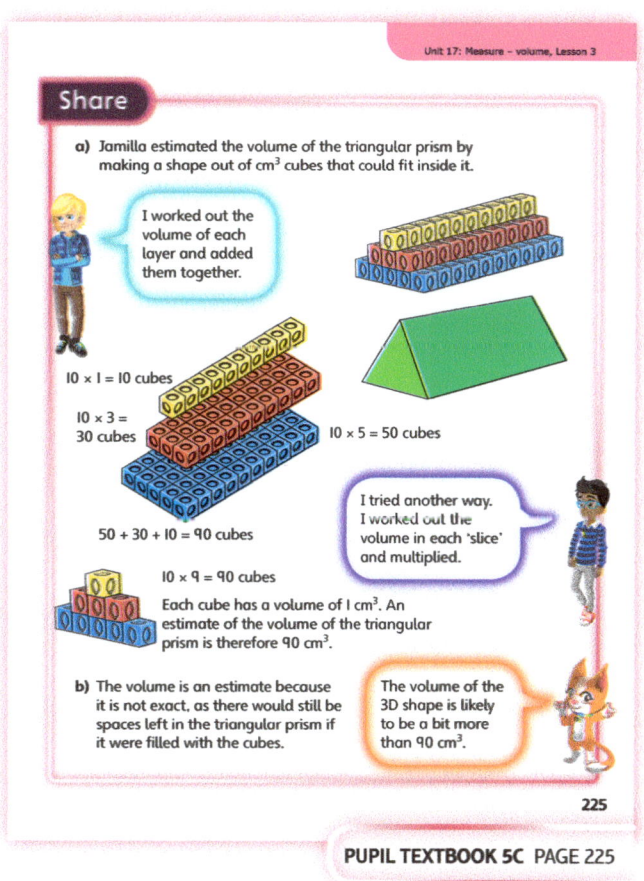

PUPIL TEXTBOOK 5C PAGE 225

Think together

WAYS OF WORKING Whole class teacher led (I do, We do, You do)

ASK

- Question ❶: *Are there any cubes that you cannot see in the model? How can you work out how many cubes there are in each model without actually making the model? How can you make sure that you have counted all the cubes in each model?*
- Question ❷: *Is it possible to use cubes to estimate any object in the classroom? What methods can you use? How can you ensure your estimation is accurate?*

IN FOCUS Question ❶ helps children to see how to use cm³ cubes to model a 3D shape approximately. Establish that the cube in part c) can be represented almost exactly using cubes, so this estimate for the volume of the cube is likely to be very close to the true volume. Discuss the benefits and limitations of the models in parts a) and b). Establish that both these models are likely to give an under estimate for the true volume. Ask: *Can you make a different model for each shape which will give an over estimate for the volume?* Challenging children to do this first will help them build visual connections between shapes and their volumes.

STRENGTHEN Provide children with cubes that they can use to model the objects. Alternatively, provide boxes in which the objects can fit. Ask: *How many cubes did you use to make the model? Is your model bigger or smaller than the actual object? What is the volume of the box? If the object fits in the box, will its volume be less or more than the volume of the box?*

DEEPEN Question ❸ is designed to challenge children into thinking more deeply about what it means to use estimation to compare the volume of different shapes and objects. It is important that children are able to apply their prior knowledge of estimation. To extend their understanding, ask: *What would happen if you used bigger cubes to make each model? Will that affect your estimation or your comparison of the volumes?*

ASSESSMENT CHECKPOINT At this point, children should be working with greater confidence, using the different methods they have been taught to estimate the volume of the shapes and objects. Children should be able to provide accurate estimates and compare their estimates by volume.

ANSWERS

Question ❶ a): 30 cm³

Question ❶ b): 30 cm³

Question ❶ c): 27 cm³; the most accurate estimate would be the estimate of the cube because if it was filled with cm³ cubes there wouldn't be any spaces.

Question ❷: Children's answers will vary depending on the object chosen. They should make a model of the object using centimetre cubes and then count the cubes to find the volume in cm³.

Question ❸ a): Make models of the different balls using centimetre cubes and then count the cubes to find the volume in cm³.

Question ❸ b): Reena is incorrect. If the football is 3 times higher than the tennis ball then the volume will be 3 × 3 × 3 = 27 times greater.

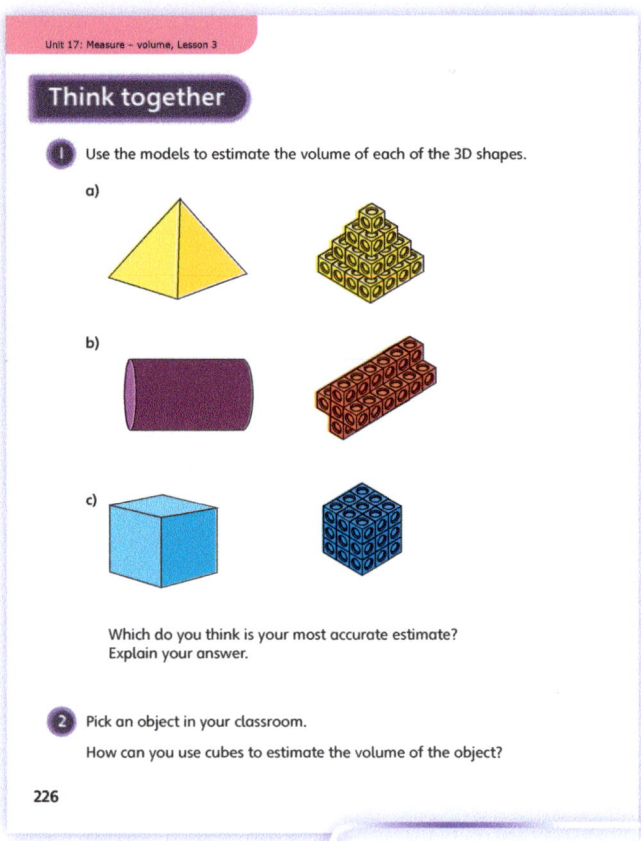

PUPIL TEXTBOOK 5C PAGE 226

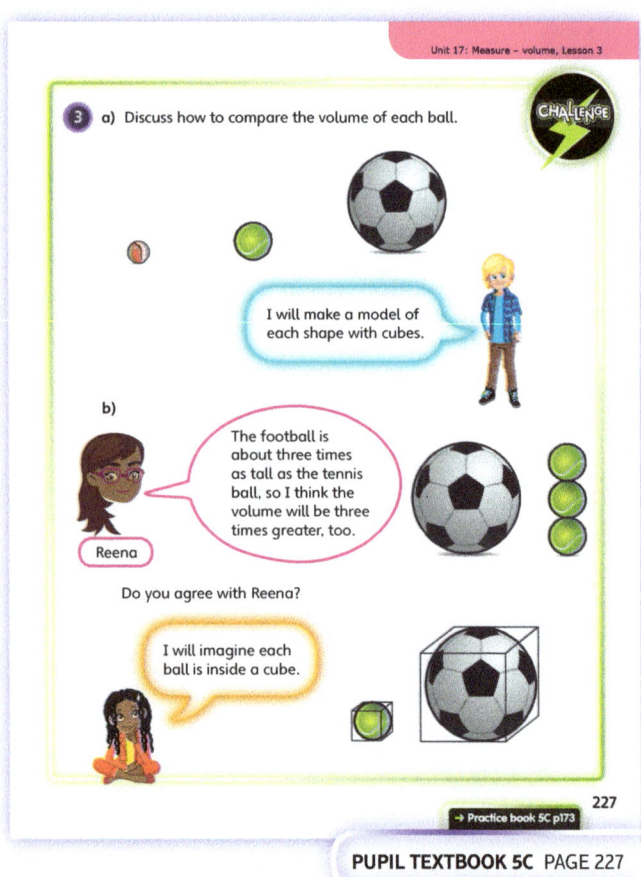

PUPIL TEXTBOOK 5C PAGE 227

Unit 17: Measure – volume, Lesson 3

Practice

WAYS OF WORKING Independent thinking

IN FOCUS Children are asked to consider what the volumes of a hemisphere, sphere, cylinder and cuboid are in question ③ – concepts they have not had to deal with before this lesson. Ask: *How many different ways can you use to estimate the volume of these shapes? Does it matter that they have curved faces?*

STRENGTHEN Provide children with cm^3 cubes and modelling clay. Ask them to make a sphere and a cylinder. Ask: *Can you predict how many cubes the volume is?* Ask children to change the shape into a cube or cuboid, then compare it to the volume of the cubes. Ask: *How many cubes make each shape? What is your estimation of the volume?* Although this is not the method children will be encouraged to use all the time, it is a useful way to strengthen their understanding of how to estimate by 'changing' the shapes into cubes and cuboids first.

DEEPEN Question ⑥ is a challenge requiring several steps. Lead children through the process. Ask: *How do you find objects with a volume of less than $10\ cm^3$? What do $10\ cm^3$ cubes look like?* The question is designed to get children to consider appropriate methods. Thereafter, they are required to use their chosen method to complete the table. Ask: *Will you need up to 1,000 cubes to estimate the volume of large objects? What method can you use? Do you need to split the shape into layers?*

ASSESSMENT CHECKPOINT At this point in the lesson, children should be confident when estimating the volume of 3D shapes and objects. They should display secure knowledge of a strategy to help them estimate, and should be able to use their estimations to compare the volume of the objects. They should be able to recognise that their calculations are estimates of the volumes and explain why this is.

ANSWERS Answers to the **Practice** part of the lesson can be found in the *Power Maths* online subscription.

Reflect

WAYS OF WORKING Independent thinking

IN FOCUS This **Reflect** question is used to help children look back on the strategies they have used in the lesson while estimating the volume of their hand. Some children may suggest drawing around the hand and then using cubes to change the drawing into a 3D shape. Other children may place cubes on their hand and count how many cubes their hand can hold. Whichever method they use, children should demonstrate their understanding of how estimation is used to calculate the volume of different objects.

ASSESSMENT CHECKPOINT Identify those children who employ a correct methodology, demonstrating a confident understanding of how to use cubes to estimate the volume of a hand.

ANSWERS Answers for the **Reflect** part of the lesson can be found in the *Power Maths* online subscription.

After the lesson

- How did you challenge children to think more deeply about estimating the volume of a shape?
- Do children now understand the concept of estimating the volume of any shape?

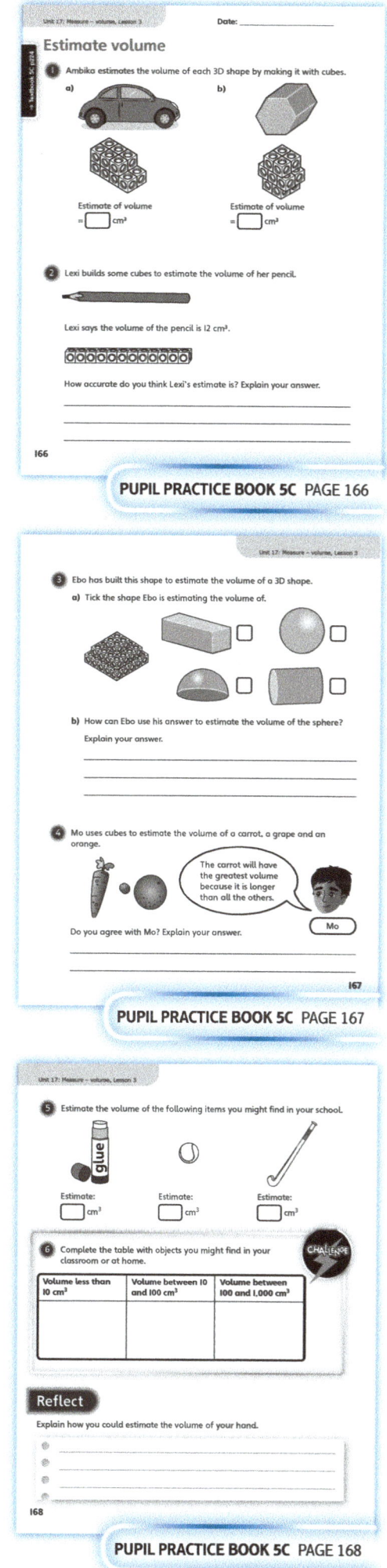

PUPIL PRACTICE BOOK 5C PAGE 166

PUPIL PRACTICE BOOK 5C PAGE 167

PUPIL PRACTICE BOOK 5C PAGE 168

263

End of unit check

> Don't forget the unit assessment grid in your *Power Maths* online subscription.

WAYS OF WORKING Group work adult led

IN FOCUS

- Question 1 assesses children's ability to find the volume of a solid.
- Question 2 assesses children's ability to find the volume of a shape by counting the cubes within it.
- Question 3 assesses children's ability to find and compare the volume of different solids.
- Question 4 assesses children's ability to use the provided dimensions of a cuboid to replicate a shape and find its volume.
- Questions 5 and 6 assess children's ability to estimate the volume of different shapes using different methods. In question 5, children may suggest different methods for estimating the volume of the triangular prism. They might suggest making a shape out of cm³ cubes that could fit inside it. However, this will result in an under estimate. They may also make a cuboid from cm³ cubes that is the same length but slightly bigger than the triangular prism. This will give an over estimate for the volume.

ANSWERS AND COMMENTARY

Children who have mastered the concepts in this unit will understand the concept of volume being a measure of the amount of space a shape takes up. They will be able to measure and estimate the volume of different 3D shapes using cm³ cubes. Children will also be able to order objects by their volume.

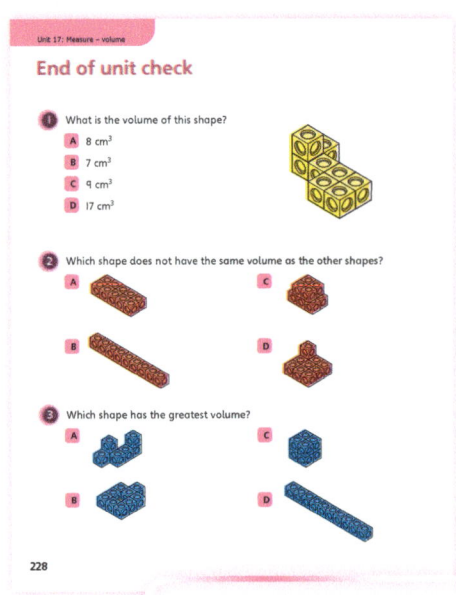

PUPIL TEXTBOOK 5C PAGE 228

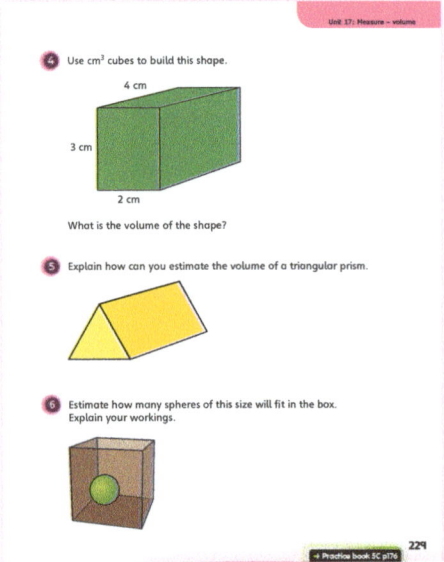

PUPIL TEXTBOOK 5C PAGE 229

Q	A	WRONG ANSWERS AND MISCONCEPTIONS	STRENGTHENING UNDERSTANDING
1	A	B or C suggest that children are unable to calculate volumes of shapes with 'hidden cubes'. D suggests that they have confused volume with area.	Strengthen understanding of the general concept of volume by providing children with opportunities to use concrete resources (such as plastic 3D shapes and cubes) to measure shapes or estimate the number of cubes that fit inside a shape. Allow children to make the shapes themselves so that they can understand what volume actually means and how they are using what they know to find it.
2	B	A, C and D suggest that children have either miscounted or compared volumes incorrectly.	
3	D	A, B and C suggest that children have miscounted cubes.	
4	24 cm³	A wrong answer may indicate that children have not built the shape correctly, or have not counted the number of cubes in each layer correctly, or have not added the results.	
5	See In focus above	Children may not demonstrate a confident understanding of how to use cubes to estimate volume.	
6	8	Children may suggest the answer is 4, since 4 spheres will sit on the bottom of the box.	

Unit 17: Measure – volume

My journal

WAYS OF WORKING Independent thinking

ANSWERS AND COMMENTARY Possible answers are:

- Split the shapes horizontally into three different layers, then add the layers. Children have access to cm³ cubes and can make each of the layers in the shape. They can then add the number of cubes in each layer to find the volume of the whole shape.

- Split the shape into three equal slices. The volume of the shape = the number of cubes in one of the slices multiplied by 3.

If children need support to explain how to calculate the volume, encourage them to make the shape and ask: *How many cubes can you see? How many cubes are there in each of the layers?*

Power check

WAYS OF WORKING Independent thinking

ASK
- What did you know about volume before you began this unit? What do you know now?
- Do you think you would be able to find the volume of any solid made of cubes on your own?

Power puzzle

WAYS OF WORKING Pair work or small groups

IN FOCUS Use this **Power puzzle** to see if children can explore different ways of estimating the volume of a football and the volume of a classroom. Children should be able to use their knowledge of estimation and the properties of a cube to predict the number of footballs that could fit into their classroom.

ANSWERS AND COMMENTARY Answers will depend on the size of each classroom. Possible approaches to finding the most accurate estimate are:
- Children fit the football in a box. They estimate the number of boxes that can fit into the classroom.
- Children should think about how many footballs they can fit along the length and width of the classroom floor. They can then multiply this together to find the number of footballs that would fit on the floor. They could then multiply this by the number of layers of footballs that could fit in the classroom.

Completing the **Power puzzle** shows that children understand what volume means and can estimate the volume of shapes of different dimensions.

After the unit

- How did children respond to the materials and approaches used to estimate volume? Were they adequately (or excessively) challenged by these exercises?

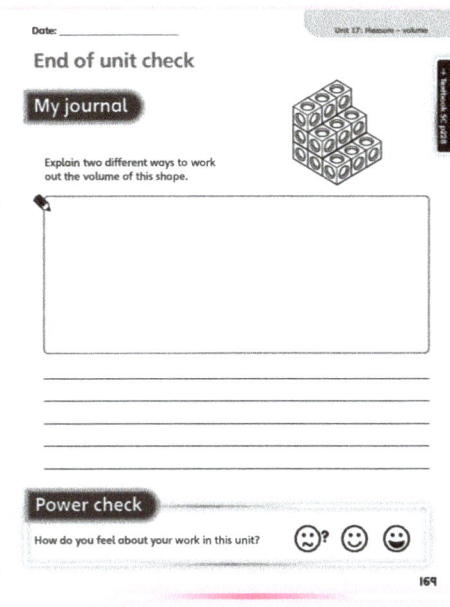

PUPIL PRACTICE BOOK 5C PAGE 169

PUPIL PRACTICE BOOK 5C PAGE 170

Strengthen and **Deepen** activities for this unit can be found in the *Power Maths* online subscription.

265

Published by Pearson Education Limited, 80 Strand, London, WC2R 0RL.

www.pearsonschools.co.uk

Text © Pearson Education Limited 2018, 2023
Edited by Pearson and Florence Production Ltd
First edition edited by Pearson, Little Grey Cells Publishing Services and Haremi Ltd
Designed and typeset by Pearson and PDQ Digital Media Solutions Ltd
First edition designed and typeset by Kamae Design
Original illustrations © Pearson Education Limited 2018, 2023
Illustrated by Laura Arias, John Batten, Fran and David Brylewski, Diego Diaz, Nigel Dobbyn and Nadene Naude at Beehive Illustration; Emily Skinner at Graham-Cameron Illustration; and Kamae Design
Cover design by Pearson Education Ltd
Back cover illustration © Diego Diaz and Nadene Naude at Beehive Illustration

Series editor: Tony Staneff; Lead author: Josh Lury
Authors (first edition): Liu Jian, Josh Lury, Catherine Casey, Belle Cottingham, Zhou Da, Zhang Dan, Zhu Dejiang, Wei Huinv, Hou Huiying, Zhang Jing, Huang Lihua, Yin Lili, Liu Qimeng, Paul Wrangles and Zhu Yuhong
Consultants (first edition): Professor Liu Jian and Professor Zhang Dan

The rights of Tony Staneff and Josh Lury to be identified as authors of this work have been asserted by them in accordance with the Copyright, Designs and Patents Act 1988.

This publication is protected by copyright, and permission should be obtained from the publisher prior to any prohibited reproduction, storage in a retrieval system, or transmission in any form or by any means, electronic, mechanical, photocopying, recording, or otherwise. For information regarding permissions, request forms and the appropriate contacts, please visit https://www.pearson.com/us/contact-us/permissions.html Pearson Education Limited Rights and Permissions Department.

First published 2018
This edition first published 2023

27 26 25 24 23
10 9 8 7 6 5 4 3 2 1

British Library Cataloguing in Publication Data
A catalogue record for this book is available from the British Library

ISBN 978 1 292 45061 2

Copyright notice
All rights reserved. No part of this publication may be reproduced in any form or by any means (including photocopying or storing it in any medium by electronic means and whether or not transiently or incidentally to some other use of this publication) without the written permission of the copyright owner, except in accordance with the provisions of the Copyright, Designs and Patents Act 1988 or under the terms of a licence issued by the Copyright Licensing Agency, Barnards Inn, 86 Fetter Lane, London EC4A 1EN (http://www.cla.co.uk). Applications for the copyright owner's written permission should be addressed to the publisher.

Printed in the UK by Ashford Press Ltd

For Power Maths online resources, go to:
www.activelearnprimary.co.uk

Note from the publisher
Pearson has robust editorial processes, including answer and fact checks, to ensure the accuracy of the content in this publication, and every effort is made to ensure this publication is free of errors. We are, however, only human, and occasionally errors do occur. Pearson is not liable for any misunderstandings that arise as a result of errors in this publication, but it is our priority to ensure that the content is accurate. If you spot an error, please do contact us at resourcescorrections@pearson.com so we can make sure it is corrected.